Intervention in the Brain

Basic Bioethics
Arthur Caplan, editor

A complete list of the books in the Basic Bioethics series appears at the back of this book.

Intervention in the Brain

Politics, Policy, and Ethics

Robert H. Blank

The MIT Press
Cambridge, Massachusetts
London, England

MIT Press books may be purchased at special quantity discounts for business or sales promotional use. For information, please email special_sales@mitpress.mit.edu or write to Special Sales Department, The MIT Press, 55 Hayward Street, Cambridge, MA 02142.

This book was set in Sabon by Toppan Best-set Premedia Limited, Hong Kong.
Printed on recycled paper and bound in the United States of America.

Library of Congress Cataloging-in-Publication Data

Blank, Robert H.
Intervention in the brain : politics, policy, and ethics / Robert H. Blank.
    p. ;   cm. — (Basic bioethics)
Includes bibliographical references and index.
ISBN 978-0-262-01891-3 (hardcover : alk. paper)
I. Title. II. Series: Basic bioethics.
[DNLM: 1. Brain. 2. Bioethical Issues. 3. Neurosciences—ethics. WL 300]
LC classification not assigned
612.8′2—dc23
2012036419

10 9 8 7 6 5 4 3 2 1

Contents

Series Foreword

Glenn McGee and I developed the Basic Bioethics series and collaborated as series coeditors from 1998 to 2008. In fall 2008 and spring 2009 the series was reconstituted, with a new editorial board, under my sole editorship. I am pleased to present the thirty-sixth book in the series.

The Basic Bioethics series makes innovative works in bioethics available to a broad audience and introduces seminal scholarly manuscripts, state-of-the-art reference works, and textbooks. Topics engaged include the philosophy of medicine, advancing genetics and biotechnology, end-of-life care, health and social policy, and the empirical study of biomedical life. Interdisciplinary work is encouraged.

Arthur Caplan
Basic Bioethics Series Editorial Board

Joseph J. Fins
Rosamond Rhodes
Nadia N. Sawicki
Jan Helge Solbakk

1

The Human Brain: An Introduction

Like genetics and stem cell research, neuroscience[1] promises to be a highly controversial political issue. However, while the political ramifications of genetic research have been well documented and widely analyzed, less systematic scrutiny has been given to the political implications of neuroscience. The array of techniques and strategies for intervening in and imaging of the brain is expanding rapidly, and these techniques will be joined in the future by even more extraordinary capabilities. In addition to their use in treating neural diseases and disorders, these innovations promise increasingly precise and effective means of predicting, modifying, and controlling behavior. The advent of neuroimaging alone has led to an enormous proliferation of scholarship: according to one estimate, over the past five years an average of one thousand peer-reviewed scholarly articles on theoretical and applied neuroimaging have been published each month (Snead 2008). However, while advances in neuroscience and technology are impressive, much of the popular literature tends to oversimplify and exaggerate the claims of presumed efficacy, thus heightening the fears of those persons who see it as a threat and creating unsubstantiated hope in those persons who might benefit from it. While this situation helps bring attention to neuroscience, it also intensifies the debate and hardens divergent predispositions on how neuroscience findings should be interpreted and used.

The 1990s "Decade of the Brain" stimulated neuroscience research on many fronts and resulted in considerable advances in the science. Unfortunately, we have been slow to develop a political dialogue to anticipate and deal with the vast implications of the brain sciences. Simply put, our political and social institutions have not kept pace with these advances.

1. Although there are nuances, the terms *brain sciences* and *neuroscience* are used interchangeably in this book.

At their base, political issues center on how we interpret the implications of these developments, a particular problem in light of the technical complexity of the subject and the speculative nature of much of the evidence to date. The questions discussed here go to the heart of the major problems facing the world today and challenge basic suppositions across the political-ideological spectrum, as well as the Standard Social Science Model of political behavior (Alford and Hibbing 2004) . As a result, the findings of neuroscience are open to ideological manipulation from many directions designed to put the most expedient spin on their meaning.

The brain has long been the subject of considerable conjecture, myth, and misconception. Throughout history, it has remained a mysterious and enigmatic entity: hidden within the skull it has been a dark territory, little understood. Major technological developments in imaging the brain, combined with leaps in knowledge about its functioning in the last several decades, have, however, appreciably expanded our understanding of its role. The evolving neuroscience perspective promises to help explain much about the biological basis of human behavior, consciousness, memory, language, and other attributes that make us what we are. Combined with research in molecular biology and other life sciences, neuroscience provides the key to understanding the foundations of human capacity, as well as mental and behavioral dysfunction.

Although we know considerably more about the brain than we did even several years ago, in many ways our knowledge is still rudimentary. There remain many competing theories even as research to identify the anatomical connections and to understand the biochemical, molecular, and genetic mechanisms of the brain accelerates. The extent of the brain's full capabilities is unknown, but it is clearly the most complex living structure known in the universe. This single organ controls all body activities, ranging from heart rate and sexual function to emotion, learning, and memory, and is thought to influence the immune system's response to disease and to govern how well people respond to medical treatments. "Ultimately, the brain shapes our thoughts, hopes, dreams, and imaginations. In short, the brain is what makes us human" (Society for Neuroscience 2006, 4). In light of the heightened potential for ever more precise and effective means to understand and potentially control behavior, it is time to broaden the dialogue over how to use these new tools for understanding the human condition.

This chapter provides a brief overview of brain anatomy and function and discusses the modular brain concept as it relates to human behavior.

It also examines neural communication and the role of neurotransmitters in brain activity. This background material elucidates the intricacy of the brain and provides a foundation for the analysis of the various modes of brain intervention in later chapters. Although the focus of this book is the brain, not genetics, obviously they are intricately connected, and discussions of genetics are included where appropriate throughout the book. Therefore, this chapter also summarizes current understanding of the interaction between genetics and the brain and presents a brain-centered paradigm of political behavior.

Anatomy of the Brain

One cannot appreciate the complexity of neuroscience without at least a rudimentary understanding of the structure of the human brain. Moreover, the uniqueness of the human brain is crucial to explaining its functioning and the problems that arise when it fails to function normally. The human brain is a three-pound maze of nerve and tissue comprising the mushroom-shaped cerebral cortex and the brain stem and is composed of hundreds of nuclei, groups of cells sharing the same anatomical region. For every given function, such as vision, hearing, and movement, a combination of nuclei and areas of the cerebral cortex act in a coordinated and highly synchronized manner. The more we discover about the brain, the more elaborate its functioning appears and the more questions arise. The anatomy of the brain and the intricacy of the connections among nuclei need be extremely complicated to be able to combine, compare, and coordinate information from every area of the nervous system. "The brain determines the relevance of each bit of information, fits it in with all other information available, and decides what actions should or should not be taken to regulate bodily functions and to permit successful interaction with the external environment" (Office of Technology Assessment 1990, 30). The brain, then, is at the core of everything that makes up the life and existence of the organism, including human consciousness itself.

The most striking part of the human brain is the cerebral cortex. Although no category of cell or type of neural circuit is unique to the human brain, the major difference between humans and animals is the exceedingly large size of the human cerebral cortex. The differential development of our cerebral cortex is primarily accounted for by the immense increase in its surface area, which is folded to fit into a confined skull space. The efficiency of the highly convoluted surface of the cortex is remarkable. Although it is only about 2 millimeters thick, its surface area is about 1.5 square meters and contains some ten to fifteen billion neurons, four times as many glial cells, and an estimated one million billion synaptic connections.

Roughly divided into the occipital, temporal, parietal, and frontal lobes on each side, the cerebral cortex contains numerous sensory receiving, motor control, language, and associated areas, with many regions having multiple functions. The frontal lobes, for instance, are thought to be the center of the higher-order processes that give us the capacity to engage in abstract thinking, planning, and problem solving. Moreover, located at the base of the temporal lobe, the hippocampus is critical to learning and for the consolidation of recently acquired information, or short-term memory. Closely situated at the base of the frontal lobes are the basal ganglia nuclei, which are associated with a variety of functions, including voluntary motor control, procedural learning of routine behaviors, and action selection, that is, the determination of which of several possible behaviors to execute at a given time.

The human frontal lobes constitute almost 40 percent of the total cortical area and, as the last cortical areas to mature, are connected to almost every other part of the brain, including the behaviorally critical limbic system. These extensive connections form the basis for the importance of the frontal lobe as an "integrator and regulator of brain function" (Restak 1994, 96). Extensive experience with patients who have suffered frontal lobe dysfunction demonstrates how critical they are to human thought and behavior. Humans are the only species capable of anticipating the future consequences of present actions, setting up plans and goals and working toward their achievement, balancing and controlling the emotions, and maintaining a sense of ourselves as active contributors to our future well-being. All these powers are diminished with frontal lobe disease.

The upper brain is surrounded by the cerebral cortex and contains discrete structures, including the hypothalamus and the thalamus. Each of these structures is composed of a collection of nuclei with specific functions. The thalamic nuclei are primarily involved in analyzing and relaying all the sensory and motor information coming into the brain. Similarly, neurons in the hypothalamus serve as relay stations for internal regulatory systems by monitoring information arriving from the autonomic nervous system and controlling behaviors such as eating, drinking, and sexual activity. This dime-sized region of the brain, comprised of ten or so nuclei on each side, controls and modulates sexual behavior, regulates the endocrine system, and controls the secretory function of the pituitary gland.

Connecting the separate left and right cerebral hemispheres is the corpus callosum, a large bundle of nerve fibers that facilitates communication between them. As discussed in detail in chapter 4, it has been posited that the corpus callosum in women is on average larger than in men, thus resulting in greater communication between the cerebral hemispheres in women (Ngun et al. 2010). Furthermore, it is suggested that women's greater sensitivity to emotional, nonverbal communication, even what is considered "female intuition," comes from this greater connectivity between the hemispheres (Menzler et al. 2011). Another structure that connects the cerebral hemispheres is the anterior commissure, which communicates visual, olfactory, and auditory information, and it too is on average larger in women than in men.

The limbic system is a complex set of structures located on each side of the thalamus just under the cerebrum. Areas that are typically included in the limbic system fall into two categories. Some of these are subcortical structures, while others are actually portions of the cerebral cortex. Cortical regions that are involved in the limbic system include the hippocampus as well as the insular cortex, orbital frontal cortex, subcallosal gyrus, cingulate gyrus, and parahippocampal gyrus. The subcortical portions of the limbic system include the olfactory bulb, hypothalamus, amygdala, the septal nuclei, and some thalamic nuclei, including the anterior nucleus and possibly the dorsomedial nucleus.

A key component of the limbic system that is critical for understanding behavior is the amygdala (plural amygdale; there is one in each hemisphere). The amygdala is a collection of thirteen nuclei (including the accessory basal, basal nucleus, central nucleus, intercalated cells, lateral nucleus, medial nucleus, and cortical nucleus) in an almond-shaped area in the medial temporal lobe, right and left. Although it comprises barely 0.3 percent of the volume of the brain, it is a vital piece of a system that initially evolved to detect dangers in the environment and modulate subsequent responses that can profoundly influence behavior (Schumann, Bauman, and Amaral 2011). As will be illustrated in later chapters, the amygdala plays a key role in an array of emotional and social functions, as well as in emotion-cognition interactions. There is also clear evidence that there are substantial individual differences in the amygdala that might contribute to many psychiatric illnesses and forms of aggressive behavior (Adolphs 2010). Despite progress in uncovering its importance, however, the amygdala remains a fairly mysterious brain structure, and its functional roles remain difficult to characterize, stemming in part from the diversity and complexity of the amygdala's functions and its complex interactions with other regions (Hamann 2011).

Beneath the upper brain, the brain stem regulates essential bodily functions, such as heart rate, blood pressure, and respiratory activity. Although the brain stem contains structures that regulate autonomic functions (those not under our conscious control), this region is also an important junction for the control of deliberate movement. All the nerves that connect the spinal cord and the brain pass through the medulla at the lower end of the brain stem. At the top of the brain stem, the pons serves as a bridge between the lower brain stem and the midbrain. Extending downward from the brain stem and divided into thirty-one segments is the spinal cord, which is composed of a central core of cells surrounded by pathways of axons. Leaving and entering the spinal cord are nerves that bring sensory data into the central nervous system (CNS) and transmit motor and activating information to the peripheral nervous system (PNS).

Highly integrated with the CNS, the endocrine system is responsible for the production of hormones. Hormones are chemical messengers that are released into the bloodstream by endocrine glands. Among other functions, they affect physical growth and development, sexual functioning, and emotional responses. The control center of the endocrine system is located in the hypothalamus. Beneath this region of the brain is the pituitary gland. The pituitary responds to signals from the hypothalamus by producing a variety of hormones that modulate the hormone secretion of other glands. The pituitary gland also produces several hormones with more general effects, such as human growth hormone and dopamine, the latter of which inhibits the release of prolactin and acts as one of many neurotransmitters. Imbalances in hormone secretions can cause both physical and mental disorders.

Neural Communication

There are two major classes of cells in the nervous system. The first type, neurons or nerve cells, conduct all information processing through networks with other neurons. Neurons consist of a cell body with long extensions called dendrites. Also projecting out of the cell body is a single fiber, called an axon, which can extend for great distances to provide linkages with other neurons. All nerves in the body are bundled axons of many neurons conveying information to and from the CNS. Glial cells, the second type of nervous system cell, support neurons by carrying on critical chemical and physiological reactions and produce a variety of substances needed for normal neurological functioning. Some produce myelin, the fatty insulating material that forms a sheath around axons and speeds the conduction of electrical impulses. Other glial cells help regulate the biochemical environment of the nervous system and provide structural support for neurons.

Information in the nervous system is conveyed by chains of neurons, usually traveling in one direction, from the dendrites through the cell body and along the axon. Information is coded as an electrical-chemical message passed from one neuron to the next when its axon connects with a dendrite or cell body of another neuron at a synapse. The two neurons never actually make physical contact but remain separated by a small gap called the synaptic space. Furthermore, because the end of the axon and the dendrites of each neuron have many branches, each axon can send messages to and receive messages from thousands of other neurons simultaneously. It is estimated that a cubic millimeter of cerebral cortex contains as many as a billion synapses by which neurons signal each another (Cheshire 2007, 137). In all, there are approximately 160 trillion synapses in the adult human cerebral cortex, with intricately precise networks of highly differentiated neurons integrating the signals flowing from synapse to synapse. According to Cheshire (2007), the brain's internal communication network of synapses underlies its capacity to interpret, reflect on, and interact with the external world, and, despite the immense number of dynamically interconnected neurons, it rarely descends into internal anarchy.

Messages are sent when a neuron generates a nerve impulse, thereby firing an electrical message along the axon toward the synapse. When the electrical message gets to the synapse, it causes the release of a chemical, one of scores of known neurotransmitters, from the axon into the synaptic space. These neurotransmitters are released from tiny vesicles in the axon terminals, disperse across the intrasynaptic space, and bind to receptors on the surface of the target neuron. At their target, they function like molecular keys to unlock a channel in the membrane of the target dendrite and allow sodium ions to enter. A neurotransmitter, therefore, acts as a messenger crossing the space and binding to the membrane of the next neuron. When this occurs the charge between the inner and outer surface of the membrane is altered and the target neuron is briefly depolarized. The binding of the neurotransmitter to the neuron membrane either initiates, or facilitates, or hinders an electrical message in that neuron, all happening in the space of a fraction of a second. Among the key neurotransmitters that have been linked to various types of behavior are dopamine, serotonin, epinephrine, norepinephrene, endorphin, and gamma-aminobutyric acid (GABA).

The most widely distributed neurotransmitter is serotonin, which has been found to be a crucial modulator of emotional behavior, including anxiety and stress responses, impulsivity, and aggression (Siegel and Douard 2011). According to Dayan and Huys (2009), its "fingerprints are on the scene of depression, anxiety, panic, aggression, dominance, obsessions, punishment, analgesia, behavioral inhibition, rhythmic motor activity, feeding, and more, in organisms from invertebrates to humans, and yet it has never quite been convincingly convicted of any single compelling influence" (96).

According to Lesch (2007), the diversity of these functions is due to the capacity of serotonin to orchestrate the activity and interaction of several other neurotransmitter systems involving numerous brain structures, including the anterior cingulate cortex, amygdala, hippocampus, ventral striatum, and hypothalamus. Serotonergic signaling pathways integrate not only basic physiological functions but also the elementary tasks of sensory processing, cognition, emotion regulation, and motor activity. Moreover, serotonin is an important regulator of early brain development and adult neuroplasticity, and, because it shapes various brain systems during development, it sets the stage for functions that regulate emotions throughout life.

The Brain, the Mind, and Consciousness

Despite the rapid expansion in our understanding of brain structure and function, we still are at a relatively early stage in developing a unifying theory of how the brain operates. As a result, vigorous controversy surrounds neuroscience in at least four large areas: the mind-brain distinction; the organization of the brain; the impact of genetics on the brain, and the influence of the brain on human behavior (see Crawford 2008; Tallis 2010). Current arguments over how the brain operates and how it relates to the mind and to consciousness are not of recent origin but rather extend back to the foundations of Western philosophy, when Aristotle rejected Plato's contention that a rational soul had its seat in the brain and advanced the cardiocentric view, in which the brain is merely a cooling system for the body. Long after a universal rejection of

Aristotle's heart-as-mind position, Descartes' dualistic theory of a separate brain and mind became the theory of choice of scholars. Despite an emerging view today that the mind is but a construct that has outlived its usefulness and that the mind is what the brain does, dualistic theory still has many supporters, who believe we cannot reduce thought and consciousness to biochemical interactions at the synapses. Another assumption of many philosophers was that at birth, the brain is a blank slate or tabula rasa. This assumption gave society broad sway in teaching the citizen and shaping reality. Hobbes and Locke are among those philosophers who based their theories on this conception of the brain, a conception that is now being challenged by evolutionary psychology, behavioral genetics, and neuroscience research.

A related long-standing controversy over the nature of the brain pits the holistic theories against those that emphasize localization and specialization of function. Under holistic doctrine, the brain functions as an indivisible whole, not reducible to its component parts. The holistic theory was dominant until the mid-nineteenth century. Although Gall's phrenology was soon discredited, his characterization of a highly partitioned brain inspired empirical research on localization of cerebral functions. Paul Broca's celebrated finding that aphasia (loss of speech) was associated with lesions in the left frontal lobe revealed an asymmetry between the two hemispheres, thereby contradicting holistic doctrine. Several years later, Carl Wernicke found that variations of aphasia were localized to lesions in the temporal lobe of the left hemisphere. With the demonstration that language was not a single function but had at least three components carried out in different regions of the cortex, the rush for empirical research for localization intensified. Interestingly, the current representation of a modular brain actually combines some elements of earlier localization and holistic conceptions of the brain.

The Modular Brain

According to the holistic theories postulated by Descartes and others, there is at some level a master site within the brain where all the separate components converge. This notion of a master control site, although intuitively attractive because it represents the "I" as a single entity, is not supported by current knowledge of how the brain operates. The brain does not act as, and in fact is incapable of acting as, a single integrated whole. Instead, very specific functions of the brain are highly localized, and these localized units (often termed modules) are linked together in a complex structure. For example, the neurons that allow our vision to differentiate straight lines differ from those that delineate curves. The modular brain theory, however, transcends simple localization of function. Despite a division of labor, the brain has evolved structures that link these components together in predictable ways. Although specific functions are localized, all neurons and nuclei communicate with other modules, and multiple connections all operate simultaneously in parallel. This means there is no cortical terminus, no master site or seat of consciousness. No one area holds sway over all others. The separate modules do not all report to a single executive center. For instance, there is no one emotion center; rather, the genesis and expression of emotions take place in a constellation of groups of neurons, or modules, which Changeux terms "integration foci" (1997, 21).

The parallel processing capacity of the brain most likely has evolved as a survival mechanism to compensate for constraints of the brain structure. By computer standards, neurons act very slowly. Electrophysiological impulses in the brain travel at about 100 meters per second, about one-millionth the speed that electrical impulses cover computer circuits. However, the brain is able to offset this relative slowness by using very many neurons simultaneously and in parallel and by arranging the system in a rough hierarchical manner. Thus, any activity engages a wide network involving many mutually interactive processes occurring concurrently in numerous parts of the brain. It makes sense, however, that when groups of neurons interact to carry out a specific function, they tend to be localized in a single region of the brain. Neurons in close proximity tend to receive similar input. Moreover, because of this proximity and the relatively short connections to nearby neurons, they allow for rapid interaction. From this standpoint, localization means that a division of labor is made most efficient by concentrating interconnected modules in one general region.

The fact that the performance of simple functions is localized does not, however, negate the possibility that overall strategies for performing an integrative operation cannot be effectuated by combining different simple functions. Kosslyn and Koenig (1992) distinguish an integrative function (e.g., language) from the simpler component functions in arguing that we can indeed have it both ways. Some functions are localized, but the brain also works as a whole to produced integrated functions that are not localized. The tasks of neuroscience research, then, are to characterize what the functions are, which parts of the brain carry out each function, and how the functions work together. Early localization notions distinguished between the midbrain's role in autonomic responses, the limbic system's role in emotions and behavior, and the neocortex's role in complex information processing, verbal language, and complex memory. Research discussed in later chapters demonstrates that this presumption is oversimplified. Under the modular brain theory, the brain is a complex system of linked structures, with an emphasis on "linked." Localized assemblages of neurons must communicate with each other or the entire system breaks down. Moreover, in so doing they actually change their structure and their interrelations. The brain, under this theory, is an active, ever-changing organ that requires constant communication among distant neurons and nuclei in order to integrate the activity of these specialized modules.

This modular conception of the brain, in conjunction with its adaptability in bypassing problematic linkages and its parallel processing, helps explain why some persons who sustain damage to one area of the brain do not exhibit the full extent of disability expected (e.g., Gabrielle Gifford). Occasionally a person with seemingly normal brain function is found on autopsy to have had only one functioning hemisphere. Furthermore, it appears that even when such a major deficit occurs early in neural development, the brain's circuits are able to compensate for the absent hemisphere. If there were complete localization, of course, such substitution could not occur because damage to one area would preclude the execution of that function. Although incapacity is probable in cases of major brain damage, in general, the brain exhibits remarkable flexibility. According to Restak (1994), this reveals that the brain is opportunistic—it uses structures that may have evolved in one context to carry out different sorts of tasks. Although specialization is a critical feature of the brain, then, there is a sharing of some functions by more than one system. The brain is modular, but only to a degree. Although the holistic theory is no longer apposite, the brain is more complex than a localization theory alone would suggest.

Memory

Memory is central to human existence. At a most fundamental level, our personal sense of identity is found in our memory, our capacity to remember past experiences and build on them. Memory is central to learning and to our ability to function at the most basic human levels. It is not surprising, therefore, that memory has been the focus of considerable neuroscience research, much of it funded to study the mechanisms of memory loss from dementias that deprive the subjects of personal identity. These studies of memory strongly reflect the modified modular theory of the brain. There is conclusive evidence that memory is not a single entity but rather a process comprised of many indispensable components. Memory cannot be found in any single structure or location in the brain, although its components have been localized with increasing preciseness. As Kotulak (1996, 20) puts it, "Thus, there are no pictures stored in the brain, as was once thought. There are patterns of connections, as changeable as they are numerous, that, when triggered, can reassemble the molecular parts that make up a memory. Each brain cell has the capacity to store fragments of many memories, ready to be called up when a particular network of connections is activated."

To date, many different dimensions and channels of memory storage have been isolated. First, there are two separate channels of memory storage concentrated in different parts of the brain. Specific recall is based in the temporal lobe and its connections to the limbic system. In contrast, habit formation, through which we remember how to perform skills, is a more diffuse system located primarily in the striatum. There is also evidence that the hippocampal system is involved in episodic memory, which over time (weeks or months) is transferred to the neocortex (Kandel and Hawkins 1992). In addition, there is a complementary relationship between two types of memory. Associative memory acquires facts and figures and holds them in long-term storage. However, such knowledge is of no value unless it can be brought to the fore by working memory, itself a combination of different types of short-term memory. Working memory, which is the basic element in language, learning, thinking, and behavior, allows for short-term activation and storage of symbolic information and permits manipulation of that information (Goldman-Rakic 1997). There is evidence that it is carried out in the prefrontal lobes, which also perform executive functions such as problem solving, planning, and organizing, which require working memory.

Furthermore, there also have been identified separate neuronal circuits for spatial, object, and verbal working memories, although it remains debatable whether there is one region that acts as a central processor for all working memory information. Goldman-Rakic (1997) believes there is not one such center but instead parallel systems, each with its own central processor. As we shall see later, recent brain imaging studies have found that working memories for facial features and locations reside in separate regions of the prefrontal cortex and in separate sensory areas. It is expected that as research on memory expands, we will find even more divisions of labor by specific modules of neurons.

Consciousness

Perhaps the most mysterious aspect of the mind is consciousness, or self-awareness, which can take many forms, from experiencing pain to planning for the future. Often the mind has been equated with consciousness, in the sense of that which makes a human a human. Consciousness provides us with the continuity of self-awareness across our life. We are conscious not only of things but also of our feelings about them. We can even be conscious of our own feeling of being conscious about something. Harth (1993) terms consciousness the "most challenging phenomenon exhibited by the brain" (133). But what is consciousness, a state of the mind or the activity of neurons?

Neuroscience research severely tests traditional ideas about the unity or indissolubility of our mental life. As noted earlier, consciousness makes it appear that a single individual is the recipient of all sensations, perceptions and feelings and the originator of all thoughts, but according to Dennett (1991) this apparent unity of the I and its self-awareness is largely an illusion, For Harth (1993), "There is in the brain no single stage on which the multiple events picked up by our senses are displayed together" (133). Rather, consciousness is a process, a kind of global regulatory system dealing with mental objects and computations using those objects. Interestingly, and of critical importance when behavior is examined in later chapters, most operations of the brain take place outside our conscious awareness, being carried out by a combination of genetic instructions and learned reactions to sensory inputs. Evidence suggests that we are unaware of much of our brain's activity. In fact, full awareness would be an impediment to our functioning, even though at the highest levels of consciousness we experience a self-conscious controller who wills, remembers, decides, and feels emotion.

Although many commentators equate consciousness with awareness, Restak (1994) disagrees. While consciousness implies awareness, we can exhibit some lower levels of awareness without being fully conscious. However, because consciousness requires a vivid awareness of oneself as the experiencer, a prerequisite to consciousness is some level of brain activity and sensory awareness. According to Restak, "Consciousness must be understood as a very special 'emergent' property of the human brain. It is not an indispensable quality, since as we have seen the vast majority of the brain's activities do not involve consciousness" (135). Although not all mental activities are accompanied by consciousness, then, consciousness cannot take place without such more basic activities.

Although there appears to be little argument with the assumption that consciousness requires brain activity, there remains discord as to whether we can ever explain consciousness solely by the workings of the brain (Häyry 2010). Tallis (2010), for instance, contends that the pervasive yet mistaken idea that neuroscience fully accounts for behavior is neuroscientism, an exercise in science-based faith. According to Tallis, the failure to distinguish consciousness from neural activity "corrodes our self-understanding in two significant ways. If we are just our brains, and our brains are just evolved organs designed to optimize our odds of survival . . . then we are merely beasts like any other, equally beholden. . . . Similarly, if we are just our brains, and our brains are just material objects, then we, and our lives, are merely way stations in the great causal net that is the universe, stretching from the Big Bang to the Big Crunch" (3). While neuroscience can reveal some of the most important conditions that are *necessary* for behavior and awareness, it is unable to provide a satisfactory account of the conditions that are *sufficient* according to critics (Crawford 2008).

Similarly, while Churchland (1995) concludes that the *state of consciousness* is primarily a biological phenomenon, she contends that the *contents* of consciousness are "profoundly influenced" by the social environment. This distinction between a state and content appears to be supported by neurological evidence. It has been discovered that while the content of consciousness, as with memory, is found in the cerebral cortex, the maintenance and regulation of a conscious state are centered in the reticular formation region of the midbrain, which serves as an activation system for wakefulness. Because consciousness cannot occur without wakefulness, it is dependent on the activity of one of the most primal parts of the brain. However, since consciousness also requires content and a relationship to that content, it must be the product of the interrelated activity of the neocortex and the reticular activating system, again manifesting the modular brain in action.

Before the death knell is sounded for dualism, then, it is important to note that there has not been a decisive resolution of the mind-brain question. Even if the mind is the expression of the activity of the brain and the two are in actuality inseparable, this does not mean that it is useless to separate them for analytical purposes. While mental phenomena arise from the brain, mental experience also affects the brain, as demonstrated by many examples of environmental influences on brain plasticity (Crews 2008). Scott (1995) concludes that it is not necessary to choose between materialism and dualism. Both can be accepted with certain reservations: "We must construct consciousness from the relevant physics *and* biochemistry *and* electrophysiology *and* neuronal assemblies *and* cultural configurations *and* mental states that science cannot yet explain" (159–160).

However, the more that neuroscience explains how the brain works, the more difficult will be the task of the dualists who demand an immaterial mind. With the rapid developments in our understanding of the mechanics of the brain, themselves products of the imagination of human minds, consciousness will lose some of the mystery that has surrounded it since the time of Plato. Although this is viewed as a threat by those who believe that we lose something special and private when we debunk the mind as separate from the brain, their fear of this shift to a modified materialism is probably premature because no matter how much we advance in neuroscience it is unlikely the debate will disappear. Similarly, despite activities in artificial intelligence, information theory and cognitive science that would reduce the mind to the workings of the computer, it is improbable that the mysteries of the human mind will be explained or replicated by even the most sophisticated computers imagined by the minds of humans. In the words of Miller (1992): "Consciousness may be implemented by neurobiological processes—how else?—but the language of neurobiology does not and cannot convey what it's *like* to be conscious" (180). Thus, it is unlikely that the philosophical debate surrounding the mind-brain relationship and human consciousness will abate, despite growing evidence of the importance of physiochemical factors for the behavior outlined in this book.

Policy Implications

What difference does it make whether the holistic, localized, or modular theory of the brain is most accurate? From a policy standpoint as opposed to a philosophical one, does it matter whether the brain and mind are indistinguishable or separate? Although on the one hand it would seem that none of this should matter, the resilience of the controversies over these topics demonstrates that they have significant policy ramifications.

First, throughout Western history, research on the brain has confronted opposition from both the left and the right ends of the ideological spectrum. On the right, this research is viewed as threatening the concept of the immaterial soul that is the foundation of much religious doctrine. Even in a nonreligious context, the claim that the "I" is simply a product of a vast network of nerve cell connections challenges fundamental beliefs of individual responsibility, autonomy, and free will. Therefore, Western theories of justice and their application in public law must be reevaluated in light of neuroscience findings (see chapter 5). From the opposite ideological direction, the findings of brain research have always raised fears of the impact of new discoveries: of social and behavioral control. The demise of the conception of the brain as a tabula rasa also has implications for education, social, and legal policies. Critics equate perceived threats of a psychochemical determinism with earlier theories of genetic determinism. Specific areas of brain research have also elicited opposition from liberals, who fear that its findings will be used as weapons of oppression (see chapter 7).

A second, more pragmatic reason why these questions matter is that they have much to do with shaping the direction of research goals and funding. Research priorities follow accepted theories of science. The frictions between cognitive scientists and neurobiologists, or more specifically between those who would substitute computer chips for neurons and those who study neurons, are evidence of the competitive nature of brain research. In the end, how we view the brain can be self-fulfilling if research is skewed too far in any one direction.

A third reason why these questions matter for policy is that the way we view the brain has significant implications for treatment strategies. For instance, are addictions and mental disorders matters of a spiritual or cultural mind, or are they primarily manifestations of biochemical imbalances or neurological deficiencies? Can localized treatment protocols be effective, or does a modular brain require multifaceted approaches? Are informed consent and individual responsibility meaningful concepts in light of our knowledge regarding free will and personal autonomy? These practical questions will always be addressed within our conventional standards of what the brain is, how it functions, and its relationship to the mind. The next section adds a complicating dimension to this equation—the role of genetics.

The Brain and Genetics

One of the most auspicious and dynamic areas of research is neurogenetics, the interface between genetics and the brain. Although this field is not nearly as developed in terms of potential clinical applications as other areas of neuroscience, it promises wide applications both as basic science and, particularly, as treatment options (Brief and Illes 2010). The amalgamation of molecular biology and neuroscience is bound to challenge prevailing notions of what it is to be a human and, possibly, to redefine the debate over equality and inequality of humans. As an example, despite strong resistance in many quarters, there is now little doubt that genes play a significant role in antisocial behavior (Raine 2008). The question of whether there is a genetic basis has been replaced by the question of how much of antisocial behavior is influenced by genes. While not all studies show significant effects, a review of over 100 twin and adoption analyses provides evidence that about 50 percent of the variance in antisocial behavior is attributable to genetic influences (Moffitt 2005). From this base, the field is now moving to the more important question, as formulated by Raine (2008): "*Which* genes predispose to which kinds of antisocial behavior?" (323).

More broadly, given its role in human adaptability and survival, it would be surprising if traits that result from variation in brain function were not influenced in part by genes (O'Donovan and Owen 2009). As illustrated by the articles on human intelligence (Deary, Johnson, and Houlihan 2009) and on aggression (Craig and Halton 2009), there is now convincing evidence that genes are involved, and the same is true for many other cognitive and behavioral traits such as addiction (Gelernter and Kranzler 2009). Although the presumption that genes contribute to variation in normal cognitive and behavioral traits remains controversial, genetic methods offer a possible route toward a better understanding of variation in brain function.

The Brain-Gene Connection

Given the complexity of the workings of the brain as compared to the human genome, it is clear that no simple one-to-one relationship exists between them. It is now estimated that the human genome contains approximately 25,000 genes, most of which are common to many species. Thus, even the differential expression of all genes would fail to explain the remarkable diversity of neuronal connections and the vast range of human behavior. However, this disparity in complexity does not negate the significant role that genes play in determining the boundaries and framework of the functioning brain. As noted by Changeux (1997), a relatively small number of genes is sufficient to control the division, migration, and differentiation of the neurons shaping the neocortex. Genes provide a template for neural functioning, a template that is completed by the environment and the experience of each individual.

In addition to prescribing the generic template for the human species, genes provide the foundations for variation among individuals in terms of neural configuration and capacity. This impact is most obvious when dysfunctions occur, either because of deleterious genes or because of chromosomal abnormalities such as those found in Huntington's disease, Tay-Sachs disease, or fragile X chromosome syndrome. Our enlightened understanding of human genetics flowing from the Human Genome Project has uncovered many direct linkages among genes and the brain that are likely to proliferate in the coming decade (see chapter 9 especially).

Not surprisingly, many potential ties between genes and behavior, such as the genetic bases of addiction, aggression, and risk-taking personalities, are already the center of considerable research attention. No matter how much knowledge of the functioning of genes emerges, however, ultimately the power of the genes will not be sufficient to explain the details of neuronal organization, the precise form of every nerve cell, and the exact number and geometry of the synapses of any individual's brain. If the differential expression of genes is incapable of explaining the diversity and specificity of an individual's neural connections, however, what is?

One intriguing theory of the gene-brain linkage has been offered by Changeux (1997). His epigenetic theory of selective stabilization is consistent with current knowledge of neuronal development and with our understanding of human genetic variability. He contends that this epigenetic process does not require a modification of the genetic material because it acts not on a single cell but rather on a higher level of groups of nerve cells. He argues that the genetic "envelop" opens to more individual variability as we move up the evolutionary chain to humans. His theory of selective stabilization assumes that the genetic influence is critical up to the point where the number of neurons peaks soon after birth. Counterintuitively, what follows at this stage for Changeux is a "growth" process based not on a neuronal building process but rather on regression as some neurons in each category die because of redundancy and some of the terminal branches or axons and dendrites of surviving cells disintegrate.

Using language acquisition and hemispheric lateralization to support his theory, Changeux (1997) points out that learning a language is accompanied by a loss of perceptual capacity, by an attrition of spontaneous sounds and syllables. Similarly, he argues that at a certain critical moment, similar if not identical neuronal structures exist in both hemispheres but are lost selectively from the right or left hemisphere early in childhood. For Changeux, then, the word *growth* should be understood in the sense of the "lengthening and branching of nerve fibers, which eventually connect the cell bodies to each other (and to their targets) after the cells are differentiated and in place," not growth in number of brain cells (212). Under this theory, to learn is to stabilize preexisting synaptic combinations and eliminate the surplus. Therefore, activity can be effective only if the neurons and their basic connections already exist before the interaction with the outside world.

Whether the process of learning and growth is based on selective stabilization, as argued by Changeux, or on the gradual building of new neural connections throughout life, the role of genetics is not deterministic for either specific neural connections or behavior. Despite these limits, however, genes, as a reflection of our evolutionary past, do exert a powerful influence on the brain and are critical to our understanding of how it works. Rapid advances in the knowledge of molecular biology and its applications of direct relevance to the brain are likely to complement corresponding developments in neuroscience. Although the genome cannot explain all the intricacies of the brain, they cannot be explained without a better understanding of the role of genetics.

One of the most critical areas in our unfolding knowledge of brain function relates to neurotransmitters, the chemical message carriers between nerve cells. Researchers have identified more than fifty substances involved in neurotransmission and have made considerable progress in understanding how they work. Ultimately, scientists hope to find ways of supplementing deficient neurotransmitters and blocking the effects of neurotransmitters that exceed the brain's needs, thus developing the capacity to restore proper chemical equilibrium in brain and body. Neurogeneticists anticipate developing the capability to isolate and analyze specific genetic defects in the neurotransmitter system and have already cloned and sequenced the majority of the genes necessary for the functions of the key neurotransmitters, dopamine, serotonin, and norepinephrine (Comings 1996, 84).

One focal point of the Human Genome Project was to identify genes that prevent normal brain development or that produce progressive brain degeneration. Identification of genetic markers for an expanding array of the neurological disorders continues, although to date our capacity to alter and express the genes remains limited. Given the progress in our understanding both of genetic disease and of brain functioning, however, it is expected that many new treatment approaches will be forthcoming.

Although single genes that make individuals susceptible to major mental disorders might eventually be found, it is likely that the mechanisms are more complicated and that the causes are multifactorial. Extensive data, however, indicate a significant genetic component to schizophrenia, mood disorders, and anxiety disorders. The probability of having schizophrenia, for instance, when both parents are affected is 46 percent, exceedingly higher than the estimated 1 percent in the population who are schizophrenic. Already genes have been identified that are associated with anxiety-related traits and excitable or novelty-seeking personality, and these might provide leads for more complicated mental disorders. Moreover, while gene therapy is now focused on identifying specific genetic factors in neurological diseases or disorders, pressures for gene enhancement are likely to follow.

Policy Issues in Genetics and the Brain

Neuroscience, in combination with molecular biology, then, provides the means to answer questions regarding the human condition that have been debated for millennia. However, it also raises difficult ethical and policy issues. For instance, finding a genetic-biochemical explanation for violent and aggressive behavior could lead to new paths for treatment or to new means of behavioral control. The line between these applications is a gray one and depends on how the condition is defined by the medical community and society. The faces of inequality, therefore, are likely to change as more knowledge about the genetic bases of brain activity and functioning accumulates and our capacity for intervention expands. Questions about the genetic bases of criminality have already engendered intense controversy (see chapter 5). Although most observers agree that it is unlikely that genetics will prove to have proximate linkages to criminal behavior, even the notion of genetically based criminal tendencies poses severe constitutional issues. Experience with XYY testing and research in the late 1970s demonstrated the explosiveness of any research into the genetic-brain linkages to criminality or antisocial behavior. Other contentious areas of discovery in genetics and neurosciences include the respective roles of genes and brains in shaping sex differences and sexual orientation, addictions, personality, and ideology. Therefore, where appropriate, material on behavioral genetics will be included in substantive coverage in later chapters.

Research into genetics and the brain promises to accentuate the already acrimonious political debates over human nature, personal identity, freedom, and equality. Traditionally, differences both in genetic complement and in behavior have tended to be defined as diseases, disorders, and conditions to be treated. How society responds to these rapid technological advances and the knowledge that spawns them depends to a large extent on our conceptions of equality and inequality. There is historical evidence, reflected most clearly in the eugenics movement, to suggest that some in society will embrace such findings as proof of inequality. In other words, neurogenetics is likely to challenge the fundamental principles of a democratic society in the minds of some observers. Furthermore, in light of this research, conventional models of behavior will need to be reevaluated and modified.

The Brain Paradigm

Despite lingering debate at the extremes by adherents of either biological or environmental deterministic models of human behavior, most observers agree that some combination of nature and nurture is crucial. Often the disagreement centers on the proportional contribution of each factor, that is, on what percentage of variation can be explained by genetics and what by the social and cultural environment. Often these positions fail to appreciate the dynamic, interactive nature of the genetic-environment relationship in their quest to explain the influence in either-or terms. This neglect becomes even more obvious when the brain is put into the equation.

The interactive paradigm holds that the genes and the environment are reciprocally related and therefore can influence one another over time. More important, the genotype and the environment, along with more deep-seated evolutionary factors, act in combination to produce a specific phenotypic expression that defines the individual (Caspi et al. 2010). (The basic interactive model of genotype, phenotype, and the environment is sketched in figure 1.1.) Although this joint action serves

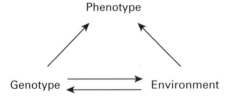

Figure 1.1
Interactive Paradigm

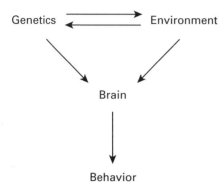

Figure 1.2
Brain Paradigm

to explain individual variation, it is not possible to legitimately generalize individual variation to population differences despite sexual, racial, and ethnic patterns in both genotype and social environment.

While this interactive model places nature and nurture in perspective, it fails to include the critical linkage factor, the brain. In order to explain behavior or capabilities, one cannot minimize the role of the brain or view it as an empty organism. According to this expanded model, the brain is the key mediator of both genetics and the environment for the individual (figure 1.2). As will be seen in later chapters, brain chemistry is increasingly being found to be critical to our understanding of behavioral patterns, personality, and a range of individual capabilities. Neuroscience, therefore, offers an indispensable tool with which to explain why we are what we are and how we might make improvements on what we are. The brain provides a focus for analyzing the rich combination of genetic, environmental, and, ultimately, neural factors that define what we are. Unlocking the secrets of the brain is the key to explaining not only why we differ from other species but also to understanding variation, as well as similarities, among humans.

Conversely, any study of the brain's structure, function, and role without a consideration of the genetic bases and environmental influences that shape its attributes would be misleading. The brain requires constant stimulation by the environment to develop properly. Without sensory input and intellectual challenges of a positive nature, full potential cannot be approached. As discussed in chapter 4, studies of infants demonstrate that the plasticity of the developing brain is remarkable, and that without both an adequate genetic base and proper environmental stimulation the brain's growth will be constrained. As Crews (2008) writes,

The early stages of life, beginning before birth and, in mammals up to weaning, are the time of maximal neuronal plasticity. Although the individual's capacity to respond to environmental change or insult with heritable phenotypic variation at a later stage is possible, it is during this early period that hormones and geno-type predispose an individual's responses to future experiences throughout the life cycle as well as the susceptibility to developing disorders. (344)

Although there is no question that cumulative experiences throughout life history interact with genetic predispositions to shape the individual's behavior, recent evidence suggests that events in past generations may also influence how an individual responds to events in his or her own life history. The new field of epigenetics studies how the environment can affect the genome of the individual during its development as well as the development of the individual's descendants, all without changing the DNA sequence. Although epigenetics is highly speculative, it demon-strates that, while the brain may be the great mediator, it too is dependent both on the genotype of the individual and on the environment within which the individual operates.

According to Hammock (2011), the brain is an experience-expectant organ that serves to maintain homeostatic balance through behavioral tasks that it has been selected to perform (the "four Fs": feeding, fleeing, fighting, and mating). To this end, mammalian brains have added layers and layers of control to these behaviors, correlating with larger brains that require longer postnatal development before reaching full maturity. This extra experience-expectant developmental time allows us to gain skills to manage and manipulate our environment, including the social environment. The significant variation in this ability across individuals, as well as across time within a given individual, comes from individual differences in brain structure and function, which in turn derive from genes, the environment, and their interaction throughout development and into maturity (Hammock 2011, 116).

Conclusions

The interface between genetic and neuroscience research is one of the most fascinating and dramatic areas of biomedical research. The possi-bilities for using the fruits of molecular biology to treat and possibly avert neurological disorders appear tremendous. In combination with an array of assisted reproductive techniques and pre-implantation and prenatal diagnostic procedures, there is considerable hope for improvement of the health of our progeny. However, these presumed benefits carry with them risks, not only to individual patients or subjects but also to future generations. Rigorous social research is necessary to analyze the value changes that accompany this rapidly expanding arsenal of technological interventions emerging from brain-gene research. Are the potentially threatening value shifts toward eugenics real or imagined? If there is substance to the fears expressed by critics, what policies can be implemented to minimize their materialization? What can be done to ensure that these technological advances are applied within a social environment that is not conducive to eugenic motivations? Furthermore, what policies, if any, are needed to protect patients and subjects involved in this research and guarantee their informed consent?

The next chapter turns attention to the impressive array of brain intervention techniques and drugs and the emerging area of neuroimaging technologies that have given us the capacity to "see" what is happening in the brain in real time. In a short few decades, the game has changed, and neuroscience has come into its own. As summarized here, in combination with behavioral genetics, it promises to change the way we view humanity. The remainder of this book attempts to map out this research and elucidate the social and political implications that accompany the new neuroscience.

2

Brain Intervention: State of the Art

This chapter describes the state of the art of a wide range of brain interventions and discusses the policy implications they raise. Most dramatic is direct brain intervention, including electroconvulsive therapy, electronic and magnetic stimulation of the brain, stereotaxic surgery, and neural implants that electrically stimulate, block, or record impulses from single neurons or groups of neurons. Similarly, drugs to modulate or enhance specific behavioral traits have proliferated and new generations of antipsychotic, anti-anxiety, and, especially, antidepressant drugs that inhibit serotonin reuptake are now among the most widely used pharmaceuticals. Moreover, nootropics for neuroenhancement are becoming more widely used and represent a growing market. This chapter also examines neurogenetics (research areas involve supplementing or blocking neurotransmitters, identifying genes that prevent normal brain development or produce progressive brain degeneration, replacing deleted or defective genes, and removing inoperable brain tumors using viral vectors) and virtual reality, a computer-mediated, multisensory technology designed to bring the mind into an alternative reality. Finally, this chapter discusses the broad array of noninvasive imaging technologies that allow researchers to examine brain structure and function in real time and coordinate these observations with behavioral tasks.

Historically, experimental and clinical interventions in the brain have elicited controversy from many directions. Illuminating are the issues surrounding early innovations of frontal lobotomies, electroshock therapy, and abuses of psychotropic drugs. Anxieties over brain intervention reflect the complexity of the object examined, the brain, and the ramifications of any evidentiary path proceeding from study of the brain. Findings in this field do not simply accumulate; rather, they interact, synergistically and convergently, bringing us ever nearer to the very seat of instinct, intellect, emotion, behavior, and responsibility. The potential

Table 2.1
Techniques of Brain Intervention

Direct brain techniques
Electroconvulsive therapy (ECT)
Transcranial magnetic stimulation (TMS)
Electronic brain stimulation (ESB)
Brain implants
Deep brain stimulation (DBS)
Vagus nerve stimulation (VNS)
Transcutaneous electrical nerve stimulation (TENS)
Transcranial direct current stimulation (tDCS)
Magnetic seizure therapy (MST)
Psychosurgery

Pharmaceuticals and biologics
Antipsychotic
Antidepressant
Anti-anxiety
Hormonal treatment
Nootropics (performance enhancers)

Virtual reality

Neuroimaging

Neural grafting

Neurogenetics

---

for progressively more reliable means to predict, modify, and control behavior raises intriguingly vexing questions along the political-philosophical line, as well as pressingly practical questions along bioethical, jurisprudential, and science policy lines (Blank 1999, 2006). To date, interest in the bioethics community, where an active neuroethics movement has emerged, has far exceeded interest in the policy-making and political science communities (see Giordano and Gordijn 2010; Racine 2010; Farah 2005, 2010; Rose 2010).

As illustrated in table 2.1, brain intervention techniques run the gamut from physical techniques that act directly on the brain to psychotropic drugs and biologics, to noninvasive virtual reality and neural imaging, to neural grafting and neurogenetics research. This chapter focuses on interventions that are already in use, but the discussion also includes emerging areas that demonstrate the swiftly growing arsenal of brain intervention technologies and the remarkable future possibilities.

Physical Intervention in the Brain

Direct electrical or physical interventions in the brain raise many ethical and policy issues and will continue to elicit strong opposition even when they are deemed safe and effective. The patient—any human— is a complex psychosomatic individual, and even if such intervention offers relief from a debilitating psychiatric condition, the deeper worry is that we are not curing a disorder but rather transforming a person into someone who is more acceptable to the rest of us (Gillett 2011). Because of the unique nature of the brain and its relationships to autonomy and personality, these interventions are viewed by many persons as assaults on personhood and autonomy. Direct brain intervention in particular also raises questions about the line between therapy and experimentation because of the uniqueness of each brain. Furthermore, questions of consent are especially troublesome where the intervention is potentially irreversible or the patient is deemed ill enough to undergo such a procedure. Finally, these techniques all raise the potential for abuse and the specter of behavioral control. These ethical aspects are discussed in detail in chapter 3.

Electroconvulsive Treatment (ECT)

Anyone who has seen *One Flew over the Cuckoo's Nest* would be hard-pressed to forget the dramatic ECT scenes where Jack Nicholson was shocked into submission. While actual use is not as dramatic, it does entail administration of a series of 70 to 130 volts of alternating current to the nervous system for one-tenth to one-half second, producing spasmodic muscular contractions. Although condemnation of ECT has not subsided in recent years, and unfavorable studies of ECT and other somatic treatments continue to portray it as a crude form of therapy at best, in 1990 the American Psychiatric Association (APA) termed the treatment safe and very effective for certain severe mental illnesses, including major depression, bipolar disorders, and schizophrenia, and ECT continues to enjoy FDA approval (APA 1990). According to Kragh (2010), ECT is experiencing a comeback because its negative historical portrayal has been accompanied by a series of books advocating its benefits in psychiatry, in part the result of major refinements in the procedure. Although patients in the past were awake, patients today are given a short-acting anesthetic and an injection of a strong muscle relaxant before the current is applied. Minutes after completion, the patient awakens, remembering nothing about the therapy and experiencing no pain, though most do experience short periods of disorientation. The usual ECT regimen entails four to twelve treatments given every two days over a period of several weeks. An estimated 100,000 persons undergo ECT each year in the United States (Mathew, Amiel, and Sackeim 2005).

According to Rasmussen (2011), ECT has been in use for many decades and has a well-established track record of being the most effective treatment for depression. Moreover, in their study of patients and their relatives, Grover and colleagues (2011) found that the actual experience of ECT has a positive impact on perceptions regarding the treatment among patients and their relatives. The relatives of ECT recipients had significantly more positive attitudes toward the treatment, whereas the relatives of the non-ECT group were often either ambivalent or critical of ECT (67). However, although numerous studies have found that subjective memory improves shortly after ECT, questioning about global impact resulted in more negative views about its cognitive effects, concordance with objective cognitive measures, and differences among treatment conditions (Brakemeier et al. 2011). According to Gupta, Hood, and Chaplin (2011), although continuation and maintenance ECT have been used for the prophylactic treatment of recurrent depression, they are poorly researched and not recommended by the National Institute of Health and Clinical Excellence in the UK.

As noted by Fink (2011), effective replacements for ECT have long been sought as alternatives for patients with depressive mood disorders, but, on the basis of his extensive review, he argues that ECT is still most effective in the relief of psychotic depression, catatonia, and delirious mania, and that no neurostimulation method challenges this dominance. He concludes that the question today is not to seek a replacement for ECT but to seek a better understanding of the mechanism of ECT and how seizures restore impaired moods and thoughts. With that knowledge, it might be possible to someday replace ECT, perhaps with an effective intervention with less optic dissonance and lesser side effects.

Although ECT normally relieves symptoms of depression rapidly, it is not without complications, and psychiatrists using it have a "serious obligation" to obtain the informed consent of the patient. Memory dysfunction can affect both the capacity to recall material learned before the procedure (retrograde) and the capacity to learn new material (anterograde). In most cases, anterograde memory gradually improves after treatment. Likewise retrograde memory capacity usually reaches full recovery within seven months after treatment, although in many cases subtle memory losses continue beyond that time. In very rare cases, such persisting losses are comprehensive and debilitating (Bootzin, Acocella, and Alloy 1993). Also, because many patients are fearful of ECT, it can and has been abused as a means of behavioral control when used as a threat to coerce cooperation.

At least thirty-seven states in the United States restrict the conditions under which ECT can be used, and according to Petersen (1994), regulations controlling ECT in many states are so stringent that many of the most direly ill patients now have little or no chance of receiving it despite its approval by the APA and overwhelming evidence of its safety and efficacy. Not only does suppressive regulation of ECT deprive patients of treatment, but it raises a number of policy questions regarding the health interests of the community, self-determination, competency, and privacy. For Petersen (1994), while the occurrence of involuntary treatment represents a tiny portion of ECT use, it has led to sensationalization and polarization and has "denied many patients a free choice under the guise of protective legislation" (30). Contrarily, critics of ECT argue that it represents an obvious form of psychiatric assault on mental patients, and they exploit strong negative images like that presented by Hollywood.

ECT devices come under the purview of the FDA's Neurological Devices Panel, where new devices are placed into one of three risk-based categories. In January 2011, the FDA's panel for ECT voted to place ECT machines in Class 3 for all but one indication (Goodman 2011). If the FDA accepts the panel's recommendation, the agency will require testing for all uses except "catatonia," which was recommended for Class 2, requiring less stringent testing. ECT devices were grandfathered into Class 3 (the highest risk-based classification for devices) in 1976 by the Medical Devices Amendments. Until now, ECT devices have been regulated through the premarket notification (510(k)) regulatory pathway, which requires a showing of substantial equivalence to a legally marketed device and is usually reserved for intermediate- and low-risk devices. In 2009 the Government Accountability Office recommended that the FDA require that all such grandfathered devices be either reclassified into Class 1 or Class 2 or undergo premarket approval. If the FDA rules in favor of the panel, premarket approval of ECT will require prospective controlled clinical trials, which will cost many millions of dollars. Alternatively, FDA officials could rule that the existing scientific evidence base for the efficacy and safety of ECT are sufficient for a determination of device classification, meaning that no new trials would be required (Kellner 2011). Only time, and politics, will tell.

Transcranial Magnetic Stimulation (TMS)

TMS is one of many neuromodulation techniques that work by changing the neural plasticity of the brain through either magnetic or electrical stimulation. In TMS, a magnetic coil running at thousands of volts is positioned outside the head, producing electrical surges inside the brain that are measured with external electrodes (Fox 2011). Recently, repetitive transcranial magnetic stimulation (rTMS) joined ECT as an FDA-approved clinical treatment for depression and other neuropsychiatric and neurological disorders. Furthermore, because it is noninvasive, rTMS can also be used in healthy volunteers as a probe to study neural physiology, connectivity, and transmission and advance our understanding of normal brain function (Rasmussen 2011). As a therapeutic intervention, rTMS is used to modulate neural functioning in the treatment of neuropsychiatric disease states such as major depressive disorder (Husain and Lisanby 2011). Moreover, rTMS has been found to have the potential to decrease negative symptoms of schizophrenia, to improve social functioning in patients with autism spectrum disorder (ASD), and to be beneficial for the treatment of refractory neuropathic pain and depression following traumatic brain injury. Finally, Endicott and colleagues (2011) suggest that deep rTMS may serve to remediate aspects of cortical dysfunction and provide a potential new avenue for the development of a treatment for impaired social relating and interpersonal understanding in ASD.

Although two large-scale clinical controlled trials found rTMS applied over the dorsolateral prefrontal cortex to have significant antidepressant effect with modest response and remission rates, Husain and Lisanby (2011) contend that rTMS is not a panacea for depression or a replacement for other forms of interventions, particularly ECT. For the authors, many questions remain, including what type of rTMS to use (e.g., single pulse, paired pulse, or repetitive pulses), how often to use it, and for what neuropsychiatric diseases (e.g., depression, schizophrenia, or anxiety disorders). Fink (2011) concludes that rTMS it is of very limited benefit and not a suitable replacement for ECT because, while the powerful magnetic field generated by the current lasts only 100 to 200 microseconds, the induced currents can cause observable and nonobservable behavior changes. One study found that, although rare, seizure can lead to brain damage and death (Heckman and Happel 2006). Although in clinical trials, one must take into consideration the probability and the magnitude of such adverse events, the use of rTMS in nonmedical applications raises concern. Although it is not physically invasive in the way that a surgical procedure is, the radiation it produces has an effect on the body similar to that of ECT.

Transcranial direct current stimulation (tDCS) is a related noninvasive neuromodulating technique that relies on electrical as opposed to magnetic stimulation. Since the 1940s, tDCS has been used to treat a myriad of medical conditions, including depression, seizures, obesity, anorexia, stroke, and aphasia. Moreover, tDCS is popular because of its versatility and portability. It is delivered by a small battery-powered device approved by the FDA in 2004 for iontophoresis (a technique in which a small electric charge is used to deliver a medicine or other chemical through the skin—basically, an injection without the needle), which delivers a very weak electric current to the surface of the brain. It works by sending constant, low direct current through electrodes placed in the region of interest, thereby inducing intracerebral current flow. This current flow then either increases or decreases the neuronal excitability in the specific area being stimulated, thus leading to alteration of brain function. Recent studies in stroke rehabilitation strategies have shown that tDCS improves one's ability to learn a simple coordination exercise, with improvement still apparent three months after the experiment ended. The method has also received attention for its potential to enhance the minds of healthy people by improving working memory, word association, and complex problem solving (Fox 2011).

Luber and colleagues (2009) concluded that tools for noninvasive stimulation of the brain, such as TMS and tDCS, have led to new insights into brain-behavior relationships through their ability to directly alter cortical activity. Such techniques also are useful tools for establishing causal relationships between behavioral and brain imaging measures. As a result, there has been interest in whether they might represent novel technologies for deception detection by altering a person's ability to engage brain networks involved in conscious deceit. The investigation of deceptive behavior using noninvasive brain stimulation, however, is at an early stage and remains to be validated.

Deep Brain Stimulation (DBS)

Although extracranial brain stimulation by magnetic fields or electric current appears to provide a relatively safe way to treat depression and possibly boost creativity, elevate mood, and strengthen cognitive skills, the use of intracranial electrodes is considerably more risky (Häyry 2010). However, while medication, psychotherapy, and other treatments are effective for many patients with psychiatric disorders, a substantial number of patients experience incomplete resolution of symptoms, and relapse rates are high. In the search for better treatments, interest has focused on focal neuromodulation, especially DBS, which is one of the most invasive of available techniques (Holtzheimer 2011; for a history, see Hariz, Blomstedt, and Zrinzo 2010). The procedure involves the implantation of a stimulation system composed of a lead and extension wire and a neurostimulator that is similar in technology to a cardiac pacemaker. The neurostimulator delivers continuous electrical impulses to the tiny electrodes at the tip of the lead wire. Wires are implanted bilaterally if the patient experiences bilateral symptoms. With the exception of the intracranial portion of the lead wire, all other hardware components are implanted subcutaneously (Farris and Giroux 2011).

Although DBS does not induce a permanent beneficial change, it has been shown to produce emotional tranquility and evoke a feeling of euphoria that relieves the psychological pain caused by severe anxiety and depression. Although the underlying principles and neural mechanisms of DBS are not yet fully understood, research has shown that DBS directly changes brain activity in a controlled manner (Kringelbach, Green, and Aziz 2011). Also, while evidence to date suggests that DBS causes no permanent anatomical damage to the brain and that its effects are reversible, its yet experimental nature warrants caution in therapeutic applications. According to Benabid (2007), technological development will further enhance and refine the effects of high-frequency stimulation and allow the extension of this method to new targets and new indications. Similarly, Clausen (2011) concluded that increased computational power and ongoing miniaturization of microtechnological components will enable the development of advanced diagnostic and therapeutic tools.

To date, more than 50,000 patients have received DBS implants in the United States. Farris and Giroux (2011) warn that these numbers are growing much faster than the number of experienced or qualified DBS clinicians and that patients who are not evaluated properly before implantation may experience unexpected or poor outcomes. DBS is an FDA-approved surgical treatment option for select patients with tremor, Parkinson's disease, dystonia, and obsessive-compulsive disorder (Aguilar et al. 2011). It also has been used for the treatment of various psychiatric disorders such as depression, Gilles de la Tourette syndrome, alcoholism, and Alzheimer's disease (Skuban et al. 2011; Christen and Müller 2011, Sen et al. 2010), as well as epilepsy, motor disorders (Bell, Mathieu, and Racine 2009), chronic pain, violent aggressiveness, and chronic insomnia (Abbott 2009c). The only current FDA-approved DBS device is manu-factured by Medtronic (Minneapolis, MN). Four models are available that vary in both size and function (Farris and Giroux 2011). However, with a better understanding of the linkage between specific nuclei and areas of the brain and debilitating disorders, electrical stimulation systems are likely to proliferate.

Depressed patients resistant to other therapies can be helped long term by DBS, with effects still apparent after six years, indicating that it does actually seem to modify the disease, something that no other treatment has accomplished (Abbott 2011). According to Schermer (2011), opera-tion techniques such as MRI-guided stereotactic surgery have improved, thus making the DBS intervention much safer. Moreover, the effects of DBS are typically reversible and the stimulation can be terminated if it proves ineffective or causes too many adverse effects. While DBS should be performed only on otherwise treatment-resistant patients and only with informed consent (Synofzik and Schlaepfer 2008), it still raises important ethical issues. Synofzik and Schlaepfer (2011) contend that the development of ethical criteria is especially crucial in light of the surge of clinical and commercial interest in DBS for psychiatric indi-cations, the high vulnerability of psychiatric patients, and the lack of extensive short- and long-term data about effectiveness.

Schmetz and Heinemann (2010) and Takagi (2009) are concerned particularly about the possible effects of DBS on the patient's personality and self-determination since it has been shown that personality changes may occur either as unintended side effects or as an intended therapeutic goal. Johansson and colleagues (2011) and Kraemer (2011), for instance, raise the issues of authenticity and alienation that might follow brain interventions when patients no longer feel like themselves (this issue is discussed in chapter 3). Glannon (2009), too, is concerned that while DBS can modulate overactive or underactive regions of the brain and, thereby, improve motor function, it can also cause changes in a patient's thoughts and personality, and he discusses the trade-offs between the physiological benefit of this technique and the potential psychological harm. Kuhn and colleagues (2009) call for transparent and well-defined regulations for the protection of vulnerable psychiatric patients, as well as appropriate support for increased therapeutic research.

Schermer (2011) argues that the most important issues with regard to DBS treatment are balancing risks and benefits and ensuring respect for the autonomous wishes of the patient. One especially controversial issue surrounding DBS is its use in children. Considering the investigational nature of DBS for psychiatric disorders, Focquaert (2011) finds it "unsettling, to say the least" that it is performed to treat dystonia in children as young as seven, even though timely intervention in childhood dystonia is important to prevent irreversible damage, obtain optimal treatment outcomes, and prevent long-term social costs due to social isolation (Isaias et al. 2008; Mehrkens et al. 2009; Clausen 2011). In contrast, Lipsman, Ellis, and Lozano (2010) conclude that DBS in children should be considered a valid and effective treatment option, though only in highly specific and carefully selected cases, and Marks et al. (2009) contend that while pediatric use of DBS has been limited, with only a few experienced centers worldwide, experience will help refine the indications and techniques for applying this invasive technology to children.

Recently, on the basis of limited animal models, isolated case studies, and evidence from some uncontrolled studies of the neurosurgical treatment of opioid addiction, DBS has been advocated for trial as a treatment of addiction. According to Carter and Hall (2011), these proposals are premature and raise major ethical issues. Unlike Parkinson's disease, addiction is not a disorder that involves an inexorable path to severe disability and death. Furthermore, it is more amenable to pharmacological and psychotherapeutic treatment. Major impediments to effective addiction treatment include lack of access to effective treatment; stigmatization, which discourages addicted individuals from seeking treatment; and punitively operated programs, which can undermine effectiveness. The addition of an expensive neurosurgical treatment costing approximately U.S. $50,000 (with maintenance costs of approximately U.S. $10,000 every few years) will worsen this situation by concentrating scarce health resources on a small number of well-to-do patients (Carter et al. 2010). Even if DBS proves to be safe and effective, Carter and Hall (2011) believe that the evidence base to support trials of DBS in addiction is weak and the opportunity costs make such trials a low priority for public funding. Despite this, addicted persons who fail to achieve abstinence may be desperate enough to undergo such an invasive treatment since history shows that the desperation for a "cure" of addiction can lead to the use of risky medical procedures before they have been rigorously tested (Carter et al. 2010).

As noted by Glannon (2010), while DBS can control and relieve many symptoms associated with a range of neurological and psychiatric disorders, the invasive procedures of implanting and stimulating the electrodes entail significant risks, including intracerebral hemorrhage, infection, cognitive disturbances such as impulsive behavior, and affective disturbances such as mania. Moreover, it is unknown whether continuous electrical stimulation of the brain might reshape synaptic connectivity and permanently alter neural circuits in deleterious ways. As Glannon (2010) notes, "The risk of these effects indicates that DBS should be used only when a patient's condition is refractory to all other interventions and when there is a high probability that the technique will benefit the patient and improve his or her quality of life. If a patient's quality of life is poor and all other treatment options have been exhausted, then the likelihood of benefit can justify physicians' exposing patients to some risk" (104). Moreover, patients with Alzheimer's disease in particular may be limited in certain cognitive dimensions that could endanger their ability to completely understand all the implications connected with DBS (Skuban et al. 2011).

Psychosurgery

Neurosurgery is a highly credible and accepted therapeutic procedure employed to remove intracranial tumors and treat brain trauma, artery diseases, severe pain, spasms, and epilepsy. Although any physical interventions in the brain have their risks, advances in imaging and microsurgery have made neurosurgical operations safer (Häyry 2010). However, when the same techniques are used to treat mental or behavioral disorders (psychosurgery), they are exceedingly controversial since they permanently destroy particular regions of the brain. Psychosurgery rests on the assumption that behavioral disorders have an organic base and that to treat the disorder the organic pathology must be corrected by physical intervention. The primary ethical justification for psychosurgery often is that it represents a last-chance option to help the patient. It is argued that, despite insufficient evidence of efficacy and safety of a procedure, in those situations where the professionals and those properly consulted favor the attempt to treat, knowing that it involves invasive procedures on the brain, it is better than therapeutic nihilism (Gillett 2011). This assumption is highly questionable for critics of psychosurgery, and the history of psychosurgery is not one to inspire confidence.

A case in point is the frontal lobotomy (frontal leucotomy), which entailed detaching the frontal lobes from the rest of the brain. Proposed as an operation of last resort, the lobotomy quickly became the first step toward creating a manageable personality. Even problem children were lobotomized (Gillett 2011). Although several techniques were used to lobotomize approximately 70,000 persons in the late 1940s and early 1950s, Walter Freeman achieved critical acclaim for his use of transorbital lobotomy. This technique is casually summarized by Freeman in its gruesome details:

I have also been trying out a sort of half-way stage between electroshock and prefrontal lobotomy on some of the patients. This consists of knocking them out with a shock and while they are under the "anesthetic" thrusting an ice pick up between the eyeball and eyelid through the roof of the orbit actually into the frontal lobe of the brain and making the lateral cut by swinging the thing from side to side. I have done with patients on both sides and another on one side without running into any complications, except a very black eye in one case. There may be trouble later on but it seemed fairly easy, although definitely a disagreeable thing to watch. (Cited in Valenstein 1986, 203)

Interestingly, the pejorative term "ice pick surgery" is accurate. Although Freeman eventually developed the "transorbital leucotome" and then the "orbitoclast" as the instruments designed to withstand the tremendous pressures necessary to crush through the orbit before severing the brain, the first transorbital lobotomy was performed in his office using an ice pick with the name Uline Ice Co. on its handle. Although by 1960 the practice of lobotomy was curtailed, primarily because of the availability of psychoactive drugs as an inexpensive alternative, in its wake it left many seriously brain-damaged persons and tainted the idea of psychosurgery to the present day (Gillett 2011).

Although a large proportion of lobotomies were performed in state hospitals on indigent patients, a substantial number were performed on well-to-do women in private hospitals and physicians' offices. The transorbital procedure left only small scars on the eyelids and could be performed in a matter of minutes. To some extent, frontal lobotomies took on a fadlike atmosphere. The fact that such a repugnant and intrusive technique won not only widespread acceptance from the medical profession as a whole and a Nobel Prize for its inventor, Egas Moniz, but also broad public and mass media approval as a miracle cure underscores an unbridled faith in a technological fix. Despite early evidence of postoperative infections and autopsies showing that large areas of brains were utterly destroyed, as well as the emotionless, inhuman quality of many lobotomized patients, thousands of individuals still willingly underwent or submitted family members for the procedure (Häyry 2010).

The demise of lobotomies did not end psychosurgery, but today, sophisticated stereotaxic instruments facilitate more precise placement of miniature electrodes to specific brain targets using geometric coordinates under imaging, thus allowing for the destruction of relatively small areas of brain tissue. Electrolytic lesioning or selective cutting of nerve fibers is conducted after the target region is localized by establishing coordinates using anatomical landmarks and x-rays. In one procedure, called stereotaxic subcaudate tractotomy, a small area of the brain is destroyed by radioactive particles inserted through small ceramic rods. The particular site varies with the nature of the disturbance. "For depressed patients, it is in the frontal lobe; for aggressive patients, it is the amygdala" (Bootzin, Acocella, and Alloy 1993, 565). There have also been some applications using radiation, cryoprobes (freezing), or focused ultrasonic beams to destroy the tissue. Voytek and colleagues (2010) discuss the use of a decompressive hemicraniectomy, a surgical procedure in which a portion of the calvaria is removed for several months, during which time the scalp overlies the brain without intervening bone.

Fountas and Smith (2007) estimate that only about a thousand amygdaloidotomies and hypothalamotomies have been performed worldwide. They conclude that "the questionable long-term benefit of these procedures, the dubious accuracy of the target localization . . . and advances in psychopharmacology" have led them to become "procedures of merely historical interest" (710). During the 1960s and 1970s there were positive reports on the use of psychosurgery for the treatment of addiction. Nearly all of these studies reported reasonable success in preventing subsequent drug use with little or no cognitive impairment and were seen by some authors as warranting neurosurgical lesioning as a "first-line treatment" for addiction (Carter, Racine, and Hall 2010). The neurosurgical treatment of addiction, however, fell out of favor during the 1970s when it became apparent that such procedures for other psychiatric disorders were not as effective as first thought and risked significant long-term cognitive deficits. In 1976 the National Commission for the Protection of Human Subjects criticized the use of psychosurgery for social or institutional control, although it acknowledged that, as then practiced, it might help some patients. Although it recommended prior screening of each proposed surgery by an independent institutional review board, its recommendations were never translated into federal legislation or regulations. Instead, action restricting psychosurgery has originated in state legislatures. Oregon, for instance, in 1982 passed a statute that strengthened its earlier restrictions and prohibited psychosurgery in the state. California also instituted regulations that all but ended the practice.

Ironically, concurrent with the imposition of legal restraints, extensions of earlier studies reported to the National Commission were favorable to the use of psychosurgery for certain classes of patients. Two independent research teams (Teuber et al. 1977; Mirsky and Orzack 1977) reported that the quality of life of respectively 70 percent and 80 percent of patients improved significantly after undergoing psychosurgery. Furthermore, they found no evidence of physical, emotional, or intellectual impairment caused by surgery. Likewise, Abbott (2009c) contends that new surgical procedures for epilepsy offer a particularly tantalizing opportunity for neuroscientists because they allow access to the cortex and the opportunity for very precise recording. Some insidious forms of epilepsy that do not respond to drug treatment originate in the medial temporal cortex and can be cured by surgical removal of this region. Neurosurgeons can identify a precise focus by implanting multiple electrodes around the medial temporal cortex, wait for a spontaneous seizure, and then determine the origin of the epileptic activity (Abbott 2009c). Careful patient selection and the use of surgery as a last resort only for those patients whose conditions fail to respond to other therapies seem central to any future uses of psychosurgery (Skuban et al. 2011).

Brain Implants, Brain-Machine Interfaces, and Brain-Gene Transfer

A rapidly expanding and provocative area of research involves brain (or neural) implants that attach directly to the surface of the brain or in the neocortex. The major impetus for brain implants has come from research designed to circumvent those areas of the brain that became dysfunctional after a stroke or other head injuries and implant pacemaker-like devices to treat mood disorders (Reuters 2005b). Brain implants electrically stimulate, block, or record impulses from single neurons or groups of neurons where the functional associations of these neurons are suspected. Advanced research in brain implants involves the creation of interfaces between neural systems and computer chips. Implants involving DBS and VNS are becoming routine for patients with Parkinson's disease and clinical depression. On the horizon are more exotic optogenetic implants composed of electrodes or fiber optic wires on the brain's surface that beam light pulses either to control brain cells or to reroute brain activity. Moreover, by introducing a few genes into specific neuron clusters, the cells can be made sensitive to certain wavelengths that can make them fire or shut them off (Dillow 2010). According to Meloni, Mallet, and Faucon Biguet (2010), these brain implants, using mechanical devices, represent deliberate approaches to abolish the brain-machine interface.

Brain-computer interfaces (BCIs) are characterized by direct interaction between a brain and a technical system (Hildt 2010). BCIs aim to transmit information from the outside world directly into our nervous system or from our nervous system directly to mechanical aids (Konrad and Shanks 2010). Examples of the former include auditory brain stem implants such as the cochlear implant, which transforms sound waves into electrical impulses and feeds them into auditory nerves, thereby allowing some deaf or near-deaf people to sense sound (Colletti et al. 2009); visual BCI implants, such as the visual cortex surface implant; and visual prostheses based on implants located in the lateral geniculate body of the thalamus (Pezaris and Eskandar 2009). In the latter type of BCIs, brain signals are extracted and used for communication and control of movement. This "reverse" communication is required when prosthetic limbs or robotic appliances need to be guided by electrical impulses of the brain. It can also offer support to patients with motor impairments resulting from amyotrophic lateral sclerosis, stroke, spinal cord injury, and cerebral palsy (Hildt 2010). However, according to Häyry (2010), human-machine interfaces do not always work, and they raise fears of the transformation of humans into cyborgs.

Another still experimental, but highly promising, approach for treating neurodegenerative and psychiatric disorders and, potentially, modification and/or enhancement of behavioral traits is brain-gene transfer (BGT). Basically, BGT consists of the introduction of nucleic acids, either ex vivo in cells that have been removed from their source and then genetically modified and transplanted into the patient or in vivo by direct importation of the genetic material into the targeted organ or structure (Meloni, Mallet, and Faucon Biguet 2010). In both cases, the utilization of viral vectors for carrying the genetic material into the cells permits a more efficient transfer and, according to the type of viral vector used, different modalities and duration of expression for the genetic material (Heilbronn and Weger 2010). The transfer of genetic material may restore the function of a mutated gene or may express a gene by coding a neuroprotective factor that counteracts the neurodegenerative process causing the disease. In some instances, brain implants or BGT may constitute a valid alternative to drug treatments by improving the specificity of the intervention or by overcoming the inefficacy of drug action. In fact, both significantly extend the possibility of therapeutic interventions beyond the realm of pharmacological approaches because they can be targeted to a specific brain region. In this way, their effects are restricted to the target, be it a cerebral structure or a specific gene (Meloni, Mallet, and Faucon Biguet 2010).

Drugs and the Brain

Although less dramatic than physical interventions in the brain, the development of powerful psychotropic drugs cumulatively represents a much larger impact on the population. Psychopharmacology, or the study of the drug treatment of psychological disorders, has exhibited remarkable advances in the last two decades and is the backbone of neuroscience research. New drugs are introduced each year, often with great enthusiasm and too much fanfare, for the treatment of a range of mental disorders. Moreover, new generations of drugs for enhancement of brain capabilities and treatment of neurodegenerative diseases promise to expand significantly medical intervention into the nervous system. Although some of these drugs fail to have the effects hoped for or are accompanied by undesirable side effects, overall the research in psychopharmacology has contributed immensely to our understanding of the chemistry of the nervous system and substantially expanded the range of treatment possibilities.

Despite this record of progress, psychotropic drug treatment mirrors many of the same issues raised by physical intervention. Other than the irreversible dimension of psychosurgery, the use of drugs to produce the desired mood and mental functioning makes similar assumptions about the organic base of behavioral problems. Drug treatment for behavioral and psychological problems also raises important issues as to whether it is wise to treat symptoms without addressing the underlying cause and under what circumstances individuals with severe mental disorders can be coerced into taking drugs that stabilize them but do not treat the cause (see chapter 5). In fact, chemical control has considerably more policy relevance because it enjoys such widespread and socially acceptable usage. The ease of administration and the potential for surreptitious applications make this form of control even more challenging than physical methods.

Antipsychotic Drugs

There are three major groups of psychotropic drugs used as therapeutic agents. The first and most powerful are the antipsychotic drugs or major

Table 2.2
Selected Antipsychotic Drugs

Trade Name	Generic Name
Abilify	Aripiprazole
Clozaril	Clozapine
Fanapt	Iiloperidone
Haldol	Haloperidol
Loxitane	Loxapine
Moban	Molindone
Navane	Thiothixene
Risperdal	Risperidone
Seroquel	Quetiapine
Stelazine	Trifluoperazine
Thorazine	Chlorpromazine

tranquilizers which are used to relieve symptoms of psychosis, including withdrawal, delusions, hallucinations, and confusion (table 2.2). With FDA approval and marketing of chlorpromazine (Thorazine) in 1954, the first of a group of powerful phenothiazenes was introduced for the treatment of major mental illnesses such as schizophrenia and paranoia. These drugs have sedative, hypnotic, and mood-elevating effects. Although they do nothing to cure, but rather suppress the symptoms of the disease, they are effective in maintaining equilibrium for many patients, often allowing them to be released from institutions and function quite well. Maintenance therapy generally requires prolonged use at reduced dose levels, and discontinuance of the drug results in reoccurrence of symptoms. Administration, however, is complex because effects and dosage levels vary across individuals and even for an individual patient.

Not surprisingly, these powerful antipsychotic drugs have serious potential side effects, including fatigue, apathy, blurred vision, constipation, muscle rigidity, and tremors. The most serious medical side effect of these drugs is tardive dyskinesia, a muscle disorder in which patients grimace and smack their lips incessantly. Unlike other side effects, this condition does not abate when use of the drug is withdrawn, and, like other side effects, it has proved resistant to treatment with additional drugs. Two other serious issues accompany the use of antipsychotic drugs. Because of their tranquilizing effect, they have been used in some institutions in high doses for the management of disruptive patients. This use for behavioral control purposes should not detract from the benefits of antipsychotic medication, but it warrants close monitoring of its use. The other issue, which is discussed later in more detail, is that patients released from institutional care into the community under the calming influence of the drug often stop taking the drug and must be readmitted or they create problems in the community and their disruptive behavior results in calls for coerced treatment or reinstitutionalization.

Antidepressant Drugs
The second major grouping of psychotropic drugs encompasses a broad range of antidepressant drugs that are used to elevate mood in depressed patients. This class of drugs includes amphetamines, monoamine oxidase inhibitors (MOAIs), tricyclic derivatives of imipramine (TCAs), and serotonin reuptake inhibitors (table 2.3). Although amphetamines have little clinical value in the treatment of depression because of their very short duration, they are widely known and used. They also have considerable potential for abuse and dependency.

Table 2.3
Selected Antidepressant Drugs

Trade Name	Generic Name
Anafranil (tricyclic)	Clomipramine
Aventyl (tricyclic)	Nortriptyline
Celexa (SSRI)	Citalopram
Cymbalta (SNRI)	Duloxetine
Elavil (tricyclic)	Amitriptyline
Lexapro (SSRI)	Escitalopram
Ludiomil (tricyclic)	Maprotiline
Luvox (SSRI)	Fluvoxamine
Marplan (MAOI)	Isocarboxazid
Nardil (MAOI)	Phenelzine
Norpramin (tricyclic)	Desipramine
Parnate (MAOI)	Tranylcypromine
Paxil (SSRI)	Paroxetine
Prozac (SSRI)	Fluoxetine
Sinequan (tricyclic)	Doxepin
Surmontil (tricyclic)	Trimipramine
Tofranil (tricyclic)	Imipramine
Vivactil (tricyclic)	Protriptyline
Wellbutrin	Bupropion
Zoloft (SSRI)	Sertraline

Until the introduction to the market of Prozac in 1987, the most commonly prescribed antidepressants were the TCAs, which produce an increase in the neurotransmitters norepinephrine and serotonin in the nervous system by blocking their reabsorption by the nerve cells. Often patients who fail to respond to tricyclic antidepressants are prescribed MAOIs, but these agents can have serious side effects, including cardio-vascular and liver damage, and, in combination with certain foods and medications, they produce fatal bouts of hypertension. Both inhibitors and tricyclics have side effects similar to those of antipsychotic drugs, including drowsiness, blurred vision, nervousness, and constipation, but normally these effects are milder with antidepressants because of the lower dosages used. Another problem with these drugs is the lag time between their administration and their clinical effect. Although they increase neurotransmitter levels almost immediately, their therapeutic effects often do not appear for two or more weeks, a long time for a severely depressed patient.

In the late 1980s, a new generation of antidepressants termed selective serotonin reuptake inhibitors (SSRIs) was marketed with the promise of offering safe and effective treatment for a wide variety of disorders. In light of increasing public awareness of depression and an aggressive marketing strategy by Eli Lilly and Company, by 1990 Prozac was the most widely prescribed antidepressant in the United States. Because it was designed specifically to block serotonin uptake, it had antidepressant action equivalent to that of the older drugs but without many of their more serious side effects. Because it keys in only on serotonin, Prozac is effective for many depressed as well as obsessive-compulsive patients whose conditions do not respond to to MAOIs or TCAs.

In response to reports of antidepressant-related suicidality, in 2010, however, a group of experts met to discuss research and lay media reports about the safety and efficacy of antidepressants for treating mild to moderate depression (Nierenberg et al. 2011). The panel concluded that the data regarding the efficacy of antidepressants are complex, making interpretation of meta-analysis results difficult. Moreover, although the issue of suicidality is a genuine concern, the risk is not great enough to abandon the use of antidepressants. The panel agreed that patients who have mild or moderate depression may benefit from receiving evidence-based psychotherapy first and stressed that additional research and novel treatments are needed to improve outcomes for patients with depression. Despite these concerns, the panel found that these antidepressants are effective tools for helping numerous patients with depression achieve remission and recovery.

Related to antidepressants are mood stabilizers, which are primarily used to treat bipolar disorder by suppressing the swings between mania and depression. Mood-stabilizing drugs are also used for borderline personality disorder and schizoaffective disorder. They are often categorized as anticonvulsants or anticonvulsant mood stabilizers.

Anti-anxiety Drugs

The third major grouping of psychoactive drugs comprises the anti-anxiety drugs or minor tranquilizers (see table 2.4). One grouping is barbiturates such as phenobarbital, which, although being the least effective, having a high tendency to produce dependence and addiction, and having a low margin of safety, are often used to treat epilepsy. Another group of minor tranquilizers includes diazepoxides such as Librium and Valium. These drugs are used effectively to control muscle spasms, hysteria in acute grief situations, and compulsion and are less dangerous than barbiturates and less addictive. Minor tranquilizers are not effective in the treatment of psychoses but are of special value in the treatment of tension and anxiety associated with situational states and stress.

The long-term use of anti-anxiety drugs is discouraged because they can produce dependence and have a tendency to be abused. Moreover, anti-anxiety drugs have side effects, including fatigue, drowsiness, and impaired motor coordination. They have been linked to automobile and industrial accidents, and to falls in the elderly. Anti-anxiety drugs also can be dangerous if combined with nervous system depressants such as alcohol. Barbiturate-alcohol combinations have led to brain death in well-publicized end-of-life cases such as that of Karen Quinlan. Anti-

Table 2.4
Selected Anti-anxiety Drugs

Trade Name	Generic Name
Ativan	Lorazepam
BuSpar	Buspirone
Klonopin	Clonazepam
Librium	Chlordiazepoxide
Tranxene	Clorazepate
Valium	Diazepam
Xanax	Alprazolam

anxiety drug use also risks the effect of rebound once treatment is terminated. Rebound means that the symptoms return with redoubled force; thus the person is likely to resume taking the drug at a higher dose in order to suppress the intensified symptoms. Anti-anxiety drugs have also been criticized because they suppress anxiety, which is a sign of a more basic problem. In effect, the drug allows the user to avoid facing the problem, thereby risking it becoming even more serious, leading to the "need" for more drugs to relieve the anxiety, and so forth.

Hormonal and Other Drugs and the Brain

In addition to the psychotropic drugs, the use of hormonal treatment to control behavior has escalated since the 1970s. For instance, studies have repeatedly found that aggression is highly dependent on testosterone levels. Elevated levels of this male hormone have been implicated in crimes of sex-related violence. The administration of female hormones to control aggressive males has been successful in treating abnormal sexual preoccupation. Depo-Provera has also been used to inhibit the male sex drive. In large doses, Depo-Provera serves, in effect, as a chemical castration by shrinking the testes. As of 2009, seven states use some form of chemical castration, with weekly injections of Depo-Provera the most common method (Nicholson 2011). In a related area, another study found that estrogen may relieve symptoms of Alzheimer's disease in postmenopausal women (Hotz 1997), thus extending potential hormonal treatment directly into the brain.

Moreover, innovations in chemical intervention for the brain have moved in new direction. At the 1996 meeting of the Society for Neuroscience it was reported that the drug ampakine CX-156 restored memory in elderly men to that of men in their twenties and would soon be tested in Alzheimer's patients. Similarly, the development of new clot-busting drugs such as tissue-type plasminogen activator (tPA) promises significant benefits for many of the half million stroke victims who die or suffer permanent damage each year. If administered within three hours of ischemic stroke, tPA clears vessels and quickly restores blood flow to the brain. If initiated after that point, however, tPA can actually make the stroke worse (National Institute of Neurological Disorders and Stroke 2005). Although tPA cannot be used in hemorrhagic strokes because it aggravates brain bleeding, in combination with brain imaging technology, this and similar drugs promise considerable benefits to many persons. In addition to a new generation of pharmaceuticals, more direct delivery methods are being tested. In 1997 a catheter was implanted in the brain of a patient with amyotrophic lateral sclerosis (Lou Gehrig's disease) to deliver glial-derived neurotropic factor (GDNF) by circumventing the blood-brain barrier.

Policy Issues in Psychotropic Drug Use

Drugs are used to alter behavior because they are effective and convenient, not because of a compelling scientific consensus as to how they help patients. In other words, they are used for nonmedical reasons to accomplish other policy objectives. In a rights-oriented society, this practice raises serious constitutional questions. However, in our drug-oriented society, psychoactive drugs are embraced to help one cope with day-to-day problems. Drugs have become a quick fix for anxiety, depression, and the social stresses of modern-day existence. Moreover, neuroenhancing drugs are expected to represent a huge market in the future (see chapter 7 for a detailed discussion of neuroenhancement).

Like any other novel drugs, psychotropic drugs raise questions of safety, especially if they are used over long periods of time (Häyry 2010). One drug that has generated considerable controversy is Ritalin, a stimulant that has been found to have the opposite effect on children with attention deficit hyperactivity disorder (ADHD). Because of its ability to calm hyperactive children, Fukuyama (2002) has argued that Ritalin is being used as a "medical shortcut" by parents and teachers who want to make lives easier for themselves. Such use has come under attack in part because the effect of prolonged exposure of Ritalin on developing brains is unknown, although it has been linked to complex changes in the CNS. Questions of its effects on personality development, innovative thinking capacity, and any psychological or physiological dependence are unanswered (El-Zein et al. 2005; Herman-Stahl et al. 2006). White (2005) suggests that the use of medication to treat ADHD in school-aged children needs to "be scaled back immediately. The current status quo which reflects broad-based use is serious, hazardous, sexist (discriminatory against boys), and far from benign in terms of its social/psychological side effects" (58; for a different view, see Hughes 2005).

Moreover, there are inherent issues whenever features that have previously been considered natural become the target of medication. Certain individuals have always been shy, absentminded, overactive, forgetful, serious, gloomy, or less social. To redefine these conditions as illnesses and prescribe drugs makes them undesirable in new ways, thus affecting people's perceptions of themselves and others. Also, while some who want to be freed from these features will undergo drug treatment, others might not see chemicals as a solution or might discontinue use to avoid potential side effects. Still others might view the qualities they have as a part of their character and feel that they are under societal pressure to alter their personalities (see discussion over authenticity in chapter 3). However, people who refuse prescribed drugs to treat their "problem" are often seen as responsible for, or even guilty of, straying from the norms accepted by a community that has strong faith in a psychopharmacological fix (Häyry 2010.

Virtual Reality and the Brain

The coming decade is certain to see astonishing advances in fabrication of worlds of illusion through virtual reality (VR) technologies. Already the marketing of rather crude VR equipment is under way as part of the ongoing computerization of social interactions. Basically, VR is a computer-mediated, multisensory experience designed to fool our senses into putting us into an alternative reality. Although movies, computer games, and even radio have long provided crude "virtual" realities directed toward one or several senses, VR combines the elements of other media forms and promises to extend this synthetic environment to all the senses so that completely artificial worlds can be created that make users feel that the artificial world they appear to inhabit is real. Moreover, unlike movies and computer games, in which the viewer or player must relate to a viewpoint or character who participates in the action, "In immersive virtual reality, the senses are provided with naturalistic data in a form which requires little learning or interpretation. The experience provided to the user appears far more immediate than other forms of media and consequently may have more impact" (Foster and Meech 1995, 210).

VR research itself might be invaluable in providing models for clarifying perceptual and sensorimotor systems in the human brain (England 1995). Most of the recent initiatives for the development of VR have focused on recreational uses, including virtual sex, but medical, industrial, and, of course, military applications are extensive (see chapter 7). Other applications of VR have been to provide simulations of architectural experience to help understand neural activation during automobile driving (Spiers and Maguire 2007). These preliminary VR experiments have shown that the hippocampus is active when the subject makes navigation decisions but not when they are externally cued. Ariely and Berns (2010) suggest that taking into account hippocampal load may be a very useful tool in architectural design, for instance, in making buildings easier to navigate. Extending this idea by considering the neurobiological changes associated with aging, it might be possible to design buildings and retirement communities that mitigate the memory loss associated with Alzheimer's disease.

Notwithstanding the considerable enthusiasm over the prospects for VR, little attention has been given to the possible dangers inherent in placing individuals into virtual realities. Although proponents argue that the person voluntarily enters VR and retains control, in effect, VR gives the computer complete control over input into the human senses, thereby altering experience, emotion, and, ultimately, thought. As VR progresses to include more senses and a more complete perceptual field, the harder it may become for some persons to distinguish the real world from the artificial one. At present, little is understood about the extent to which VR might interfere with normal psychological processes, but it is possible that such interference could place at-risk individuals in mental or emotional peril. Even persons with minor neuroses or perceptual problems might find that their sensations and reactions are exaggerated in VR and their "residual memories and learning" may become "distorted upon returning to the real world" (Cartwright 1994, 24).

Like ECT and frontal lobotomies of the past, VR also raises ethical questions and concerns over behavioral control abuses. Even in the absence of such abuses, however, treatment by VR raises questions. While the first targets of medical researchers have focused on VR therapy for phobias and fears, its applications are being extended to a wider range of mental and psychological disorders. One problem is that once a patient is in VR, the motivation to change might be reduced or eliminated. Because VR allows one to experience what is comforting and thus become a means of avoiding anxiety and distress, it might also result in social withdrawal, particularly for those who find reality threatening. VR might also have adverse effects because it contradicts or short-circuits basic psychological principles that are taken for granted in the real world but do not exist or operate differently in the virtual world. For instance, while contact with the world around us is often used as an indicator of mental health and psychological adjustment, VR represents a deliberate manipulation of the senses to produce a hallucinatory state, thus producing a "very fine line" between some kinds of VR experiences and certain schizophrenia-type states (Cartwright 1994, 24). VR also produces an environment in which persons can create parallel lives, often with a completely different set of physical, social, and emotional attributes.

Ironically, although VR is potentially a formidable communication medium, like video gaming today, it could lead to isolation of individuals (Burdea and Coiffet 1994).

Other areas where VR confuses psychological principles are the deliberate creation of altered states of reality, disembodiment and rematerialization into a virtual body. and the prospect of electronically projecting one's ego-center into a virtual space beyond the real body. Although this decentering of self might be illuminating, it could also be "destabilizing and destructive" (Cartwright 1994, 25). Recent evidence of problems raised by immersion of some persons in the Internet should raise warning signs regarding VR, which promises to be an even more seductive experience. While VR could be a positive substitute to drugs in certain pathological cases, it also could create a "whole new class of depersonalized addicts" (Burdea and Coiffet 1994, 254).

Brain Imaging Techniques

Antecedent to and concurrent with advances in brain intervention have been advances in our understanding of structural and functional links to human behavior (Tingley 2011). Noninvasive functional imaging is making analysis of the brain an obligatory intermediary between genotype and behavior (Biswal et al. 2010; Hurley and Taber 2008). Until recently, our understanding of brain structure and function was restricted, based on inference from trauma, disease, and autopsy; extrapolation from animal studies, or the outcomes of very risky "therapeutic" physical intrusions in patients. Although animal models continue to provide a valuable source of information for neuroscience, they inherently have limits, and at some stage human models are crucial. Safe, practical ways to study the living human brain functionally began to evolve long ago with electroencephalography, followed in the mid-twentieth century by radioisotope scanning. Today, techniques that provide vivid images of live brains promise to greatly enhance our understanding of the relationship between the anatomy of the brain and psychological functioning. The increasingly sophisticated use of x-rays, radioactive tracers, and radio waves, combined with rapid advances in computerization, allows the noninvasive, safe investigation of the structure and functioning of the brain (Wahlund and Kristiansson 2009).

Broadly speaking, brain (or neuro-) imaging techniques can be divided into two categories (table 2.5). Structural or anatomical neuroimaging is limited to observation of the brain's architecture, whereas functional

Table 2.5
Brain Imaging Technologies

Computed axial tomography (CAT, CT)	Images brain structure.
Dynamic computed tomography (DCT)	Traces blood flow throughout the brain structure.
Echo planar MRI (EPI)	Produces enhanced MRI data using multiple high-power, high-speed oscillating field gradients, and advanced image processing.
Electroencephalography (EEG)	Measures electrical activity or lack thereof in specific regions of the brain. Can measure event-related potentials (ERPs) to observe changes in electrical activity over time that follow a specific stimulus.
Functional MRI (fMRI)	Measures increases in blood oxygenation that reflect heightened blood flow to active brain areas (or absence thereof) under resting and activated conditions.
Magnetic resonance imaging (MRI)	Provides high-resolution three-dimensional images of brain structure and activity caused by molecular changes in the brain when exposed to a strong magnetic field.
Magnetic resonance spectroscopy (MRS)	Extension of MRI that examines molecular composition and metabolic processes instead of anatomy.
Magnetoencephalography (MEG)	Reveals real-time resolution measurement of small magnetic field patterns in specific regions of the brain. When combined with fMRI, it provides precise information on the location and timing of activation.
Positron emission tomography (PET)	Creates images that map the distribution of radioactively labeled substances to measure cerebral blood flow, oxygen consumption, glucose utilization, and neurochemical activity.
Single-photon emission computed tomography (SPECT)	Can be designed to attach to specific receptors for precise mapping of brain activity or lack thereof.
Three-vessel angiography/ digital subtraction angiography (DSA)	Provides detailed patterns of blood flow or lack thereof in specific areas of the brain.
Transcranial magnetic stimulation (TMS)	Noninvasive technique used to induce, interrupt, or modulate brain activity in specific regions.
Voxel-based lesion-symptom mapping (VLSM)	Technique that produces maps that illuminate the relationship between the severity of a behavioral deficit and the voxels that contribute to that deficit.

neuroimaging involves the construction of computerized images that measure the brain's activity (Snead 2008). In the 1990s, axial tomography became computerized and then adapted from x-radiation to magnetic resonance (MRI) and positron emission (PET) scanning. PET is a nuclear medical imaging technique that produces a three-dimensional image or map by creating computerized images of the distribution of radioactively labeled substances in the brain following injection into the blood or absorption through inhalation. As the radioactive substances move through the brain, investigators are able to visualize regional cerebral blood flow and glucose utilization, as well as neurochemical activity. The more active a region, the more blood will flow through it and the more glucose it will use. The system detects pairs of gamma rays emitted indirectly by a positron-emitting radioisotope, which is introduced into the body on a metabolically active molecule.

MRI can detect molecular changes in the brain when it is exposed to a strong magnetic field. MRI provides clear and detailed images of brain activity and is used to detect structural abnormalities, changes in the volume of brain tissue, and the enlargement of cerebral ventricles in patients. The activity within particular regions of the brain can be analyzed to determine damage or malfunction and to correlate it with behavioral manifestations. Other currently available functional neuroimaging techniques based on hemodynamic principles are functional magnetic resonance imaging (fMRI), single-photon emission computed tomography (SPECT), and near-infrared spectroscopy (NIRS). The structure of the brain also can be studied by computed tomography (CT), which uses computers to combine a series of x-rays to provide a layered picture of the brain. These techniques measure abnormal activity in specific brain regions or the brain as a whole, or to detect the normal asymmetry of activity between the two sides of the brain. Also, because PET can use labeled drugs that attach to specific receptors, it is possible to identify the number and the distribution of receptor populations. For the first time in history, therefore, we have noninvasive techniques that facilitate precise mapping of normal brain activity and identifying variations from it that are related to specific behavioral manifestations.

Functional MRI measures the increases in blood oxygenation that reflect a heightened blood flow to active brain areas. It does not measure brain activity directly but rather utilizes a strong magnetic field to detect local deoxygenation of hemoglobin in the brain. When a person performs a cognitive or behavioral task, neuronal networks in specific brain regions are activated to initiate, coordinate, and sustain this task. Images are then constructed based on blood-oxygenation-level-dependent (BOLD) contrast, which is an indirect measure of a series of processes. These sets of neurons require large amounts of energy in the form of adenosine triphosphate (ATP) to sustain their metabolic activity. Because the brain does not store its own energy, it must make ATP from the oxidation of glucose. Thus, when an area of the brain is activated, its metabolism increases and more blood flows to that area, bringing the necessary glucose and oxygenated hemoglobin in and forcing the deoxygenated hemoglobin out. Because the magnetic properties of deoxygenated hemoglobin vary from those of oxygenated hemoglobin, this process illuminates what parts of the brain are most active and therefore display different signal intensities on MRI. These changes rest on the fact that oxygenated blood contains increased amounts of iron-rich hemoglobin. Because iron has magnetic properties, it can be detected by the MRI machine's magnet. The more oxygenated blood in a particular area, the more iron-containing hemoglobin is present, thus the brighter the signal on the fMRI brain scan. A more recent innovation is the machine learning classifier analysis of fMRI data (Pereira et al. 2010). This highly advanced software analyzes the data for "pattern localization" in the brain picture and "pattern characterization" to interpret these patterns (Shamoo 2010).

Generally, fMRI is more readily available to researchers and clinicians and provides greater detail and resolution than comparable imaging methods. It also enables the integration of anatomical (structural), neural, and molecular information in a single session. The gleaning of data on brain activity, along with information about the anatomy and the neural and metabolic status of the region, therefore, can be achieved within seconds. In this way, fMRI surpasses all other noninvasive modalities of brain imaging.

These discrete imaging systems are supplemented by a new generation of three-dimensional spatial imaging systems. Stereotactic imaging combines a series of two-dimensional scans into a three-dimensional virtual object. For instance, the BrainSCAN Radiosurgery System provides three-dimensional imaging by correlating anatomical information from preexisting MRI studies with diagnostic data from CT and angiography by means of automatic image fusion. Future advances in software are likely to match hardware improvements and provide even more remarkable and precise imaging of the brain. In addition, the development of computer programs that can alter or rearrange anatomical brain images from MRI and PET to match a standardized brain map makes it possible to compare the anatomy and function of different brains. In 2003, for example, the Brain Atlas, comprised of digitalized high-definition structural maps collected from MRI studies of more than 7,000 subjects, was published on the Internet (The Whole Brain Atlas 2012). Layered over the anatomical maps are brain functions such as memory, emotion, and language. Users can view individual pictures, composite pictures of subgroups by age, race, or gender, or the composite of all the subjects (Beasley 2003).

A second category of imaging techniques is based on electrophysiological principles instead of blood flow and includes electroencephalography (EEG), magnetoencephalography (MEG), and transcranial magnetic stimulation (TMS). One of the latest advances in imaging, MEG allows real-time resolution measurement of small magnetic field patterns in specific regions of the brain. In contrast to other imaging systems, MEG can measure millisecond-long changes in magnetic fields created by the brain's electric currents. By presenting stimuli at varying rates, the examiner can determine how long neural activation is sustained. When combined with fMRI, this technique provides precise information on the location and timing of activation. Moreover, newer techniques specifically directed at the brain include measurement of electrical activity through enhancement of conventional EEGs by computer analysis. Electrical activity can be measured while the patient is performing particular cognitive or sensory tasks or at rest, permitting investigators to observe changes in brain responses. Using knowledge of normal ranges, researchers are able to identify variations linked to particular mental disorders or behavioral problems.

In addition to these imaging technologies, other methods have been developed to study brain activity. Chance and colleagues (1998), for instance, introduced near-infrared spectroscopy (fNIR), which is portable and less expensive than imaging techniques and does not require immobilization of the subject. A related technology, touted by its inventor Lawrence Farwell as forensic brain fingerprinting, is the multifaceted electroencephalographic response analysis (MERA), which measures a specific electrical brain wave known as a P300 (Brain Fingerprinting Laboratories 2003). A P300 wave is emitted by the brain within a fraction of a second when an individual recognizes an incoming stimulus as significant. In contrast, irrelevant stimuli do not trigger a P300 wave. Thus, by carefully presenting a mixture of significant and insignificant stimuli to a person, memory can be "detected." The issues surrounding this and other "lie detection" technologies are examined in more detail in chapter 6.

Moreover, through its illumination of changes in local circulation and metabolism, neuroimaging has alerted us to the ongoing and costly intrinsic activity in brain systems that most likely represents the largest fraction of the brain's functional activity (Raichle and Mintun 2006). Although most imaging techniques, particularly fMRI, suffer from an inability to link their measurements to serotonin concentrations or release (Dayan and Huys 2009), through the development and characterization of novel radiotracers, we are now able to visualize and quantify in vivo many of the key molecular sites, including serotonin receptors, reuptake transporters, and enzymes responsible for serotonin metabolism. Furthermore, while molecular imaging of the serotonergic system is not yet in use in clinical practice, according to Parsey (2010), significant advances are expected in the next decade.

Although science has not yet been able to produce an imaging technology that can be used to image motivation, responsibility, or propensity for behavior, ever-expanding advances are raising expectations for more detailed mapping of brain structure and function. Among these discoveries is volumetric analysis, a technique for imaging morphology and size of regions of interest. Other advances include voxel-based morphometry, which will permit a systematic review of changes in the gray and white matter across the whole brain; cortical surface analysis; magnetic resonance spectroscopy; diffusion weighted imaging; and diffusion tensor imaging. Investigators are also researching ways to combine different imaging technologies, such as quantitative EEG and magnetic encephalography, with fMRI. Finally, transcortical magnetic stimulation can produce interactive maps of brain function (Khoshbin and Khoshbin 2007).

The Debate over Brain Imaging

The most popular technology in behavioral research used to locate increases in neural activity in the brain is fMRI, which, when used in combination with PET and other techniques, can give us accurate images of brain function and correlate them with behavior. According to Bandettini (2009), fMRI has become the tool of choice for the cognitive neuroscience community and is essential to researchers interested in understanding the functional correlates of behavior and disease in populations and individuals. Functional MRI has grown largely because of its noninvasiveness, relative ease of implementation, high spatial and temporal resolution, and signal fidelity, which is in general highly reproducible and consistent.

Imaging techniques, however, are not without limitations and detractors at several levels. For instance, while the MRI procedure sounds benign and rarely injures patients or test subjects, an MRI scanner is a powerful medical device, capable of causing serious injury or death if operated carelessly. More important, MRI of the brain carries psychological risk. On the one hand, as will be discussed in chapter 3, incidental findings of clinically significant conditions in volunteer research subjects raise a host of ethical concerns. On the other hand, clinically irrelevant MRI findings sometimes lead to needless and dangerous interventions. Moreover, identification as a participant in neuroimaging studies of cognitive impairment, such as research involving Alzheimer's disease or dementia, could in theory place a subject at risk for discrimination in insurance underwriting or employment (Kulynych 2007).

A second, more technical criticism directed at fMRI is that it is impossible to determine whether blood flow or metabolism has increased because of processes that excite or inhibit neural activity. Although direct neuronal response to a stimulus can occur on the order of milliseconds, increases in blood flow, or the hemodynamic response to this increased neural activity response (the proxy for neural activity measured by fMRI), has a one- to two-second lag, which is important in determining the temporal resolution of fMRI (Greely and Illes 2007). Because the neural activity of interest occurs on a time scale that is orders of magnitude faster than the hemodynamic, brain scans are not really images of cognition in process, thus raising "a profoundly disconcerting problem" for the users of neuroimaging in that the cumulative measure of brain metabolism is neither theoretically nor empirically linked to the momentary details of the neural network at the microlevel. As McDermott (2009) points out, "any given moment of activity does not offer much information about the circuits of activity that preceded or followed the image that is captured in time on the brain scan" (578). Similarly, Crawford (2008) concludes that an fMRI image may serve to indicate mentation, but because of the time-scale difference it does not preserve the machine states that encode mentation. "With such signs, we do not have a picture of a mechanism. We have a sign that there is a mechanism. But the discovery that there is a mechanism is no discovery at all, unless one was previously a dualist" (71).

Moreover, the measured differences in activity are small (often only a few percentage points) compared to the constant metabolism flux throughout the brain. In most imaging studies even the statistically significant differences in brain structure and function are very subtle (Dossey 2010). It is also important to realize that blood flow can be affected by many factors, including the properties of the red blood cells, the integrity of the blood vessels, and the strength of the heart muscle, as well as the age, health, and fitness level of the individual. For example, the velocity of cerebral blood flow decreases with age, and similar differences may be induced by pathological conditions. Some medications can also modify blood flow. Fluctuations in any of these variables could affect the signal measured and the interpretation of that signal. Greely and Illes (2007) note that because complete medical examinations are not routinely given to subjects recruited for fMRI studies as healthy controls, it is important to take this potential variability into account when interpreting and comparing fMRI data.

Shibasaki (2008) suggests that because the hemodynamic response correlated with neuronal activity tends to be more widespread in space and lasts longer in time compared with the neuronal activity, and since each technique has its own characteristic features, especially in terms of spatial and temporal resolution, it is important to adopt the most appropriate technique for solving each question, and it is useful to combine two techniques, either simultaneously or in separate sessions. In a multimodal approach, the combined use of EEG and MEG, EEG and PET, or EEG and fMRI, is applied for the simultaneous studies, and for the separate use of two different techniques, the information obtained from fMRI is used for estimating the generator source from EEG or MEG data (fMRI-constrained source estimation). Moreover, functional connectivity among different brain areas can be studied using a single technique, such as the EEG coherence or the correlation analysis of fMRI or PET data, or by combining a stimulation technique such as TMS with neuroimaging (Shibasaki 2008).

Although Poldrack (2006) agrees that neuroimaging techniques provide a measure of local brain activity in response to cognitive tasks undertaken during scanning, he cautions against the use of "reverse inference," by which the engagement of a specific cognitive process is inferred from activation of a particular brain region. Although such inferences can be helpful for advancing our understanding of the brain and the mind and provide useful information, he argues they are not deductively valid. He suggests that combining behavioral and fMRI results offers stronger evidence and that the mining of neuroimaging bases can also provide further insights into specific inferences from the data. Thus, reverse inferences are useful in suggesting novel hypotheses to be tested in subsequent experiments, but we must be aware of their limitations (Poldrack 2006).

Methodologies used in brain imaging also have been attacked because the accuracy of the correlations reported in many studies is not warranted by the statistical methods used. Critics charge that the neural correlates seen in fMRI studies are exaggerated and that many fMRI researchers have engaged in "cherry-picking" of their data (Dossey 2010, 278). For example, Vul and colleagues (2009) assert that fMRI researchers are producing results that defy statistical probability.

To sum up, then, we are led to conclude that a disturbingly large, and quite prominent, segment of fMRI research on emotion, personality, and social cognition is using seriously defective research methods and producing a profusion of numbers that should not be believed. Although we have focused here on studies relating to emotion, personality, and social cognition, we suspect that the questionable analysis methods discussed here are also widespread in other fields that use fMRI to study individual differences, such as cognitive neuroscience, clinical neuroscience, and neurogenetics. (286)

A basic problem is that emotions are difficult to measure quantitatively. There is no way to give a precise value to emotions such as sadness, rejection, love or truth. If one cannot value truth above an accuracy of, say, 60 percent, then a correlation of truth telling with fMRI patterns can never be more accurate than that. Yet Vul and colleagues found that researchers were reporting fMRI correlations of 80 percent or higher with behaviors and emotions. Based on their analyses, they maintain that the highest correlation that researchers can hope to achieve between a human behavior or emotion and a brain scan is 74 percent. Moreover, they found that the typical fMRI article does not provide sufficient detail to describe how the high correlations were arrived at. Thus, more than half of the investigations in this area used methods that were guaranteed to offer greatly inflated results and were "entirely spurious," even some studies that had been published in elite journals such as *Nature* and *Science*.

Similarly, Bennett and Miller (2010) have questioned the reliability of fMRI studies. They argue that brain imaging studies that claim to link facets of personality, memory, and emotion to specific brain activity are not easy to reproduce. Their exhaustive analysis questions the reliability of fMRI studies that image brain activity while people perform mental tasks. Although nobody disputes the validity of the basic method, which uses MRI scanners to monitor the brain's oxygen usage with millimeter-scale precision (Lovett 2010), Bennett and Miller, like Vul and colleagues, question many studies that have reported 80 or 90 percent correlations between specific regions of brain activity and personality traits and emotions. Moreover, as noted above, popular neuroimaging techniques may be painting a misleading picture of brain activity, because scientists using such techniques make the assumption that blood flow into a particular brain region is directly linked to the amount of activity in the cells of that region. These findings suggest that scientists who use fMRI may need to interpret their data differently, because the mismatch between neural activity and blood flow demonstrated is "extreme" (Smith 2009).

Not surprisingly, the critiques have stirred great controversy and elicited a defense by numerous researchers. Lieberman, Berkman, and Wager (2009), for instance, defend social neuroscience and argue that although they accept that some correlations are overstated, the problem is not nearly as bad as critics contend. They also disagree with the implied claim that the overstated correlations have distorted the understanding of social neuroscience research and object to the critics' focus on social neuroscience, given that the same statistical issues arise in all sorts of brain imaging studies. Finally, they point out specific areas where they feel that critics have mischaracterized their data analysis methods. Similarly, Lindquist and Gelman (2009) suggest that the way forward is to go beyond the correlation and the multiple-comparisons framework that causes so much confusion. Rather than correcting for problems arising from multiple significance tests, perhaps it is more appropriate to represent all relevant research questions as parameters in one coherent multilevel model.

Although Weber and Thompson-Schill (2010) agree that popular portrayals of fMRI tend to attribute too much inferential power to the technique, they are concerned that some of its detractors seem to invent weaknesses of the technique. They note that the popularity of the use of brain stimulation techniques in cognitive neuroscience is rising, and researchers often justify using these techniques by invoking the inability of functional imaging to address causal matters (Iacoboni 2009). For example, Uddin and colleagues (2006) state that "functional magnetic resonance imaging (fMRI) provides only correlational information about the relationship between a given brain area and a particular cognitive task. Causal relationships between brain and behavior can be tested with [transcranial magnetic stimulation (TMS)]." But Weber and Thompson-

Schill (2010) counter that fMRI is not merely correlational. Imaging studies provide causal information, although not causal certainty, about the influence of brain activity on behavior. Moreover, the influence on brain activity of experimentally manipulated variables, such as stimuli and task instructions, is unarguably causal. While it is essential that cognitive neuroscientists be clear about the comparative strengths of non-imaging studies such as TMS, it should not be at the expense of imaging techniques.

Furthermore, it is argued that the presentation of simple, attention-grabbing pictures showing activity in the brain tempts one to jump to conclusions about the power and possibilities of brain imaging while they actually conceal enormous complexities. "Brain images are particularly vulnerable to misuse because they are so visually attractive. This visual power can easily result in misunderstanding what the images show and what they mean" (Khoshman and Khoshman 2010, 171). Tingley (2011) adds that obtaining a clean fMRI of neural processes is difficult even with million dollar equipment. Although fMRI scans are highly processed representations of an indirect measure of neural activity, they are often described as if they were direct snapshots of the mind in action (Garnett et al. 2011). What appears as a striking color picture of brain activity, one that might even show apparent changes in brain activity over time, is a visualization of a probabilistic model of brain activity, and this model is a complex, mathematical abstraction. "Like observatories pointed inward, modern brain scanners routinely capture breathtaking images of the gyral swirls and neuronal clusters that underlie human cerebral nature. These brain portraits are composed of three-dimensional voxels, or volume picture elements, digitally displayed in shades of grey. Knowledge in neuroscience can be measured in degrees of resolution" (Cheshire 2007, 135).

The presumption of a modular view of the brain, which assumes that it has unique centers for each mental function and that these centers can be revealed with fMRI, has led critics to call brain imaging in general, and fMRI in particular, the new phrenology (Dossey 2010). Whereas old phrenologists felt the shape of the skull, believing they could find out things about one's mind and character, the new phrenologists, so the criticism goes, measure changes in the brain's metabolism for the same purpose and with little more scientific validity (Árnason 2010). Crawford (2008) notes that when applied to medical diagnosis of, for instance a brain tumor, brain imaging is "straightforward and indubitable," but the use of neuroimaging in psychology is a fundamentally different kind of enterprise that depends on the premise that mental processes can be analyzed into separate and distinct faculties, components, or modules, and further that these modules are instantiated, or realized, in localized brain regions. Current evidence, however, suggests this premise is wrong and that neither mental functions nor the physical systems that realize them are decomposable into independent modules. Most complex psychological or behavioral concepts do not map to a single center in the brain (Crawford 2008, 65).

Brain imaging provides vast quantities of data in the form of voxels from the entire brain. Neuroscientists sometimes focus on an area of interest by searching for voxels that are activated when subjects perform different tasks in an experiment, for example, looking at a face or an inanimate object. But fMRI data are intrinsically very noisy, producing many "false voxels." Problems arise when researchers use the same data to select a particular brain region and then to quantify the experimental effects there by asking how much more strongly the region responds to a face than to an inanimate object (Abbott 2009a). Cheshire (2007, 135) contends that implicit in every voxelous reconstruction of the brain is the idea that the brain is virtually, if not essentially, reducible to matter. Although reductionism can clarify, it can also mislead. Vibrant voxels may elucidate pertinent facts, but exclusive attention to them may overlook important truths.

Another concern is that the development of functional neuroimaging has led to the neglect of brain structure in favor of a focus on mapping functional responses associated with performance of specific behaviors, despite the fact that brain anatomy and function are inextricably linked and together determine our behavioral responses. Compared with "gold standard" techniques for brain anatomy, such as histological techniques or tracer-based tract tracing, neuroimaging approaches are crude, and the images are difficult to interpret, and there is rarely a one-to-one relationship between a given imaging measure and an underlying anatomical feature (Johansen-Berg 2009). Thus, it is likely that neuroimaging studies of brain anatomy will always rely on information from classical neuroanatomy to guide their interpretation, as reflected in the increasing use of the combination of functional imaging techniques and neuroimaging methods for probing the gross architecture and the microstructural anatomy of the brain (Bandettini 2009).

Given these potential problems with neuroimaging, particularly as applied to human behavior, and the question of what brain images really indicate, should we take this research seriously? It is here argued that despite their limitations, neuroimaging-based studies have enormous advantages in that they can be performed on living human brains whose functional infallible responses and behavioral outputs can be studied at the same time. Neuroimaging, in particular fMRI, offers unparalleled opportunities for studying whole-brain anatomy in health and disease (Johansen-Berg 2009). If brain imaging can give us insights about human behavior, even though not infallible, it is still very important. In fact, many of the criticisms of neuroimaging pale in comparison with the known shortcomings of a dependence on verbal responses in survey research, which is the key method in social science. Moreover, brain imaging is widely available, it is noninvasive and safe, it has a relatively good resolution, and it can show possible activity in the whole brain.

According to Friend and Thayer (2011), brain imaging will make major contributions to the discipline of political science in the study of aggression and violence, prejudice, and in-group/out-group formation and in the analysis of voting behavior and partisan preference. However, while it is clear that measurement of psychological traits using neuroimaging is in principle possible, it is not clear whether such measurement is a likely development in the near term or a more remote but still realistic goal for the future, or whether it remains effectively a science fiction scenario. On the basis of their in-depth review, however, Farah and colleagues (2009) conclude that functional neuroimaging can provide more predictive biological correlates of psychological traits than genotyping. This is not surprising, in principle, because brain function is a causal step closer to behavior than are genes.

Imaging Genetics

Advances in molecular biology, brain imaging, genetic epidemiology, and developmental psychopathology provide a unique opportunity to explore the interplay of genes, brain, and behavior within a translational research framework (Viding, Williamson, and Hariri 2006). The innovative field of imaging genetics uses neuroimaging methods to assess the impact of genetic variation on the brain (Kempf and Meyer-Lindenberg 2006). Generally, multiple imaging methods are used together to achieve an optimal characterization of structural-functional parameters in large groups of individuals, whose genotypes are then statistically related to these data across the subjects, a form of genetic association study. Although this technique is in its infancy, the emerging literature shows that it can be useful in identifying neural processes involved in mediating the effect of genetic polymorphisms on mental disease risk, thus moving us closer to understanding the complexities of the specific mechanisms involved in the etiology of psychiatric disease. To date, studies have dealt with genes involved in risk for schizophrenia, Alzheimer's disease, depression, anxiety, and violence, but extensions to genes linked to genetically based behavioral tendencies will likely follow in the near future as our understanding of the interplay between genetics and the brain improves.

For instance, an important consideration in experience-dependent plasticity research is the potential contribution of genes to behavioral and neural variation. Imaging genetics has begun to identify associations among common genetic variants, human cognitive functions, and their neural correlates (Losin, Dapretto, and Iacoboni 2010). Until now, identified genetic polymorphisms associated with cognitive and neural effects have been primarily those related to monoamine neurotransmitter receptors and transporters. For instance, genetic variation in the catechol-O-methyltransferase gene (COMT) has been linked to variation in working memory and prefrontal cortex function, while genetic variation in the serotonin transporter gene has been tied to variation in anxiety-related behavior and cingulate cortex–amygdala interactions (Losin, Dapretto, and Iacoboni 2010). Moreover, such polymorphisms have been found in higher frequencies in certain regional groups, for instance East Asians, than others, thus stressing the importance of considering the potential effects of such regional genetic variation on cross-cultural behavioral and neural differences.

Imaging genetics provides an enormous amount of functional-structural data on gene effects in living brain, but the sheer quantity of potential phenotypes raises concerns about false discovery (Meyer-Lindenberg et al. 2008). Moreover, while it provides a powerful new approach to the study of genes, brain, and behavior, imaging genetics' full potential can be realized only by expanding the scope and scale of the experimental protocols. Although single gene effects on brain function can be documented in small samples, the contributions of multiple genes acting in response to variable environmental pressures is ultimately necessary for the development of truly predictive markers that account for the majority of variance in any given phenotype. "Ultimately, we anticipate that such a mechanistic understanding will allow for the early identification of individuals at greater risk for emotional regulatory problems that can have long-term health related implications" (Hariri, Drabant, and Weinberger 2006, 895). Tairyan and Illes (2009) detail the wide range of ethical concerns accentuated by imaging genetics.

Conclusions

Whether the intervention technique utilizes chemical, electrical, surgical, or computer techniques, issues of safety, efficacy, and consent are inherent in any attempt to modify the structure and functioning of the brain. Moreover, broader social issues arise because most interventions are directed at suppressing symptoms, not treating the root cause of the disorder. This is especially problematic when the human subjects are children or other vulnerable persons who are unable to exercise full informed consent or when the long-term effects of the intervention are unknown or irreversible.

The discussion here demonstrates the urgent need for more systematic, anticipatory analysis of the social consequences of the rapid diffusion of these innovations. Because of our heavy dependence on technological solutions to health and social problems and the prominence of the medical model of health, it is always difficult to curtail or slow the proliferation of the latest wonder drug or procedure. Active marketing and publicity often encourage use before the risks of the intervention are fully understood, as evidenced by frontal lobotomies in the 1950s and Prozac in the 1990s. Thus, it is even more urgent that the policy issues they raise are thoroughly analyzed with alacrity. Like genetic and neuroscience research in general, these intervention techniques have significant social implications that warrant analysis by policy and social scientists as well as by the medical and scientific communities. Moreover, as stated by O'Donovan and Owen (2009), while brain disorders are "undeniably challenging targets for biomedical research, they are also amongst the most important, the brain being central to every aspect of human life" (91). Chapter 3 examines in detail the ethical and policy implications that accompany this broad array of interventions in the brain.

3
Neuroethics and Neuropolicy

This chapter first summarizes the range of ethical and related policy issues that accompany various forms of brain intervention. It then examines in more detail the movement of neuroethics issues to the policy domain and discusses how this complicates the context by bringing to the fore political considerations and divisions that place the dialogue and resolution of these issues in the milieu of interest-group politics. Because of the huge economic, social, and personal stakes surrounding neuroscience, this is unavoidable. As noted in chapter 2, historically, experimental and clinical interventions in the brain have elicited intense debate from many directions. Although new advances promise considerable benefits, as with genetic technologies, neuroscience research and technologies challenge social values concerning personal autonomy and rights and for some observers raise the specter of mind control and fears of a Big Brother society. As a result, difficult policy questions are opened that must be dealt with in the coming decades.

The chapter concludes by examining the range of policy options available to deal with the fruits of neuroscience, and offers a brief look at the process by which policy will be made.

Ethical Issues in Brain Intervention

Unlike brain policy, where there has been a dearth of action, there is a vast literature and considerable international scholarly activity in neuroethics. General ethical issues include the identification of new and ethnic diseases, the possibility of stigma and labeling, questions of privacy and personal autonomy, response sensitivity, what actions should be prompted by incidental findings, how to ensure clinical validity, and issues associated with prediction of future states and intervention decisions. Thus, with every intervention and every observation in every human brain, we confront a multitude of ethical issues. Even ECT, used more or less successfully since the 1930s, remains "one of the most controversial treatments in medicine" (Carney and Geddes 2003, 43). Less widely appreciated, though, is how easily traditional topics in clinical and research ethics merge into new topics in political ethics and, therefore, policy. The following represent some of the major issues that breach the lines between ethics and politics.

Informed Consent

The mandate to obtain informed consent holds that a person contemplating medical treatment or participation in a research protocol must be apprised of its risks and benefits and must understand its implications. The underlying notion from an ethical standpoint assumes that the subject is capable of making such a decision free from constraints, both personal and social. Explicit in these assumptions is the supposition that the subject is mentally competent to understand what he or she is consenting to and the consequences of the decision. With brain intervention, however, consent, though conscientiously elicited, is unlikely to be informed fully for any first intervention and may be less competently offered, if better informed, for subsequent interventions. Consent in such cases has to come from the very organ that needs evaluation, assistance, or repair or that presents itself, once or repeatedly, for experimentation (Müller and Walter 2010). On the one hand, the question arises as to whether fully informed consent is possible; on the other, paternalistic imputation of consent may seem sufficiently sensible to weaken safeguards and trigger fears of potential abuse.

Skuban and colleagues et al. (2011), for instance, are highly skeptical as to whether patients with occasionally debilitating psychiatric diseases are capable of giving fully and freely their consent to a comparatively experimental procedure such as deep brain stimulation (DBS). Patients with Alzheimer's disease especially are limited in certain cognitive dimensions, and this restriction could endanger their ability to understand all the implications connected with the intervention. Similarly, cognitive impairment is a common finding in patients with depression, and this dysfunction may result in deficits of attention, perception, concentration, and memory, thereby leading to a significant ambivalence of the patient (Clark, Chamberlain, and Sahakian 2009). Moreover, patients suffering from substance abuse frequently are impaired with respect to tasks that involve highly goal-directed behavior and thus might be constricted in their free decision-making process (Boutrel, Cannella, and De Lecea

2010). Furthermore, if certain information obtained by neuroimaging is sensitive and personal, thus belonging to a person's sphere of privacy, the dissemination of data itself can be harmful even if no (other) detrimental consequences ensue (Häyry 2010).

The issue of consent is especially problematic in relation to more experimental procedures such as neurogenetic and neurotransplantation technologies. Given the heightened, compound uncertainties these interventions raise, Giordano (2010) urges the development of some "system of metrics" that would provide a level of power necessary and sufficient for clinicians to ascertain benefit-risk probabilities and inform patients about them while simultaneously preserving the patient's autonomy and right to accept or refuse any such interventions based on clinical advice and the patient's own values and goals. Therefore, at a minimum the notion of informed consent is problematic when any physical, electrical, or chemical intervention in the brain is contemplated.

Experimentation or Therapy?

As noted by Skuban and colleagues (2011), many interventions in the brain, even those widely used in clinical settings, such as DBS, at their base are at least partly experimental in each application. Therefore, while the line dividing experimentation from therapy is often a fine one in medicine generally, it nearly disappears here in the sense that each person's brain is unique in ways that affect identity, memories, judgments, and preferences. Because the brain defines the person, even minor alterations might shade a definition of self or, more observably, an assessment of character. Moreover, the complexity of the human brain ensures that side effects can never be fully predictable or controlled. However, as in other areas of medicine, there has been a tendency to move quickly from experimental to therapeutic status, and although this is understandable in light of the often urgent need and vulnerability of patients who undergo these techniques, their very vulnerability necessitates a cautious approach.

This concern over risks and uncertainties is especially critical when the techniques are used in nonmedical, especially commercial, settings. According to Goldberg (2007), no technology is absolutely safe, and even noninvasive procedures such as MRI have some risk. Caution is particularly imperative when an intervention is irreversible, as is the case with psychosurgery, but it is also warranted when the short- or long-term consequences may be too subtle to notice, too idiosyncratic to define, or too safely within the range of normal to be called adverse. For instance, even as DBS has proliferated as an accepted therapy, concern over its long-term effects persists (Schlaepfer, Lisanby, and Pallanti 2010). The same holds for drugs such as Ritalin and Prozac. Concern is deepened when the intervention involves that treatment of behavioral disorders that lack a proven organic origin or the modification of questionably disordered behaviors that are troublesome principally to families, societies, governments, or insurers.

Autonomy, Authenticity, and Mind Control

Any form of intrusion in the brain conjures the image of mind control, thus potentially undermining individual responsibility and the scope of culpability. As the center of personal autonomy and identity, the brain enjoys special status, and modifying it even slightly raises concerns of manipulation. In parallel fashion, the incremental elucidation of neurological factors in human behavior, especially aberrant behaviors, may greatly complicate moral, civil, and jurisprudential arguments (Gazzaniga 2005). These potential issues are exacerbated when persons are compelled to undergo such interventions by the courts, insurance companies, or employers. Similar issues of privacy and autonomy arise with the potential use of brain imaging studies undertaken in the course of medical treatment by insurers and employers.

Given the sensitive nature of the brain as the core of autonomy, questions of privacy also are endemic when one intervenes in the brain for whatever reason. Neuroimaging may, even if imperfectly, acquire personal information and disclose facts about a person that the person prefers to keep private. Moreover, as with tissue samples used for genotyping, brain images can be obtained with consent for one purpose but later analyzed for other purposes, particularly when correlations between brain function and psychological traits are obtained in the context of tasks that lack an obvious relation to the trait being measured (Farah et al. 2009). For example, extraversion and unconscious racial attitudes are both correlated with brain activity evoked by simply viewing pictures of faces (see chapter 4). Hence, in such studies, subjects could be told that the purpose of the scan was related to face perception, thereby making it relatively easy to obtain information about a person without his or her knowledge. Similarly, a substantial minority of drug-dependent persons exhibit a characteristic pattern of brain activation in response to drug cues that could enable identification of them as drug dependent (Farah 2002). Moreover, characteristic patterns of brain activity in childhood and adolescence may predict an increased risk for addiction later in adult life. According to Greely (2005), this possibility raises the same ethical issues as are raised by testing for alleles that predict an increased risk for serious neurological disease.

A related issue concerns authenticity, or the self's sense of its own uniqueness and individuality, and the desire to be true to this self (Singh 2005; Newson and Ashcroft 2005). For instance, psychotropic drugs may diminish a "real" self, or transform the self, if given on the assumption of a self that is identifiable, coherent, and stable. This concern has been extended to physical interventions such as DBS (Kraemer 2011). Erler (2011) is concerned that "memory editing" poses a significant threat to authenticity. According to Johansson and colleagues (2011), authenticity urges us to live in accordance with our given nature, meaning "that which we are" has a privileged position. Therefore, diversions from our given nature are morally problematic, particularly those that distance a person from his or her true self. Although an individualist emphasis can encourage an elemental vision of the self, notions of authenticity do not necessarily map onto a real core identity. Instead, the language of authenticity may more accurately reflect the desire to provide a moral justification for particular medical decisions.

For instance, Singh (2005) found that the dilemmas mothers face regarding their use of Ritalin for their boys on weekends revolve around a dialectic of authenticity and personal freedom: Who is the real boy, and can he be free to be who he really is when he must be chemically controlled in order to be free? Interestingly, while Singh found that many mothers saw their son on medication as the authentic self where he felt best about himself, other mothers refused to medicate their sons on weekends, implying that a boy's behavior is part of who he really is and to deny or restrain that part of him through medication is to subject him to a suspect medical explanatory model. This antitherapeutic narrative sees the unmedicated boy as the authentic boy and justifies withholding medication on the weekends by mothers who want their sons to be themselves, or to know who they really are when they free from the confines of a school setting (Singh 2005; Appelbaum 2005).

A related issue of concern is that neuroscience risks medicalizing what many see as essentially psychosocial phenomena, and thereby attempts to change our nature through intervention in the brain (Adolphs 2003). For instance, the idea that personality traits such as shyness and anxiety have a biological base and represent candidates for neural intervention raises questions of boundaries. In contrast, Ashcroft and Gui (2005) suggests that contemporary work in the neurosciences might present a more fluid and socialized idea of human nature, in which many aspects of behavior are shaped by both biological and social factors, and human social cooperation, and even ethical behavior, is naturalized and seen as having a biological basis.

Stigmatization

To whatever extent judgments of character are supplanted by judgments of genetics or neurons, stigmatization will adapt, moving from an evaluative realm in which intuition serves imperfectly to one in which it serves only to harm. People with mental disorders have always been treated differently than those with somatic diseases. Furthermore, discrimination against the mentally ill, including budgeting for their care, is widespread even among well-meaning people. The very presence of brain intervention techniques, therefore, can heighten stigmatization and discrimination.

In a coming era of neurosciences insight, such discrimination might lessen in old ways while worsening in new ones the balance of the two trends, in a way hard to estimate prospectively. "New knowledge can also reinforce, reconstruct, or undermine widely held cultural ideas about what it is to be normal or abnormal. This can challenge forms of discrimination and stigma, or can alternatively, create new forms of social division, which form the basis of inequality" (Ashcroft and Gui 2005, 27). The potential to pigeonhole and discriminate on the basis of test results could lead to negative consequences, including the development of a "neuroscientific underclass" denied access to societal benefits on the basis of their neuroscience test results (Garland 2004). Complicating matters, Meyer-Lindenberg and colleagues (2008) suggest that frequent false positive results are probable in neuroimaging.

Incidental Findings

Conducting research that uses MRIs on apparently healthy subjects results in a substantial number of findings that may be of clinical significance. ... In other words, it turns out that many research subjects may be sick without even knowing it. But the question remains: how should the research community react? To date, there is no uniform answer. (Goldberg 2007, 236)

The profusion of imaging techniques for research on human behavior has recast a perennial clinical epistemological problem: the incidental finding (IF). Wolf, Lawrenz, and colleagues (2008) define an IF as "a finding concerning an individual research participant that has potential health or reproductive importance and is discovered in the course of conducting research but is beyond the aims of the study" (219). Although fMRI and other imaging technologies have great potential for unraveling the secrets of the brain and brain development, they come at the potential cost of revealing IFs (Nelson 2008; Illes and Bird 2006; Illes and Chin 2008; Miller, Mello, and Joffe 2008).

As stated by Wolf, Paradise, and Caga-anan (2008), "IFs may seem at first blush a minor or peripheral concern. On closer scrutiny, however, IFs raise fundamental questions" (363). From a clinical standpoint, IFs include unexpected masses, aneurysms, evidence of current or past trauma, or anatomical evidence of dementia. Although Wolf, Paradise, and Caga-anan (2008) admit that social and behavioral IFs are harder to ascertain than physiological and genetic ones, they suggest that all raise similar issues. For example, what should be done with observed signs of alcohol abuse in an adolescent subject in an fMRI study of cognition unrelated to alcohol use, or troubling signs of suicidality in a study unrelated to that phenomenon? Their survey of the literature found an IF on 13 to 84 percent of brain fMRI or MRI scans, with 1.2 percent needing immediate referral, 0.4 to 14 percent urgent referral, and 1.8 to 43 percent routine referral (Wolf, Lawrenz, et al. 2008, 221). Similarly, in 151 studies they reviewed, Illes and Chin (2008) found IFs requiring referral occurred in 6.6 percent of subjects and concluded, "Now with advanced methods such as magnetic resonance imaging (MRI), and greater attention to ethical issues in research, the Pandora's Box of incidental findings has been opened and the need for guidance and resolution has become clear" (303). How such regulations should look is, nonetheless, highly controversial. "The truth is that no-one knows how to handle these difficult questions" (Wolf, Lawrenz, et al. 2008, 216).

According to Heinrichs (2010), two questions are critical with regard to IFs. First, how can the research subject's right not to know be guaranteed, and second, should a diagnostic check of scans by a neurologist or radiologist become an obligatory part of neuroscientific research protocols? There is a widely shared consensus that it is a fundamental right of patients to decide whether they want to know about diagnostic findings or not. However, within the context of brain research and neuroimaging, the right not to know can become highly problematical. While Miller, Mello, and Joffe (2008) argue that this preference surely should be honored when a subject has explicitly indicated that she does not want to receive incidental findings, it is doubtful whether a strict adherence to the right not to know is always feasible. Wolf, Lawrenz, and colleagues (2008, 233) contend that potential research subjects should be informed they have a right not to know and that researchers should respect this right unless concealing an IF could be life-threatening or comparably dangerous for the research subject or third persons, for instance, if the subject is unfit to drive. Research subjects should be fully aware that in such cases their right to informational self-determination might be outweighed by the right to their and others' psychophysical integrity.

Although IFs are common to all neuroimaging methodologies, MRI research has received the most discussion because it is more commonly used to study normal populations and often is conducted by persons not trained to interpret the scan clinically. IFs often appear when a researcher or student reads a scan and notices something unusual. According to Royal and Peterson (2008), the "point of greatest controversy over how to handle IFs is whether scans of all research participants should be reviewed in an active search for pathology and, if so, whether this search should be performed by a radiologist" (305). Obviously, the cost of this obligation could be significant, especially for social or behavioral studies using fMRI. In their recommendations, however, Wolf, Lawrenz, and colleagues (2008) state that the "cost of compensating the consultant for IF verification and evaluation should be built into the research budget" as either direct costs of the project or an infrastructure cost (237). Milstein (2008) adds that in cases where IFs are found, qualified physicians should be available to explain to the subject the significance of the findings. While investigators have the obligation to minimize risks and burdens for participants and thus a duty to respond to IFs that emerge in the course of research, this "does not entail an obligation to actively seek out IFs through routine clinical review of research data or to provide follow-up clinical care" (Miller, Mello, and Joffe 2008, 279).

Distributive Justice

Intervention in the brain also raises questions of distributive justice, mainly involving the access to services, many of which are costly on a per-case basis. Although resource allocation discussions to date have been limited in neuroscience, they are critical because while resources are finite, demands and expectations fueled by new technologies have few bounds. While it might be premature to speculate about the relative costs and benefits of yet undeveloped procedures, it is logical to assume that cumulatively their cost will be high. As such, these interventions raise questions as to who should get priority for access to these procedures, whether insurance companies must pay for what might be experimental treatments, and the extent to which a government can or should regulate their use. Will access be equitable and coverage universal, and, if so, how will it be funded? Or will it be available to the affluent and denied to persons who lack sufficient resources?

This question of access becomes especially important when endowment of social advantage becomes possible through enhancement of brain function (Wolpe 2003). Could such endowment ever be fair if not available on an equitable basis? Here the comments of Mehlman and Botkin (1998) regarding genetic technology are highly relevant. They make a persuasive case that access to the benefits of these technologies is bound to be inequitable in the United States because the traditional market-oriented, third-party payer system leaves out many people. Giordano (2010), for instance, concludes that high-tech medical interventions are often only partially covered, and in some cases, not covered at all by the majority of health insurance plans. Moreover, with over 45 million Americans currently uninsured, there is a possibility that state-of-the-art neurotechnological interventions will become a form of boutique neurology (Wolf 2008).

On a broader level, what priority should the search for knowledge about the brain and the ever- expanding applications of this knowledge have vis-à-vis other strategies and health care services? What benefits would such research and applications hold for the population as a whole compared to other spending options? On the one hand, over the last several decades there has been a dangerous proclivity to develop

Box 3.1
Allocation and Use of Ritalin

There is evidence of the importance of social and economic factors when individuals decide to or give Ritalin (or combination of wide variety of drugs) to their children. One study found the ratio of treatment between whites and blacks in elementary school was 2:1, but by middle school it had increased to 2.6:1, and by high school to 5.6:1 (Safer and Malever 2000). In other words, as children enter high school, the diagnosis of ADHD becomes overwhelmingly a "white thing." According to Griggins (2005), this disparity in treatment rates between rich and poor and whites and blacks means not only that affluent whites are receiving cognition-enhancing drugs more than minorities but also that, because a diagnosis of ADHD is also recognized as a legal disability, whites are afforded certain rights and academic accommodations more often than minorities.

and widely diffuse expensive curative techniques without first critically assessing their overall contribution to health outcomes. On the other hand, the availability of effective and inexpensive imaging studies might help provide valuable information for disease prevention and health promotion because it would make feasible the examination of a large population of individuals at heightened risk for diseases that could be reduced by early intervention. What priority should these interventions have, then, particularly when used to treat behavioral problems?

Emerging Policy Issues in Brain Intervention

As discussed in chapter 2, the arsenal of techniques and strategies for intervention in the brain is expanding rapidly and will be augmented in the future by even more notable technologies. Although many physicians and patients welcome these broadened opportunities for diagnosing and treating mental disorders, their advent is not without concern. This concern is directed not only at the potential risks and unanticipated side effects that accompany any interventions in such a complex organ as the brain but also at potential abuses of the powers of control that they bring. While new generations of psychoactive drugs and invasive procedures, as well as technologies such as virtual reality, are introduced and diffused with enthusiasm, the lessons of our past experience with cruder forms of intervention in the brain cannot be ignored.

The Right to Refuse Treatment
A large part of the legacy of earlier interventions in the brain, such as ECT and frontal lobotomies, is that what is carried out voluntarily on some groups in society can easily be used with coercion on others. Because these interventions can be used to control or alter behavior as well as to treat disorders as defined, at times subjectively, by society and the medical profession, the motivations behind each application must be scrutinized. As discussed earlier, informed consent is especially problematic when dealing with brain intervention, in part because of the experimental nature of using even well-tested techniques on an individual patient's brain and the irreversible nature of some of these techniques. Special problems arise if the patient agrees to the recommended treatment when his or her competency is questionable, but normally they can be minimized by substantiating consent from a proxy decision maker. Similarly, if a clearly competent person refuses treatment and poses no risk to him- or herself or others, the decision not to treat should be

clear-cut. The more difficult problems arise when a person of questionable competency refuses treatment that others deem in that person's best interest or when a competent but potentially dangerous person refuses treatment. This latter situation often arises in cases of coerced treatment of diagnosed mentally ill patients who refuse to take medication that is needed to stabilize them (Ryan 2010).

The history of attempts to limit the power of the government to impose treatment on unwilling patients has involved a variety of legislative and judicial actions. Because of their traditional responsibility in health-related policy, state legislatures have been active in setting statutory limits on the involuntary treatment of mental patients. California Penal Code (Sec. 2670–2680), for instance, declares that all persons "have a fundamental right against enforced interference with their thought process, states of mind, and patterns of mentation." In the absence of a court order, no drugs, ECT, electronic brain stimulation, or other treatment that inflicts pain can be used on persons incapable of consenting. The authorities must prove to the court that the therapy will be beneficial, that its administration is supported by a compelling state interest, and that no less intrusive therapy is available. Similarly, a Florida statute (Sec. 394.459(3a) 1995) requires "express and informed consent" for treatment and sets procedural criteria for the authorization of involuntary treatment that is essential to appropriate care. Other sources of state limits include judge-based tort law and regulatory limits imposed by state administrative agencies. The result of these actions is that many states have granted civilly committed patients the right to refuse involuntary treatment with antipsychotic medication, although most states allow exceptions after due process, procedural reviews have taken place.

In interpreting the actions of state legislatures and the courts regarding a right to refuse treatment, it is important to look at the context. Although the U.S. Supreme Court has ruled against involuntary administration of antipsychotic drugs for punishment in a prison setting, it has not agreed to hear similar cases from other types of institutions or the community. By and large the courts have recognized the heightened degree of state interests in institutional settings and of the responsibility of the institution for safety. Therefore, the courts have been likely to defer to the institutional officials as long as procedural protections are met. Clearly, the right to refuse treatment for mental disorders is relative, not absolute, and is dependent on the setting, duration, extent, and effects of the treatment, as well as on the competency of the intended subject. The lines between involuntary and voluntary intervention are not bright. There are many different degrees of coercion possible since each patient poses different degrees of potential danger to self and others. Moreover, the procedures vary greatly as to safety, efficacy, and intrusiveness. It seems logical that the right to refuse a highly experimental, risky, and irreversible procedure such as psychosurgery be stronger than the right to refuse psychotherapy or drugs of proven efficacy.

Psychotropic Drugs and Children
Another issue of growing concern is the potential for overmedication of children. Although this issue extends to all prescribed drugs, including antibiotics, anti-asthmatics, and anticonvulsants, attention has focused on the use of psychotropic drugs. Because the vast majority of drug trials are conducted only on adults, approximately 80 percent of drugs are not approved specifically for children. However, once FDA approval for sale is given, physicians can prescribe a drug to anyone for any purpose. For psychotropic drugs in particular, this is highly problematic because of the lack of any data about the possible long-term effects on developing brains, even for widely prescribed drugs like Ritalin and Prozac. An additional worry is that antidepressants, unlike antibiotics and other drugs, are often taken for many years, and the proper dosage is difficult to calculate, especially for younger children.

As noted earlier in the discussion of Ritalin use, drugs prescribed to treat behavioral disorders and problems are especially controversial when given to large numbers of children. The prescription of Ritalin to over 1.5 million children daily (in addition to the many other children prescribed the stimulant Dexedrine) raises questions concerning the accurate diagnosis of attention deficit/hyperactivity disorder (ADHD), which is known to affect considerably fewer children. Even for children who are accurately diagnosed with ADHD, some observers argue that unknown long-term consequences, slowed height development, and other side effects raise serious doubts over the tendency, "when in doubt, treat with these drugs."

Most recently, attention on the potential overmedication of children has focused on the rapidly escalating prescription of antidepressants, particularly Prozac during the 1990s. In 1996, nearly 600,000 children and adolescents were prescribed one of these drugs. Interestingly, this growth in pediatric prescriptions corresponded to saturation of the adult market and a decrease in adult prescriptions during that period. Moreover, a 2003 survey found the rate at which American children were prescribed antidepressants, including Prozac, Zoloft, and Paxil, almost doubled (to 49 percent) in five years, with the steepest increase, to 64 percent, among preschool children. In response to this sharp increase and a Harvard study that found that 22 percent of children and adolescents who had been prescribed any one of the antidepressants in the selective serotonin reuptake inhibitor (SSRI) group suffered drug-induced psychiatric adverse effects within three months and 74 percent experienced "an adverse event to an SSRI over the course of their treatment" (Wilens et al. 2003, 143), in 2004 the FDA issued a public health advisory warning that the drugs might be associated with an increased risk of suicidality in both children and adults. After the warning, the number of antidepressant prescriptions dispensed to patients aged eighteen and under dropped nearly 20 percent (Rosack 2005, 1). Katz and colleagues (2008), however, point out that this decrease was accompanied by a decrease in ambulatory visits and a significant increase in suicides.

Despite this decline, it is estimated that at least four million children and adolescents suffer from depression and thus are candidates for Prozac or alternatives. Many millions more might also be targeted for use when SSRIs are prescribed for behavioral problems. However, because these drugs treat symptoms rather than causes, it is questionable what benefit they have in the long run for many of these children. One could argue that the prescription of drugs to treat serious psychological and emotional disorders in children fails to properly address the social dimensions. Why do so many children now suffer from depression? Also, for minors, the issue of consent is often challenging. Although one might assume that most parents and guardians place the child's health as predominant in consenting to drug therapy, experiences with Ritalin suggest that at least in a minority of cases the prime reason is to better manage or control the child's behavior. As noted earlier, while informed consent is always at issue when any type of intervention in the brain is involved, with children the issue is clouded even further.

Policy Implications of Brain Science Findings

As these two issues illustrate, inherent in all applications of neuroscience are challenging policy issues and trade-offs that reveal an urgent need for more systematic, anticipatory analysis of the social consequences of these game-changing innovations. Our heavy dependence on technological solutions to health and social problems makes it difficult to curtail or slow the widespread use of the latest drug or procedure. Moreover, active marketing and publicity often promote their use long before the risks of intervention are fully understood. The policy implications of the new neuroscience, however, are expansive and touch on most areas of human existence. Because of the ethical issues inherent in brain intervention, these techniques must be scrutinized for their impact on the individual and society as a whole. For instance, Illes and Racine (2005) conclude that the "interpretation of neuroimaging data is a key epistemological and ethical challenge," and to that one should add a "policy challenge" as well.

Although controversy over the brain is nothing new, the political debate concerning the emerging knowledge about the brain and new intervention techniques will deepen. These discoveries are already confronting prevailing societal values by demonstrating that much of what we are can be explained by the actions of neurons and neurotransmitters. Neuroscience findings require a reevaluation of democratic concepts of equality, individual autonomy, freedom, and responsibility. Neuroscience findings also challenge conventional perceptions of human nature and the conceptualization of the mind as a blank slate to be shaped by experience, which is a core assumption of the Standard Social Science Model (Fowler and Schreiber 2008).

Until recently, brain intervention techniques were highly intrusive and crude (e.g., frontal lobotomies and early electroconvulsive techniques) and knowledge of how the brain functions was highly speculative (e.g., phrenology). The intrusive physical character of many early procedures resulted in gruesome cases of abuse that could be exploited by opponents and generate public outrage. Although the more overt and dramatic physical risks of brain intervention have been largely eliminated, because the new techniques are both more effective and subtle, they might in reality be more problematic. First, these procedures are more difficult targets for criticism because they are less easily sensationalized. Second, because they appear less intrusive, newer interventions have intrinsic support in a value system dependent on technological solutions. Psychotropic drugs, rather than being feared, are welcomed as solutions to a wide array of personal and social problems and generally portrayed in positive terms by the mass media. They are appealing because, under the auspices of medicine, they promise happiness, health, weight loss, and, increasingly, performance enhancement, and thus any caution about overuse or misuse tends to be dismissed.

Just as the "old" eugenics was replaced by more nuanced forms based on genetic screening and diagnosis, so the older forms of mind control have given way to more subtle methods such as the use of mood-altering drugs, DSB, and virtual reality techniques, which are often embraced from an early stage. Fears of eugenics and behavior control are more easily manifested in involuntary sterilizations and coerced psychosurgery than in routine uses of brain imaging, medication, or brain stimulation. Similarly, as behavioral disorders and addictions are transformed into neurological disorders caused by biochemical anomalies, they instinctively become the province of medicine. The subject of a proposed intervention to alter his or her behavior is now a patient who must be treated with the latest procedure or drug. Thus, the diffusion of more sophisticated and seemingly less invasive intervention techniques complicates rather than removes policy concerns over their use. As a result, there are likely to be increased demands for government involvement in neuroscience applications (see, e.g., Greely and Illes 2007; Kulynych 2007).

The move of neuroscience to the policy domain alters the context by bringing to the fore political considerations and divisions and situating the resolution of these issues in the milieu of interest-group politics. With the high economic, social, and personal stakes surrounding neuroscience and its applications, this is unavoidable. The policy issues raised are reflected in the almost daily announcements of new findings in neuroscience and related scientific fields and in the development of innovative technological applications. Intervention in the brain is a particularly controversial policy area because of rapid, successive advances in knowledge and the shortened lag time between the announcement of basic research findings and their clinical application. As noted by Hyman (2007), the potential use of brain imaging "to reconstruct a person's recent experience or investigate his or her veracity" not only raises traditional bioethical questions of privacy but also engages broader communities that are not often represented in discussions of bioethics.

Although many of the specific issues raised by neuroscience applications are distinctive, fundamentally the policy dimensions are similar to those in other areas of biomedical research. At their base, there are three policy dimensions relevant to neuroscience. First, decisions must be made concerning the research and development of the technologies. Because a sizable proportion of neuroscience research, both civilian and military, is funded either directly or indirectly with public funds, it is important that public input be included at this stage. The growing prominence of forecasting and assessing the social as well as technical consequences of biomedical technologies early in research and development suggests one means of incorporating broader public interests, although it is questionable how best to design assessment processes so as to evaluate the efficacy, short- and long-term safety, and social impact of brain interventions, especially when there is a ready market for them. The need for anticipatory policy making is crucial here but, until now, has been largely absent in neuroscience.

The second policy dimension relates to the individual applications of technologies once they become feasible. Although direct governmental intrusion into individual decision making in the medical arena until recently was limited, governments have at their disposal an array of more or less explicit devices to encourage or discourage individual use, including tax incentives or disincentives, the provision of services, and education programs. Although conventional regulatory mechanisms could be utilized to protect potential users or targets of brain interventions, it is critical that their effectiveness be assessed and that they be molded to fit neuroscience. As illustrated throughout this book, neuroscience policy has conspicuous importance in contemporary politics because it challenges passionately held societal values relating to the self, privacy, discovery, justice, health, and rights.

The third dimension of neuroscience policy centers on the aggregate consequences of applications in a population. For instance, how might the use of neural imaging to type personalities affect our concept of equality of opportunity or legal responsibility? How might it be used in employment, and what social and economic impacts would this have? What impact will the diffusion of DBS and new drugs have on the treatment of the mentally ill? What impact will widespread cognitive enhancement have on society? Policy making here requires a clear conception of goals, extensive data to predict the consequences of each possible course of action, an accurate means of monitoring these consequences, and mechanisms to cope with the consequences deemed undesirable. Moreover, the government has a responsibility to ensure safety and quality control standards, as well as fair market practices.

As table 3.1 illustrates, the governmental response to neuroscience developments can take many forms. Moreover, governmental intervention can occur at any point in the process, from the earliest stages of research to the application of specific techniques in specific populations. Brain policy, then, can be permissive, affirmative, regulatory, or prohibitive. The government always has the option of taking no action, thus permitting any actions by the private sector. It can also make affirmative policies that promote or encourage certain activities, for example, by funding research or providing services to facilitate more widespread use

Table 3.1
Types of Governmental Involvement in Neuroscience

Ban or block technology
Regulate technology
Take no action
Encourage individual use
Mandate use of technology

of a particular technique or application. The question of whether the government ought to be providing such encouragement, and if so by what means, is a matter of debate. Should public funds be used to pay for expensive brain interventions when patients cannot afford them? Should private insurers be required to cover these expenses? Should we distinguish between therapeutic and enhancement uses? Affirmative policies are often redistributive and thus introduce potential conflict between the negative rights of individuals to use their resources as they see fit and the positive rights of recipients of government support who need it. Also, in some instances, the line between encouragement and coercion or mandate can be readily crossed.

The most obvious examples of the regulation of neuroscience are psychoactive drugs, including those used for cognitive enhancement (Flaskerud 2010). Although the research-and-development phase of all pharmaceuticals is highly regulated by the FDA, control of individual use is less clear, as is the potential overuse in the aggregate. A broader regulatory approach would allow a substance to be studied socially and ethically as well, by focusing attention on the various social processes involved in moving a technology along the different axes of regulation. A good example might be Ritalin, which has both official and unofficial uses. The unofficial uses attract the legal power to control illicit amphetamine use, while the official uses depend on the soft forces of parental and teacher approval for its use and the harder forces of official approval for stricter control of errant children's behavior (Ashcroft and Gui 2005). Moreover, in all instances an important regulatory device is price, including price modification through taxation or license fees.

Although regulatory policy can be framed to apply only to government-supported activities, it usually consists of sweeping rules governing all activities, in both the public and private sectors. Regulation can be used to ensure that standards of safety, efficacy, and liability are adhered to, and, unlike professional association guidelines, which also often set minimum standards, regulations have the force of law and usually include legal sanctions for violations. Greely and Illes (2007, 414), for instance, propose a rigid regulatory scheme for new lie detection technologies similar to the FDA system for controlling new drugs or biologics where anyone caught using them without approval is subject to both criminal and civil sanctions. Therefore, the sponsors of a lie detection technology would be required to prove through rigorous large-scale and inclusive trials that it was safe and applicable. Similarly, after an exhaustive review of the maze of federal regulations of MRI applications, Kulynych (2007) concluded that we need to "eliminate the duplicative functions of our multiple research bureaucracies, replacing them with a single federal office promulgating one set of regulations for all research, with the flex-ibility to propose and implement new regulations as necessary to keep pace with evolving oversight concerns" (316). Such a move, of course, would be highly politically charged in the United States.

Martin and Ashcroft (2005) present useful distinctions regarding the regulation of neurotechnology. The *institutions* of regulation range from local ethics committees, to the courts, to subnational and national bodies, both private and public. The *functions* of regulation can entail safety requirements, restriction of access to a specific class of users, deterring abuse or misuse, and so forth. The *impacts* of regulation may be desired and expected, desired and unexpected, undesired but expected, or undesired and unexpected. (Of note, the impacts of regulation may affect different segments of society differently.) The *subjects* of regula-tion are both those who are regulated and those who are affected by such regulation, while the *principles* of regulation relate to the extent to which regulatory policy is designed with specific moral or social principles in mind. Finally, there are various *styles of regulation*— centralist or democratic, formal or informal, egalitarian or libertarian in respect to distributive justice, and libertarian or communitarian in respect to criminal justice.

Last, although far less common than regulation, prohibitive policies could be implemented that would reduce the options available in neuro-science applications. The most straightforward form is laws that impose criminal sanctions on a particular research activity or application such as psychosurgery. Another type of prohibitive policy is precluding public funding for specific areas of research and development (e.g., for certain types of fetal or human embryo research) or specific services, such as DSB. It remains to be seen what, if any, areas of neuroscience are candi-dates for prohibition, but governments do have that option, just as some

Favor the Technology

Favor government involvement	Oppose government involvement
Mandate use	Absolute individual choice
Fund public research	
Favor free market	
Incentives for private research	Commercialization without
Encourage individual use	government intervention
• incentives	
• education	Professional guidelines
• free services	
Private policy	
Consumer protection	
Set standards of practice	Bioethical deliberation

Favor government involvement	**Oppose government involvement**
Monitor social consequences	No public funding
Technology assessment	
Regulate marketing practices	Fear mandates, social control,
Discourage individual use	stigmatization, Big Brother
Strict regulation	
Prohibit use	

Oppose the Technology

Figure 3.1
The Role of Government in Neuroscience

jurisdictions have limited ECT. These policies often reflect political motives or are a response to the demands of particular interest groups.

Figure 3.1 illustrates the wide range of possible policy responses. Many of these options have been used by various countries with regards to stem cell research, reproductive and genetic technologies, or earlier forms of brain intervention. They clearly demonstrate the wide divergence of policy responses available, as well as the diametrically opposed positions on the role of the government that are found within and across societies. Given the history of policy in these related fields, there appears very little likelihood of anything approaching a consensus emerging either on the role of government in neuroscience or on the preferred policies regarding specific applications.

Setting a Neuropolicy Agenda

The difficulties of neuroscience policy making can be best understood if analyzed as part of a broader policy process. There have been many useful analyses of this process, which is usually presented as a series of stages or types of action. For example, Anderson (1990) envisions the process as consisting of five stages: problem identification and agenda setting, policy formulation, policy adoption, policy implementation, and policy evaluation. In the problem identification and agenda-setting stage, an issue becomes a matter of public concern. Of the multitude of problems faced by society, only a small number receive public recognition, with even issues that are salient matters of public debate failing to trigger governmental action. To explain why this happens, Cobb and Elder (1983) identify two agendas. The *systemic* agenda consists of all issues that are commonly perceived by members of the political community as falling within the legitimate jurisdiction of existing governmental authority. In contrast, the *institutional or formal* agenda is that set of items explicitly up for the active and serious consideration of authoritative decision makers. Only those issues that are "commonly perceived by members of the political community as meriting public attention and involving matters within the legitimate jurisdiction of existing government authority" are placed on the systemic agenda (Cobb and Elder 1983). Thus, the systemic agenda consists of general categories that define which legitimate priorities merit attention by the government, while the institutional agenda consists of those problems perceived as important by decision makers that engender an effort to develop a course of action.

After an issue has reached the government's formal agenda, policy formulation begins. This is a complex process involving a range of actors inside and outside government, in which interest groups push for particular policies and attempt to influence priorities. Formulation usually includes analysis of various policy options, including inaction. Although policy adoption, which typically includes a legislative enactment or an executive directive, is usually the most salient stage of policy making, policy formulation is the stage during which the boundaries of government action are defined. Once the policy is adopted, the focus shifts to the executive branch, which is responsible for implementation. Agencies make rules, adjudicate, use their discretion to enforce the rules and laws, and maintain program operations.

As soon as a policy is implemented, it is important to evaluate its impact. Evaluation entails comparing expected and the actual performance levels to determine whether goals have been met. It is also the stage in which the impact of new technologies on the existing policy can be assessed and adjustments in policy made to accommodate them. Are new policies needed, or can existing policies simply be adapted? Anderson (1990) makes a useful distinction between policy outputs, which "are the things actually done by agencies in pursuance of policy decisions and statements," and policy outcomes, which "are the consequences, for society, intended and unintended, that stem from governmental action or inaction" (223).

To be effective, a policy must progress through all five stages; however, the newness of many neuroscience issues means that the most urgent attention must be directed toward agenda setting. Because there is always a multitude of issues competing for placement on the formal agenda, the fact that many issues arising from neuroscience seem to be matters of public concern is no guarantee that policy makers will recognize them, consider them a political priority, or put them on the formal agenda. The policy importance of a technological innovation depends both on the degree to which it provokes a public response and on how it is perceived by organized economic and political interests.

The most prominent work in the field of agenda setting has been done by Kingdon (1995), who stresses the importance of timing and suggests that moving an issue onto or higher up on the agenda involves three processes: problems, proposals, and politics. *Problems* here refers to the process of persuading policy makers to pay attention to one problem over others. The likelihood that a problem will rise on the agenda is heightened if it is perceived as serious by policy makers. *Proposals* refers to the process by which proposals are generated, debated, and adopted. Typically, this process takes patience, persistence, and supporters' willingness to try many tactics. Framing a proposal in such a way that it is seen as technically feasible, compatible with policy-maker values, reasonable in cost, and enjoying wide public support increases the chance of success. Finally, *politics* here refers to political factors that influence agendas, such as the political climate and the actions of advocacy or opposition groups. While these three processes operate independently, the actors may overlap. For Kingdon, successful agenda setting requires that at least two of the processes come together at a critical time, thus opening up a *policy window*. Policy windows, however, are not just chance opportunities; they can also be nurtured. Furthermore, under the right circumstances, they can be seized on by key political players to move an issue onto the agenda. Baumgartner and Jones (1991) add that the image of the policy problem is crucial. If it is portrayed as a technical problem rather than as a social question, experts can dominate the decision-making process. In contrast, if the ethical, social, or political implications of proposed policy assume center stage, a much broader range of participants might be involved.

Agenda setting in federal systems like the United States is further complicated because policy making can take place at the federal, state, or local levels, with multiple agendas in play at various institutions at each level. A particular problem might be perceived and acted on by decision makers in one institution at one level of government in one state but not be perceived at all, or be perceived differently, in others. Developments that would cause a particular issue to be placed on the government agenda, such as technological change, interest-group demands, or a widely publicized event, might be of limited interest geographically. A complicating factor for neuropolicy is the diversity of agendas among the myriad professional associations, business enterprises, and individual practitioners in the private sector that are active participants in neuroscience and thus have a stake in any policy. The mass media can also play a significant role in getting issues recognized by policy makers and, under the right circumstances, can be decisive in pressuring them to act quickly in response to a perceived "crisis."

Although attention here focuses on public policy making, Bonnicksen (1992) presents a private-sector policy model that defines "regularized rules and procedures in the medical setting as the desired end of decision-making" (54). She favors increased emphasis on this model because she believes that governmental action has been "unlikely, premature, and unwise" in many areas of biomedicine. "Private policies have weaknesses, but these can be partly remedied by political strategies. Public policies, in contrast, are difficult to refine or revamp if found to be erroneous or misguided. Traditional values of self-determination are perhaps more easily protected in the medical than in the political setting. At the very least, a private policy alternative warrants consideration for contentious biomedical issues" (55). It is argued here, however, that while it is important that private-sector solutions be pursued, the wide scope of issues emerging from neuroscience and their broad implications for many societal groups make them a matter of public concern and, thus, of public policy.

It should be emphasized that policy making is principally a gradual process, not manifested in quick, decisive action. Neuroscience policy, like genetic and reproductive policy, is most likely to form in fits and starts, in a fragmented, unsystematic manner. Most policy analysts agree, and evidence from neuroscience helps explain why (see chapter 9), policy making is not a textbook rational process but at best an incremental one in which analysis is limited to a small number of alternatives, the emphasis is on ills to be remedied rather than on positive goals to be met, and the analysis of future consequences is scant. Moreover, under the incremental model, new policies usually do not vary substantially from past ones, meaning that novel areas like neuroscience have a difficult time penetrating the process inertia. Additionally, as a product of bargaining, compromise, and accommodation, policy making favors the status quo, while decision makers tend to avoid considering choices that vary significantly from existing practices.

Throughout the policy process, governments have at their disposal many mechanisms for facilitating expert input. Permanent mechanisms include the use of internal bureaucratic expertise, science advisers, offices of science and technology, and science advisory councils, while temporary mechanisms may make use of task forces, ad hoc committees, commissions, consultants, conferences, hearings, issues papers, and white papers. Such mechanisms for acquiring expert opinion may be specific to a particular application of neuroscience, such as DBS, broader in scope across the range of brain intervention technology, or, as illustrated by the President's Council on Bioethics in the United States, cover a wide swath of issues.

The gathering attention to neuroscience has raised awareness of its importance to all aspects of human life, but it has not yet placed neuroscience high on the policy agenda, even within the health care arena. The enormous ramifications of neuroscience for society, however, suggest this situation will change. The political debate surrounding the expanding knowledge about the brain and new intervention techniques promises to be intense. The wide array of new intervention capacities and the tremendous costs of CNS-related health care problems, along with the coalescing view that the the mental and physical dimensions of health are inseparable, will elicit considerably more attention from policy makers in the coming decades. At the least, this analysis demonstrates the pressing need for more methodical and pro-active investigation of the social consequences of the rapid diffusion of these impressive, often dramatic, innovations as neuroscience slowly moves onto the public agenda.

Conclusions

This chapter has introduced some of the major ethical and policy issues arising from interventions in the brain. Although these issues are present in all intervention modes, they become increasingly problematic as the techniques used become more intrusive and the changes they effect potentially irreversible. The uncertainty of potential, often subtle, long-term effects of many interventions is especially difficult to predict at this stage of development, but risks both to individuals and to society at large are always present whenever we deal with the brain. Moreover, while some of the issues, such as concerns related to informed consent, stigmatization, incidental findings, and distributive justice, can be addressed by establishing protective procedures or legislation, other, more conceptual issues such as those relating to authenticity, autonomy, and cognitive or psychological enhancement will provoke increased debate as we move from the clinical to the behavioral dimension. Furthermore, because neuroscience challenges fundamental beliefs across the ideological spectrum, those on both the political left and the right are likely to interpret the findings in the most favorable light to serve their purposes and reject findings that contradict their views. The following chapters focus on this latter dimension and consider the policy implications of neuroscience research, especially neuroimaging.

4

Implications for Political Behavior: Addiction, Sex Differences, and Aggression

Neuroscience research and applications by their nature raise difficult political and policy issues and have wide-ranging implications for political behavior. Novel studies are reported almost daily in the media that raise questions about the role of the brain in explaining, among other things, altruism, love, truth telling, addiction, personality, political leanings, racial prejudice, moral reasoning, and individual responsibility (Davidson and Begley 2012). This chapter examines the ramifications of neuroscience findings for understanding political behavior in three disparate areas: addictive, risk-taking behaviors; sex differences and sexual orientation; and aggressive behavior. Each of these areas also produces distinctive, politically sensitive policy dilemmas that flow from the discussion of ethical and policy issues raised in chapter 3. In combination, they illustrate how brain research challenges long-held assumptions about human behavior and provokes political controversy.

The Brain and Addictive Behavior

Addiction is a major social and health problem that increasingly has been linked to the brain. In the U.S. alone, the five most-chronicled "hard" addictions— alcohol, drugs, tobacco, gambling and eating disorders—are estimated to cost taxpayers and businesses $590 billion annually, primarily lost productivity and government-assisted medical treatment. Moreover, this does not count the huge amounts that users personally spend on drugs, alcohol, tobacco and gambling or the major disruptions on families and society at large. Furthermore, it is estimated that almost 20 percent of deaths in the U.S. are linked to the use of tobacco, alcohol and illegal drugs with another 18 percent to obesity. (Mokdad et al. 2004)

The concept of addiction is complex, and the delineation of its defining characteristics has fostered considerable debate, but most researchers agree that it involves a compulsive pattern of use, even in the face of negative health and social consequences (Campbell 2010; Andrews et al. 2011; Taylor, Curtis, and Davis 2010). A major finding is that all abused substances, no matter what the mechanism, stimulate the brain's reward system and induce feelings of pleasure that can override basic survival activities. "All drugs of abuse stimulate brain mechanisms responsible for positive reinforcement and reward, and most of them can also alleviate unpleasant feelings" (Zangen 2010). When the brain is exposed to these addictive substances, tolerance builds up, and with excessive stimulation over time, the brain learns to adjust and its functions change.

Addiction as a Disease of the Brain

There continues to be extensive debate as to what constitutes addiction and whether behavioral compulsions should be classified as diseases. Vrecko (2010), for instance, while admitting that there are physiological components to behavioral compulsions and that effective addiction therapies may act on the bodies of patients, argues against conceptualizing behavioral compulsions as diseases and the therapeutic interventions for such compulsions as treatments. Buchman, Illes, and Reiner (2010) warn that while the language of neuroscience used to describe addiction as a brain disease may reduce attitudes such as blame, it might also yield unintended consequences by fostering discrimination commonly associated with pathology. Similarly, Keane and Hamill (2010) contend that neural explanations of addiction pathologize drug users by viewing drugs as agents of disease that inevitably produce brain dysfunction and conclude that while neuroscientific theories of addiction represent a major contribution to knowledge, they are limited in scope because they remove addiction from the social context in which its harms and dilemmas are experienced (also see Campbell 2010). Vrecko (2010) adds, however, that social analyses of addiction biology are particularly important today, insofar as addiction is increasingly understood and managed in terms of biology, and that current approaches developing within neuroscience and psychopharmacology raise important challenges to social accounts of addiction that give inadequate consideration to physiology (Kushner 2010; Courtwright 2010; Kuhar 2010).

Quenqua (2011) contends that the rethinking of addiction as a medical disease rather than a strictly psychological one began about fifteen years ago when researchers discovered through brain imaging that drug addiction resulted in identifiable physical changes to the brain. The National Institute for Drug Abuse brain disease paradigm, developed in the 1990s, asserts that prolonged substance use turns on "a switch in the brain" that permanently transforms brain mechanisms, making interventions aimed at reversing addictions extremely difficult (Kushner 2010). Thus, addiction is a chronic, relapsing brain disease with a social context, a gene-environment-stress interactive component, and significant comorbidity with other mental and physical disorders. Addictive substances, in effect, hijack the brain's reward circuit and the malfunctioning limbic system overrides the regulatory functions of frontal lobes, thus destabilizing the brain's checks and balances (Kushner 2010). Moreover, "studies of effects of drugs on receptors have shown us how drugs can change gene expression and how drugs can change the biochemical makeup of the brain" (Kuhar 2010, 33).

Although drug use often begins voluntarily and develops slowly over time, users progressively lose control with the onset of addiction. Persistent use leads to long-term changes in brain structure and function as neurons become more responsive to the biochemical changes triggered by drug consumption. Imaging studies have shown specific patterns of abnormal activity in the brains of many addicts (Courtwright 2010). Campbell (2010, 101) argues that redefining addiction as a brain disease rendered it amenable to neuroimaging technologies. Neuroscientists were recruited to the substance abuse research arena in response to the recoding of addiction, drug dependence, or substance abuse as a "chronic relapsing disorder characterized by compulsive drug seeking, a loss of control limiting intake, and emergence of a negative emotional state when access to the drug is prevented" (Koob and Le Moal 2006, 19).

Addiction: The Mounting Evidence

The two most important tracks in current research are concerned with the neural-genetic basis of addiction (Windle 2010; Kushner 2010) and the impact of addictive substances on the brain and its functioning (see Buchman, Skinner, and Illes 2010). Recently, analogous research has been initiated on eating disorders (Vocks et al. 2010) and obesity (Davis and Carter 2009; Taylor, Curtis, and Davis 2010). Although addiction can affect all organs of the body, the primary target is the brain (Kuhar 2010). Addictive substances or behaviors are linked to the brain's capacity to experience feelings of pleasure and pain, which has evolved to manage fundamental behaviors such as feeding, reproduction, and aggression. When the brain's pleasure centers are stimulated, the brain sends out signals to repeat the pleasure-producing behaviors.

According to Dupont (1997), the brain is characterized by the immediate quest for pleasure. "When it comes to many natural pleasures, the brain has built-in protections. It has powerful feedback systems to say 'enough' when it comes to natural behaviors, including aggression, feeding, and sex" (1997, 5). The brain is selfish, however, in that automatic brain mechanisms do not account for delayed gratification. Therefore, when the brain comes into contact with the addicting substance or behavior, and when this experience triggers the pleasure centers, there is a strong incentive to repeat the exposure. These feelings, of course, are constantly mediated by culture and other environmental forces that influence the behavior (Chiao and Ambady 2007). The brain is programmed to produce survival behavior, and drugs seem to utilize and seduce these circuits, thus explaining why drugs are so powerful and gain so much control over a person's behavior (Kuhar 2010, 33). The transition from recreational drug use to addiction can be seen as an escalating cycle of dysregulation in the brain reward systems, leading eventually to compulsive drug use and a loss of control (Zangen 2010).

To complicate matters, there is substantial evidence of genetic predispositions to addictive behavior and possibly vulnerability to addiction from particular substances such as alcohol (Hall, Carter, and Morley 2004). For instance, twin studies have generally yielded heritability estimates for alcohol dependence in the range of 50 to 60 percent (Gelernter and Kranzler 2009). Moreover, people who are genetically oriented toward immediate gratification or toward impulsive behavior and risk taking are also at higher risk for addiction (Anderson et al. 2011; Taylor, Curtis, and Davis 2010). Despite the importance of genetic, cultural, and social factors in explaining addiction, ultimately we must focus on the brain, not only because it is the key to unlocking the causes of addiction but also to determine how addictive drugs and behaviors affect the functioning of the brain and change the brain of the addicted person (Kuhar 2010).

Much of what we are learning about addiction and the effect of these substances on the brain comes from research utilizing PET and fMRI with addicts in craving or withdrawal (Anderson et al. 2011; Dunbar, Kushner, and Vrecko 2010). Imaging shows that when addicts feel a craving, there is a high level of activation in the mesolimbic dopamine system. In one study (Childress et al. 1999), PET was performed on patients undergoing treatment for cocaine addiction while they were exposed to cues associated with past craving episodes. The scans indicated activation of the dopamine system in the ventral tegmental area at the moment the addicts expressed intense craving. Another study

(Tanda, Pontieri, and DiChiara 1997) found that the mesolimbic dopamine system was also active in nicotine addiction, while a third (Rodriguez de Fonseca et al. 1997) found that marijuana affected the same brain circuit.

As our understanding has increased, it has become clear that addiction extends far beyond the physical need for chemicals to a wide range of activities that produce feelings of dependency in our neural networks and affect the brain circuits involved in emotions and judgments. All drugs of abuse produce their effects by traveling through the bloodstream to the brain. Once in the brain, each drug alters the function of specific brain cells. Stimulants such as cocaine act as an exciting influence on certain nuclei, while depressants such as alcohol and the narcotics act to inhibit their activity. Some drugs act by blocking the reuptake of neurotransmitters from the synapse to the sending axon, thus facilitating transmissions by prolonging the time the neurotransmitter remains in the synapse. Other drugs actually mimic particular neurotransmitters by sending their own messages and occupying the receptors. Moreover, some substances, such as alcohol, interfere with the cell membrane, while others affect the synapse, working either as agonists (activating transmission across the synapse) or antagonists (blocking the receptor sites on the dendrites).

Hommer, Bjork, and Gilman (2011) compared imaging evidence for two major hypotheses of how alterations in the brain's reward system underlie addiction: (1) the impulsivity hypothesis that addiction is characterized by excessive sensitivity to reward combined with a failure of inhibition and (2) the reward-deficiency hypothesis that addicted individuals have a reduced response to non-drug rewards that leads them to seek drugs in preference to more socially acceptable goals. They found that while PET studies of dopamine receptor density and dopamine release strongly support the reward-deficiency hypothesis, more recent fMRI studies of goal-directed behavior provide both support and contradiction for each of the hypotheses. Thus, it "is likely that aspects of brain function described by both the impulsivity and reward-deficiency hypotheses contribute to the pathophysiology of addiction" (Hommer, Bjork, and Gilman 2011, 50). One problem with addiction is that more than 80 percent of addicted individuals fail to seek treatment. Although this treatment resistance could in part reflect a failure of society to recognize addiction as a disease and the blame and repudiation placed on addicted person, Goldstein and colleagues (2009) suggest that this impairment also reflects a dysfunction of the neural circuits underlying insight, self-awareness, and appropriate social, emotional, and cognitive responses.

There is also evidence that drug addiction involves a failure of various subcomponents of the executive systems that control key cognitive modules that process reward, pain, emotion, and decision making and that the diverse patterns of drug addiction and individual variations in the transition to addiction may emerge from differential vulnerability in one or more of these subcomponents (George and Koob 2010, 232). Dolan, Bechera, and Nathan (2008) also found that individuals with substance dependence exhibit poorer executive functioning, which could partially explain poor treatment outcomes: "a substance-dependent individual with poor baseline executive functioning may have greater difficulty changing his or her behavior because of deficits in set-shifting abilities or cognitive flexibility" (818).

Whatever the precise mechanism of a specific substance, when a neurotransmitter is excessively stimulated over a long period, the brain reestablishes equilibrium by reducing the sensitivity of the affected receptors or by decreasing their number, and tolerance builds. This process, termed *down-regulation*, means that the more the brain is exposed to chemicals affecting a neurotransmitter, the less the brain responds to a specific dose. Therefore, to experience the same effect, the addict must use higher doses. A related effect, physical dependence, is manifested by withdrawal symptoms experienced when use of the substance is discontinued. Such symptoms vary by substance and reflect the cellular adaptation of the neurons of that area of the brain to the sustained presence of it. Withdrawal symptoms manifest the shock to the brain of a rapid alteration in the chemical environment. Frequently they are interpreted by the addict as a "need" to resume use of the substance.

A key to understanding the biochemical bases of addiction, then, is at the molecular and cellular levels in the mechanisms of neurotransmitters (De Biasi and Dani 2011; Kuhar 2010). Drugs acting at specific receptors actually alter gene expression and produce significant changes in the chemical makeup of the brain. For Kuhar (2010), the brain on drugs really is a different brain. Two prominent theories of addiction focus on dopamine and the endorphins. According to the former theory, most of the drugs of abuse, including alcohol, cocaine, amphetamines, and narcotics, stimulate the dopamine-producing neurons in the median forebrain bundle, the neural pathway that connects the midbrain to the forebrain (Kuhar 2010). Specifically, researchers have discovered that several areas of the brain, the ventral tegmental areas and the nucleus accumbens, have high concentrations of dopamine-containing neurons, and that all drugs of abuse, through varied mechanisms, trigger the release of relatively large amounts of dopamine into the synapses of these neurons. This increased production of dopamine contributes to the processing of rewarding sensory stimuli, creating the euphoria associated with the high, thereby reinforcing its continued use. Research has demonstrated that if dopamine production is turned off by dopamine-suppressing chemicals, the stimulating effects of the drug are blocked. For instance,

Changes in the brain arise during chronic exposure to nicotine, producing an altered brain condition that requires the continued presence of nicotine to be maintained. When nicotine is removed, a withdrawal syndrome develops. Thus, nicotine taps into diverse neural systems and an array of nicotinic acetylcholine receptor (nAChR) subtypes to influence reward, addiction, and withdrawal. (De Biasi and Dani 2011, 115)

The second theory applied specifically to opiate addiction focuses on a group of peptides, the endorphins, of which more than a dozen natural forms are known. The endorphin brain system moderates pain, promotes pleasures, and manages stress. Endorphins also act as neurohormones and can affect nerve functioning at distant sites in the nervous system through the blood. It is postulated that endorphins can explain the physiological dependence of heroin. When external opiates are taken, the brain ceases to produce natural endorphins, and as a result, the person becomes totally dependent on the drug for relief of pain or the feeling of pleasure. Termination of drug use results in withdrawal symptoms until of the brain resumes normal endorphin production (Bootzin, Acocella, and Alloy 1993, 324). Furthermore, research indicates that the opiate receptor sites can be occupied by antagonists such as naloxene and naltrexone, which are used to treat overdose and addiction. Even if the opiates get to the receptor sites first, the antagonists cover the sites, thus blocking the drug's capacity to produce a rush.

With a better understanding of the function of neurotransmitters and receptor sites and the mechanisms by which drugs influence neural activity, we should be able to ascertain why some people have heightened susceptibility to addiction, and to offer preventive treatment. Silveri and colleagues (2011), for instance, found that adolescents with a family history of alcoholism exhibit differences in brain structure and functional activation when compared with adolescents without such a family history. Moreover, corrresponding to the observation that frontal brain regions and associated reciprocal connections with limbic structures undergo the most dramatic maturational changes during adolescence, Silveri and colleagues found heightened activation in the frontolimbic search territory in these subjects. They suggest this demonstrates a significant influence of family history on brain activation during the performance of a response inhibition task, reflecting a neurobiological vulnerability that might include reduced neuronal efficiency or recruitment of additional neuronal resources, or both.

In the largest imaging study of the brains of adolescents, Whelan and colleagues (2012) found that diminished activity in specific networks, particularly one in the orbitofrontal cortex (OFC), is associated with experimentation with alcohol, drugs, and cigarettes in fourteen-year-olds, to the extent that someday it could be used as "a risk factor or biomarker for potential drug use." Similarly, in their fMRI study of adolescent smokers, Galván and colleagues (2011) found that smoking behavior and nicotine dependence were negatively related with neural function in the cortical regions, suggesting that smoking too can modulate prefrontal cortical (PFC) function. Because of the known late development of the PFC, which continues through adolescence, it is possible that smoking may influence the trajectory of brain development during this critical developmental period.

In addition to helping explain how one becomes addicted, this research provides insights into how drugs affect the brain. Studies of brain cells demonstrate that repeated exposure of the brain to addictive drugs represents a chemical assault that alters the structure of the neurons in the pleasure circuitry. Over time, these changes starve the affected cells of dopamine, thereby triggering a craving for the drugs that will reactivate the release of high concentrations of that neurotransmitter. During withdrawal, a different brain circuit in the same brain region releases a small protein, corticotropin-releasing factor (CRF). When a person suddenly stops taking the addictive substance, CRF levels rise and withdrawal symptoms appear. Moreover, a significant finding is that many drug-induced brain changes are enduring (Kuhar 2010). Addictive drugs not only modulate transmission in the OFC, they also lead to a loss of synapses, therefore exacerbating an abnormal behavioral state that is characterized by impulsiveness and delay aversion (Bennett and Hacker 2011). Moreover, cocaine users have been shown to have a 5 to 11 percent decrease in the gray matter of the OFC, and there is loss of OFC gray matter following amphetamine use as well (Crombag, Ferrario, and Robinson 2008). Recent studies have also documented structural changes in the OFC in those with schizophrenia, most likely due to a decrease in the loss of synaptic connections, not changes in the density of nerve cells (Sapara et al. 2007; Nakamura et al. 2008).

Policy Implications of Brain Research on Addiction
While this research is progressing rapidly, there is little corresponding progress in the social interpretation of the findings that raise questions about antidrug policy, the legalization of marijuana, alcohol policy, and other issues surrounding addiction. Society's treatment of nicotine and alcohol may have to be modified for consistency. The evidence of the interchangeability of substances demonstrates that antidrug policies that concentrate on specific drugs at the exclusion of others are likely to fail (Kuhar 2010). For the addict, no drug is a safe drug, only a substitute. Neuroscience research on addiction, therefore, is likely to undercut many current policy initiatives and treatment regimens, but it also offers the promise of more creative and effective solutions.

Although research on addiction raises questions about the effectiveness of antidrug policies, most governments continue to make abstinence the centerpiece of their policy. The histories of addiction indicate that official abstinence is a failed policy because addiction is a chronic relapsing-remitting syndrome. The most problematic aspect of addiction and the treatment of addicted individuals is the high risk of relapse, which persists long after any withdrawal symptoms have abated, and often for a lifetime. The presence of actual physical changes in the brain could explain why some smokers will still crave a cigarette thirty years after quitting (Quenqua 2011). The common view suggests that the basic neural foundation for relapse is that addictive behaviors produce prolonged alterations by usurping normal mechanisms of associative memory and reward (Zangen 2010). Moreover, animal studies reveal that they will self-administer the same drugs that humans abuse, suggesting that drug addiction has a biological basis throughout the animal kingdom (Kuhar 2010). Any policies that fail to address this natural drive are unlikely to succeed in the long run.

In light of the technological advances of brain imaging and the current research on the biological bases of addiction to substances, such research has expanded into other habitual behaviors, such as gambling, aggression, sex, and eating. In doing so, however, there is a danger of extending the notion of addiction to any behavior that becomes patterned because it is found to stimulate the pleasure centers. Such a move would have wide legal repercussions and threaten the idea of individual responsibility for one's own actions. As will be discussed in chapter 5, the courts will be faced with novel defenses based on scientific evidence of neuronal susceptibilities. Although the social and cultural dimensions of addictive behavior remain crucial and the causes and effects of addiction are complex, eventually, neuroscience research might transform the long-running debate over moral versus medical models of addiction by providing a detailed causal explanation of addiction in terms of brain processes (Hall, Carter, and Morley 2004, 1491; MacKenzie 2011).

Compulsive Eating and Obesity

The World Health Organization has stated that obesity is increasing at an alarming rate globally, while the 2006 International Congress on Obesity warned that an "obesity pandemic threatens to overwhelm health systems around the globe." Of particular concern is the skyrocketing rate of obesity among young children and adolescents, especially in the United States, where the percentage of overweight children has doubled in the last three decades. The tracking of obesity from childhood to adulthood is well substantiated, with obese children being much more prone to chronic diseases. It has been suggested that today's children might be the first generation to die before their parents because of health problems related to weight. According to health expert Jay Olshansky, "within the next 50 years, life expectancy at birth will decline, and it will be the direct result of the obesity epidemic that will creep through all ages like a human tsunami. . . . There has been a dramatic increase in obesity among the younger generation and it is a storm that is approaching"' (quoted in Reuters 2005a).

Obesity has many negative health, social, and policy ramifications. Mortality and morbidity rates are elevated among overweight and obese persons. Obesity is a known risk factor for diabetes, heart disease, stroke, hypertension, sleep apnea, osteoarthritis, pregnancy complications, and some forms of cancer. A person 40 percent overweight is twice as likely to die prematurely as a person of average weight. As a result, obesity and the conditions related to it account for a mounting share of health care expenditures (Reidpath et al. 2002). Additionally, obesity has huge negative economic consequences for society in terms of lost productivity and premature pension payouts, as well as indirect costs for individuals, including poor health and a reduced quality of life. Although the cause of obesity is multifaceted and varies according to the individual, it is clear that in most cases, chronic overconsumption of food plays a fundamental role. When this type of overeating becomes compulsive and out of control, it is often classified as a food addiction, a label that has caused much clinical and scientific controversy (Davis and Carter 2009).

Food addiction, which more accurately may reflect addiction to specific components of food, can be described in much the same way as other addictive behaviors (Taylor, Curtis, and Davis 2010). Davis and Carter (2009) conclude there is compelling evidence that highly palatable foods eaten in abundance have the potential to cause the same alterations in the brain as conventional substance dependence and thus justify its inclusion as an addiction disorder. Like drugs, foods induce tolerance over time, so that increased amounts are needed to reach and maintain satiety. Likewise, withdrawal symptoms, such as distress and dysphoria, often occur during dieting, and there is a high incidence of relapse. These symptoms parallel to a remarkable extent those described in the *Diagnostic and Statistical Manual of Mental Disorders,* fourth edition (DSM-IV), for substance abuse and dependence, leading some to argue that food addiction should be considered a psychiatric illness (Volkow and O'Brien 2007; Davis and Carter 2009).

In making the argument for overeating as an addictive behavior, it is clearly not appropriate to include all cases of excessive food consumption. For some individuals, overeating is a relatively passive, habitual event that occurs almost without awareness, in the form of liberal snacking and large portion sizes (Drewnowski and Darmon 2005). For others, however, it can be compulsive and excessively driven. According to Davis and Carter (2009), there is sound clinical and scientific evidence that binge eating disorder is a phenotype particularly well suited to an addiction conceptualization. Imaging studies have shown that specific areas of the brain, such as the caudate nucleus, the hippocampus, and the insula, are activated by food as well as drugs, and that both cause the release of striatal dopamine (Taylor, Curtis, and Davis 2010).

The role of dopamine in the various aspects of reward has been vigorously debated in recent years. Some have argued that mesocorticolimbic dopamine fosters the motivation to engage in rewarding behaviors more than liking the reward does. However, this position is still debatable since reward involves a myriad of emotions, including anticipation, expectation, pleasure, and memory, that are difficult to separate experimentally. Other neurotransmitter systems such as gamma-aminobutyric acid (GABA), the opioids, and serotonin are also integral to the reward process (Davis and Carter 2009). Until recently, the neurobiology of overeating largely focused on the hypothalamus, but with the increasing sophistication of neuroimaging techniques it has been found that pre-frontal systems also play a prominent role in eating and appetite. There is now a view that the reinforcing effects of palatable foods are regulated largely by the same dopamine pathways as addictive drugs (Corwin and Grigson 2009); Davis and Carter 2009).

While food shares with addictive drugs the property of increasing dopamine in brain reward pathways, it also can bypass the adaptive mechanisms of normal reward, such as the habituation that constrains the responsiveness of the brain (Di Chiara and Bassareo 2007). Additionally, endogenous opiates are activated by food, particularly sweet foods, whereas the opioid blocker naltrexone has been shown to reduce cravings (Yeomans and Gray 1997). Compounds that act as inverse agonists in the endocannabinoid system are used both to treat substance addictions and to promote weight loss (Pelchat 2009). Furthermore, similar to what happens after other addictive transfers, following gastric bypass surgery to reduce the amount that can be physically ingested, patients often experience other addictive behaviors, such as alcoholism, smoking, or compulsive spending (Sogg 2007).

All this research suggests that, for some individuals, a predisposition to food addiction might be hard-wired. This makes sense from an evolutionary perspective since it would have been highly adaptive for the consumption of food to be rewarding, especially in the case of foods rich in fat and sugar, which can be rapidly converted into energy (Taylor, Curtis, and Davis 2010). A widely accepted theory postulates that the reward pathway evolved to reinforce the motivation to engage in naturally rewarding behaviors such as eating to promote survival in times of famine (Volkow and O'Brien 2007). However, recent developments in food technologies have changed the environment through the creation and production of foods that enhance their rewarding properties but are fattening (Kessler 2009). Furthermore, a shift toward sedentary work and lifestyles means this excessive caloric intake is no longer burned off.

Not everyone who is exposed to drugs becomes an addict, and not everyone who is exposed to high-fat, high-calorie foods becomes a compulsive overeater. These differences in susceptibility can be attributed in part to a genetic predisposition and/or to brain adaptations to excessive use over time, specifically, down-regulation of the dopamine D_2 receptors linked to addictive behavior (Roberts and Koob 1997). Vulnerability may also stem from various personality traits. For example, obese individuals tend to be more sensitive to reward and punishment and to display more impulsive behaviors (Davis et al. 2008). For these individuals, the forces driving food consumption are likely to go beyond physiological hunger. Research also suggests that eating for some is a means of self-medication in response to negative emotional states, such as depression, loneliness, boredom, anger and interpersonal conflict (Davis, Strachan, and Berkson 2004).

The concept of food addiction does not negate the role of personal choice or diminish the importance of cultural and social factors, but it does provide some insight into why many individuals continue to struggle with obesity. Classifying obesity as an addiction is a strong statement and implies much more than merely a change in semantics. It indicates that screening for addiction and binge eating should become a routine part of treatment for obesity, and, in the case of gastric surgery, that such screening should be an important part of postoperative follow-up. It also helps explain the lack of success of programs that do not incorporate pharmacotherapy or behavioral strategies specifically designed to address the addictive, brain-centered component of this illness. Although medicine may not yet accept compulsive overeating as an addiction, it cannot ignore evidence highlighting the role played by biological vulnerability and environmental triggers (Taylor, Curtis, and Davis 2010).

In contrast, Ziauddeen, Farooqi, and Fletcher (2012) find the addiction model unconvincing. Moreover, while Foddy (2011) agrees that there is a compelling body of evidence for food addiction and a growing consensus among neuroscientists that people can become addicted to food, he contends that few researchers concur that obesity is a disease in the same sense as drug addiction, that is, a disease of behavior with a neurological cause. Thus, he argues that obesity ought not be considered a neurobehavioral disease comparable to drug addiction. "Once we recognize that obesity and addiction share the same underlying neurological mechanisms, and once we understand what these mechanisms are, we must conclude either that they are either both diseases of the brain, or that neither of them is a disease of any kind" (Foddy 2011, 87).

Sexual Differences and the Brain

Until recently, human sexuality was the domain of psychology; however, extensive neuroscience findings are rapidly shifting emphasis to biochemical processes in the brain. This transformation has been widely criticized by those observers who hold that nurturing, culture, and the social environment are the most powerful forces influencing sexual behavior and other behaviors and the cognitive traits that typically differentiate the sexes. Current biological and neuroscience research, however, demonstrates that variation between the sexes and in sexual orientation is inextricably linked to differing hormonal influences on brain development that lead to striking variation between the brains of males and females, both on average and in normal distributions (Luders and Toga 2010). Moreover, some emerging research shows that sex-linked genes also play a direct role in human gender development (Hines 2011, 70). Specifically, sex-determining genes that are triggered by the sex chromosome complement initiate a series of events that determine an organism's sex and lead to the differentiation of the body in sex-specific ways. Such events also contribute to many unique sex differences, including the susceptibility to different diseases (Sanchez and Vilain 2010).

Although none of these findings exclude environmental contributions to behavior, cumulatively they require a shift in balance from nurture to nature as a prime focus of inquiry (Cahill 2010; Kimura 2007). For Cahill (2010), the "striking quantity and diversity of sex-related influences on nervous system function argue that the burden of proof regarding the issue has shifted from those examining the issue in their investigations generally having to justify why, to those not doing so having to justify why not" (29). As a result, neuroscience is producing intensified debate and shaking the conventional foundations of our perceptions of sex differences and equality. For instance, Kaisera and colleagues (2009, 53) argue that sex and gender differences, especially with respect to higher cognition, are not fixed and immutable in the cerebral organization but rather are open to experience during life. They argue that even if differences between women and men are found in neurocognitive fMRI research, it does not give any indication of whether they are the origin or the result of sex or gender behavior and experience, or if they are an immutable, hard-wired condition. Furthermore, they disparage the current focus of neuroscience on sex differences, which they argue are often overstated.

Notwithstanding this debate, studies conducted using state-of-the-art neuroimaging systems provide dramatic evidence that male and female brains process information differently. According to Cahill (2006), the past decade has witnessed a "surge of findings" from animal and human research concerning sex influences in many areas of brain and behavior, including emotion, memory, vision, hearing, processing faces, pain perception, navigation, neurotransmitter levels, stress hormone action effects the brain, and disease states. Moreover, recent behavioral, neurological, and endocrinological research indicates that the effects of sex hormones on brain organization occur so early in life that, from birth, the environment acts on differently wired brains in males and females (Kimura 2007, Cahill 2010). The implications of these findings for public policy force us to conclude that "men and women have been living for the past thirty years with the absurd expectation that moral and political correctness demands gender sameness"(Nadeau 1996, 14).

The biological factors that contribute to sex-specific behaviors can be traced both to differing levels of sex hormones and to sex-specific differences in the brain (McCarthy, de Vries, and Forger 2009). Male and female fetuses differ in testosterone concentrations beginning as early as week eight of gestation, and these differences exert permanent influences on brain development and behavior (Hines 2010). Elsewhere, Hines (2011, 70) contends there is convincing evidence that prenatal exposure to testosterone influences the development of children's sex-typical toy and activity interests, and growing evidence that testosterone exposure contributes to the development of other human behaviors that reflect sex differences, such as sexual orientation, core gender identity, and many sex-related cognitive and personality characteristics. In contrast, there is no evidence that the postnatal social environment plays a vital role in gender identity or sexual orientation (Bao and Swaab 2011).

In addition to hormonal influences in the prenatal period, early infancy and puberty may also be critical periods when hormones influence human neurobehavioral organization. Neural mechanisms responsible for these hormone-induced behavioral outcomes are beginning to be identified, and the evidence suggests the importance of the hypothalamus and amygdala, as well as interhemispheric connectivity (Hines 2010). According to Cahill (2006), sex differences exist in every brain lobe, including in many cognitive regions such as the hippocampus, amygdala, and neocortex, but they can also be relatively global in nature. Furthermore, "it seems that the sexual dimorphisms uncovered so far, abundant as they may be, represent only a fraction of the sexual dimorphisms that are likely to exist in the brain" (Cahill 2006, 2). In many cases, moreover, the differences might not be evident in overt anatomical structure but in some type of functional dimension (hence the distinction above between "functional" and "structural" dimorphisms). For example, a region may differ between the sexes in aspects of its neurotransmitter function or in its genetic or metabolic response to experience. The presence, magnitude, and direction of observed sex differences depend on a number of factors, including (but not limited to) the brain structure examined (cerebral cortex, corpus callosum, etc.), the specific brain feature assessed (cortical thickness, cortical convolution, etc.), the degree of regional specificity (global gray matter volume, voxel-wise gray matter volume, etc.), and whether measurements are adjusted for individual brain size or not (Luders and Toga 2010). Ngun and colleagues (2010) suggest that even though most brain studies have focused on gross manifestations of differences in the size of specific regions or nuclei, there is mounting evidence of sex differences at a finer level, including differences in synaptic patterns and neuronal density.

There are three major areas of sexual differentiation: the internal genitalia, the external genitalia, and the brain. Although all three have genetic and hormonal foundations, attention here focuses on sex differences in the brain. The evidence comes from extensive animal studies and more recent human studies using PET and MRI. Although most attention to date has centered on two major sites, the hypothalamus and the corpus callosum, other regions of the nervous and endocrine systems also exhibit differences by sex. Huster, Westerhausen, and Herrmann (2011), for instance, found evidence of sex-related differences that reflected disparities in the degree or quality of interhemispheric interaction. Females are thought to have a decreased magnitude of interhemispheric cortical lateralization compared with males, whereas males exhibit a greater degree of cortical asymmetry for components of cognitive and perceptual processing (Draca 2010). It must be emphasized that the research findings reflect statistical averages and variations, and causality remains largely speculative, although evolutionary theories abound (Paus 2010; Ellis 2011).

"Neuroanatomical sex differences include regional volume, cell number, connectivity, morphology, physiology, neurotransmitter phenotype and molecular signaling, all of which are determined by the action of steroid hormones, particularly by estradiol in males, and are established by diverse downstream effects" (Lenz and McCarthy 2010, 2096). Male brains are on average larger than females, but this is mostly due to body size differences. More important, notable differences exist between the sexes in the size and functioning of particular brain regions. The hypothalamus is a natural target for such research because it is the regulatory center of primal activities, including feeding, drinking, blood pressure control, body temperature control, growth, and emotional responses. The hypothalamus is not only interconnected with the functions of the amygdala and hippocampus, it also controls the secretory function of the pituitary gland. The sexually dimorphic nucleus of the hypothalamus is associated with sexual behavior, neural control of the endocrine glands, and sexual orientation. When a child is two to four years old, the release of testosterone promotes cell growth and prevents cell death in this nucleus; as a result, it doubles in size in male children.

Accumulating evidence demonstrates that male and female hippocampi vary significantly in size (larger in women than in men when adjusted for total brain size), anatomical structure, neurochemical makeup, and reactivity to stressful situations (Cahill 2006). Moreover, Lopez-Larson and colleagues (2011) found greater female local connectivity in the right hippocampus and drew attention to "the importance of examining gender differences in imaging studies of healthy and clinical populations" (187). Similarly, there is mounting confirmation of the sexually dimorphic nature of the human amygdala (Cahill 2006, 4; Tranel and Bechara 2009). For example, amygdale are significantly larger in men than in women when adjusted for total brain size. Furthermore, one fMRI study that examined amygdala responses to fearful faces in men and women found significantly different patterns of responsiveness depending both on the sex of the subjects and on whether the right or left hemisphere amygdala was being studied (Kilpatrick et al. 2006).

The region termed the medial preoptic area has been found to have a vital role in male-typical sexual behavior. This region has major hormonal inputs, especially testosterone, and incorporates several small nuclei and axonal tracts. When this region is destroyed in male animals, there is a cessation or reduction of copulatory behavior. Conversely, electrical stimulation of this region has the opposite effect. Two of the nuclei in this region, INAH 2 and INAH 3, are on average larger in males than females. In contrast, female-typical sexual behavior is modulated in the ventromedial nucleus, a region slightly behind the medial preoptic area. Although it has been linked to feeding behavior, this nucleus is associated with female copulatory behavior and is strongly influenced by sex steroids. Sex-specific experiences have also been isolated in the sirprachiasmatic nucleus in the hypothalamus, which is spherical in males but elongated in females. For both males and females, an intact hypothalamus is necessary for generation of sexuality, and puberty for both sexes is under its direct control through its complex circuitry with the neocortex and the amygdala.

In their fMRI study of the effect of sex on regional brain activity, Bell and colleagues (2006) found differential patterns of activation in males and females during a variety of cognitive tasks, even though performance did not vary significantly. Similarly, according to Lenroot and Giedd (2010, 46), fMRI studies have shown sex differences in white matter development during adolescence as well as different patterns of activation without differences in performance, suggesting that male and female brains may use slightly different strategies for achieving similar cognitive abilities. Moreover, longitudinal studies have shown sex differences in the trajectory of brain development, with females reaching peak values of brain volumes earlier than males.

Compared to men, women show more white matter and fewer gray matter areas related to intelligence (Haier et al. 2005). In men, IQ/gray matter correlations are strongest in the frontal and parietal lobes, whereas the strongest correlations in women are in the frontal lobe and Broca's area. In their MRI study, Lv and colleagues (2010) found significant gender-related differences in cortical thickness in the frontal, parietal, and occipital lobes. Moreover, Menzler and colleagues (2011) found regional microstructural differences between the brains of male and female subjects that were most prominent in the thalamus, corpus callosum, and cingulum. They concluded that all differences were primarily due to differences in myelination.

Sexual dimorphisms also occur in a wide array of neurotransmitter systems, including serotonin, GABA, acetylcholine, vasopressin, opioids, and monoamines (Cahill 2006, 5). One study found that levels of monoamine oxidase were significantly higher in several brain regions in women than in men (Curtis, Bethea, and Valentino 2005). Another study showed that serotonin production was a remarkable 52 percent higher on average in men than in women, which might help clarify why women are more prone to depression—a disorder commonly treated with drugs that boost the concentration of serotonin (Luders et al. 2009). Differences in metabolic activity between the brains of males and females have also been discovered through imaging studies. One study found seventeen regions of the brain where there were statistically significant differences in brain activity between male and female subjects. Also, men on average have higher levels of activity in the temporal limbic system, a more primitive area of the brain associated with action. In contrast, women have higher levels of activity than men in the middle and posterior cingulate gyrus, areas of more recent evolution associated with symbolic action. Despite these differences, there remains substantial overlap in the sexes, however.

Although the hypothalamus is critical in explaining differences in sexual behavior, recent studies have found substantial differences by sex in other areas of the brain associated with nonsexual abilities, functions, and behaviors. The relative size of the corpus callosum, which connects the right and left hemispheres, is about 5 percent larger in women (Leonard et al. 2008). With more fibers, communication between the hemispheres in females is heightened simply because there are more routes connecting them. This could help explain why female brain function is less symmetric than males and why damage to one hemisphere in females has less effect than a comparable injury in a male (Kimura 1992, 123). Research demonstrates that in part because of the greater interaction between the hemispheres, the cognitive tasks of women tend to be localized in both hemispheres. In contrast, in males the two hemispheres act more independently, thus localizing cognitive tasks in only one hemisphere.

Language functions, for example, tend to be localized in different regions of the brain in men and women. One MRI study found that the male brain performs language tasks in the inferior frontal gyrus of the dominant hemisphere, while in females it takes place in both hemispheres (Shaywitz, Shaywitz, and Pugh 1995). Because females have a stronger concentration of left hemisphere linguistic function as well as more reliance on the right, they have a superior ability to learn complex grammatical constructions and master foreign languages. Moreover, the heightened interaction with the right hemisphere appears to enhance the range and complexity of linguistic representations in women. It might also help explain women's relative advantages in associational, expressive, and word fluency (Hyde and Linn 1998).

Although the question of why male and female brains vary continues to be speculative, as noted above, the differences have been linked to exposure to sex hormones during the prenatal period. Kimura (2007) terms these effects "organizational" because they appear to alter brain function permanently during a critical developmental period, suggesting that men's and women's brains are organized along different lines very early in life. Throughout development, however, sex hormones continue to direct differentiation (Kimura 2007). Furthermore, according to Hatemi (2011), gender (femininity-masculinity, not to be confused with sex) is largely a function of genetic and unique environmental influences, including the in utero environment, while socialization has no significant role. Savic, Garcia-Falgueras, and Swaab (2010) agree there is no evidence to suggest that social environment after birth has an effect on gender identity or sexual orientation.

The prior question as to why such developmental differences exist is even more speculative, but one evolutionary perspective has been widely disseminated. According to this theory, sex differences in cognitive patterns arose because they proved advantageous. The assumption is that our brains have remained essentially unchanged over the last 100,000 years or so and reflect a division of labor in hunter-gatherer societies that put different selection pressures on males and females. Males were responsible for hunting, which necessitated skills in long-distance navigation, the shaping and use of weapons, and spatial acuity. In contrast, women had responsibility for raising children, tending the home area, and preparing food and clothing. This responsibility required short-range navigation, fine-motor capabilities, and a perceptual discrimination sensitive to small changes in the environment, skills that are generally consistent with findings of cognitive research. Moreover, men needed to be more aggressive for hunting and defense, while women required cooperative and consensual skills in the home and community.

Behavioral Differences by Sex
Whatever the ultimate cause of sex differences in the brain, they are reflected in varied cognitive capabilities and behavioral tendencies. In addition to the language skill differences noted above, men on average perform better than women on certain spatial tasks, particularly those involving mental rotation of objects in space. They also outperform women on mathematical reasoning tests, in route navigation, and in target-directed motor skills like throwing a ball. Moreover, in general, men and women construct three-dimensional space differently. Although women are stronger at verbal reasoning in mathematics, men are stronger at abstract mathematics. Other research has demonstrated that women, on average, are more skilled at hand-eye coordination; have better sensory awareness, night vision, and wider peripheral vision; have longer attention spans; and are less likely to be either dyslexic or myopic. Weiss and colleagues (2003) found gender-specific differences in the neuropsychological processes involved in mental rotation tasks.

Several intriguing behavioral studies add to the evidence that some behavioral sex differences arise before a baby draws its first breath (Luders et al. 2009). From the age of three to eight months, girls were found to choose dolls over the toy cars and balls that boys prefer, a toy-preference behavior that cannot be explained by social pressure (Bao and Swaab 2011). In the toy studies, males—both human and primate—preferred toys that can be propelled through space and that promote active play (Hines 2010). These qualities, it seems reasonable to speculate, might relate to the behaviors useful for hunting and for securing a mate. Similarly, one might also hypothesize that females select toys that allow them to hone the skills they will one day need to nurture their young (Alexander and Hines 2002).

Similarly, extensive studies of infants have found that males are more interested in objects than in people, are more skilled at throwing objects, and are better at following objects in space. In contrast, female infants are more interested in people's faces and voices and appear to be significantly more adept at assessing mood based on visual or voice cues. Baron-Cohen (2010) found that one-year-old girls spent more time looking at their mothers than did boys of the same age. Moreover, when these babies were presented with a choice of films to watch, the girls paid more attention to a film of a face, whereas boys preferred a film featuring cars. To eliminate the possibility that these preferences might be attributable to differences in the way adults handle or play with boys and girls, Baron-Cohen (2010) and his students videotaped one-day-old infants in the maternity ward where infants saw either the friendly face of a live female student or a mobile that matched the color, size, and shape of the student's face and included a scrambled mix of her facial features. In reviewing the tapes, they found that the girls spent more time looking at the student, whereas the boys spent more time looking at the mechanical object.

The games of girls place an emphasis on cooperation and physical proximity, and they are anxious to integrate newcomers into the group play. Interaction is favored over specialized roles. Contrarily, boy's games emphasize competition and action and favor clearly defined winners and losers. On average, boys are indifferent to newcomers and accept them only if they are useful. Girls have also been found to be better auditory listeners, while boys are better spatial-visual listeners. A similar sex difference is seen in children's spontaneous drawings. Girls five to six years old tend to draw women, flowers, and butterflies in bright colors, while boys prefer to draw more technical objects—soldiers and fighting and means of transportation, in bird's-eye-view compositions using darker colors. For Bao and Swaab (2011), fetal exposure to higher levels of male hormones has lasting effects on playing behavior and artistic expression

In a brain imaging study of adults that received widespread media coverage, Singer and colleagues (2006) compared how men and women reacted when watching other people suffer pain. If the sufferer was someone they liked, areas of the brain linked to empathy and pain were activated in both sexes. If they disliked the person experiencing the pain, however, women had a diminished empathic response, but it was still there, whereas men showed instead a surge in the reward areas of the brain, with empathy completely absent. The researchers concluded that empathic responses in men to other people are not automatic, as had been assumed in the past, but are shaped by the perceived fairness of others. Moreover, men admitted to having a much higher desire for revenge than women and derived satisfaction from seeing the unfair person punished (Singer et al. 2006). In the general population, therefore, females on average have a stronger drive to empathize, while males on average have a stronger drive to systemize (Baron-Cohen 2010).

Male-typical behavior, therefore, demonstrates a strong bias toward action, heightened aggression, and command-oriented hierarchical structuring. Female-typical behavior, conversely, places more emphasis on consensus, cooperation, and interaction. While the male brain constructs reality in terms of vectors marking distance and space, and thus very segmented, the female brain tends to construct reality in terms of more extensive and interconnected cognitive and emotional contexts. As a result, females are more likely to feel a need to be included and attached, to share mutual feelings, and to receive confirmation of these feelings. Men, on the other hand, remain more distant, unattached, and independent.

One of the most studied differences between the sexes, and one that is highly dependent on prenatal androgen exposure, is aggression. As discussed below, most research attention has focused on the amygdala's contribution to behavior that has a strong emotional loading, such as aggression or fear-driven behavior. Raine and colleagues (2011) conclude that sex differences in the ventral and middle frontal gray volume also contribute to sex differences in antisocial personality disorder and crime. Interestingly, studies of girls exposed to excess androgens during the prenatal period, and who as a result have congenital adrenal hyperplasia, show them to be more aggressive than their unaffected sisters (Kimura 2007).

Health Implications of Brain-Based Sex Differences

Although the brains of men and women are similar overall, there are numerous examples of sex differences orchestrated by the interplay of genetic, hormonal, and environmental influences (Qureshi and Mehler 2010, 77) that are linked to differential susceptibilities to a range of diseases. Ultimately, the study of the biological basis of sex differences should improve health care for both men and women (Ngun et al. 2010).

For instance, a growing body of imaging research confirms what clinicians have long known, that variations in the brains of males and females make women more susceptible to certain psychiatric and neurologic disorders (Sukel 2008, Bao and Swaab 2011). Some reports show that women experience anxiety disorders at a rate double that of men and are also twice as likely to develop depression during their reproductive years (Ngun et al. 2010, 6). In contrast, Parkinson's disease is overrepresented in males and its onset is later in women (Van Den Eeden et al. 2003). And, although some have argued that these differences can be attributed to exposure to trauma or environmental factors, recent research suggests that at least some can be traced to the brain itself (Miller and Cronin-Golomb 2010). Table 4.1 lists the sex difference ratios for a selection of neurological or psychiatric diseases that more commonly affect women.

Table 4.1
Sex Ratios of Selected Neurological or Psychiatric Diseases

Disease	Women/Men (%)
Rett syndrome	100/0
Anorexia nervosa	93/7
Hypnic headache syndrome	84/16
Bulimia	75/25
Alzheimer's disease	74/26
Multiple sclerosis	67/33
Anxiety disorder	67/33
Anencephaly	67/33
Posttraumatic stress disorders	70/30
Dementia	64/36
Unipolar depression, dysthymia	63/37
Severe learning disability	38/62
Substance abuse	34/66
Amyotrophic lateral sclerosis	33/67
Stuttering	29/71
Schizophrenia	27/73
REM sleep behavioral disorder	24/76
Dyslexia	23/77
Attention deficit/hyperactivity disorder (ADHD)	20/80
Autism	20/80
Sleep apnea	18/82
REM sleep disorder	13/87
Gilles de la Tourette syndrome	10/90

There is also strong evidence that while drug addiction is more common among men, women experience higher relapse rates and faster progression of compulsive drug abuse and dependence (Ngun et al. 2010). Becker and Hu (2008) found that while there are some differences among specific classes of abused drugs, the general pattern of sex differences is similar for all drugs of abuse and is present for all phases of drug abuse. Females begin self-administering licit and illicit drugs of abuse at lower doses than do males, their use escalates more rapidly to addiction, and they are at greater risk for relapse following abstinence. For instance, based on their finding that females showed greater brain reactivity to cocaine cues than males but exhibited no differences in craving, Volkow and colleagues (2011) concluded that deactivation of brain regions from control networks (prefrontal, cingulate, inferior parietal, and thalamus) in females could increase their vulnerability to relapse, a finding that underscores the importance of gender-tailored interventions for addiction.

The Brain and Sexual Orientation
One of the most controversial findings of neuroscience centers on the role of the brain, in combination with the genes, in determining sexual orientation. According to Bao and Swaab (2011), differences in sexual orientation are programmed into our brain during early development, with no evidence that one's postnatal social environment plays a crucial role in gender identity or sexual orientation. Since the early 1990s, both structural and functional brain differences have been linked to sexual orientation. The first difference found was that the suprachiasmatic nucleus in the anterior part of the hypothalamus is twice as large in homosexual compared to heterosexual men (Swaab and Hofman 1990). In the following year, Allen and Gorski (1991) reported differences in the size of the anterior commissure, the axonal connection between the left and right hemispheres. While the major finding was that it is larger in women than in men on average, they also found that it is larger in gay men than in either straight men or women, indicating that cerebral functions are less lateralized in gay men than in straight men. Since this structure is found to be larger in heterosexual women than in heterosexual men, it could be involved in sex differences relating to cognition and language (Swaab 2008).

In a highly publicized sequel to this work, LeVay (1991) scanned the cadaver brains of gay and straight men and women assumed to be heterosexual. He focused his attention on the INAH 3 nuclei in the medial preoptic region of the hypothalamus, which is known to be sexually dimorphic and has major hormonal inputs characterized by high levels of androgen and estrogen receptors. LeVay found that on average, the INAH 3 nuclei of gay men was the same size as that of the women and between one-third and one-half the size of that in straight men. This finding suggests that gay and straight men may differ in central neuronal mechanisms that regulate sexual behavior. LeVay suggested two possibilities as to how this might come about. First, it could result from differences in levels of circulating androgens between gay and straight fetuses at the critical period for development of the INAH 3 nuclei. Or second, it could be that while levels of androgens are similar, the cellular mechanisms by which the neurons of INAH 3 respond to the hormones are different (LeVay and Hamer 1994). They concluded that, while inborn and environmental factors influence the anatomical and chemical structure of the brain, there are "intrinsic, genetically determined differences in the brain's hormone receptors or other molecular machinery that are interposed between circulating hormones and their actions on brain development" (LeVay and Hamer 1994, 127). Although not all the factors that determine sexual orientation are known, they posited that sexual orientation is "strongly influenced" by events occurring during the prenatal period when the brain is differentiating sexually under the direction of gonadal steroids.

Recently, Savic and Lindström (2008) found a possible relationship between the difference in anterior commissure size and the sex-atypical hemispheric asymmetries observed in homosexual men and women. Functional scanning revealed that the hypothalamus of homosexual men appeared less responsive to fluoxetine than that of heterosexual men, indicating different activities of the serotoninergic system (Kinnunen et al. 2004). Moreover, Savic and Lindström (2008) explored the influence of putative pheromones on sexual behavior by means of PET and found their concentrations excreted in perspiration ten times higher in men than in women. They also found that these pheromones stimulated the hypothalamus of heterosexual women and homosexual men in the same way. However, heterosexual men were not stimulated by a male scent, suggesting that pheromones may be a contributing factor in determining our choice of partner.

Summary: Brain Differences by Sex

Not surprisingly, the neuroscience findings that reveal structural and functional differences by sex raise potentially explosive political issues. Even citing the findings of these studies is judged by some to be politically incorrect. One might ask why it matters whether there are differences in the brain across the sexes. Certainly, it matters in terms of any policies we establish, particularly in developing and prescribing drugs that might operate differently in male and female brains and in developing sex-specific treatments for a host of conditions, including depression, addiction, schizophrenia, and posttraumatic stress disorder. It also has ramifications for education policy. According to brain research, it might be counterproductive to treat boys and girls the same when it comes to education and social training.

Neuroscience research, then, offers new insights into why male and female interests, abilities, and worldviews are often at odds. It demonstrates that the influence of hormones on neural development is a powerful explanatory aid for patterns that have long been highly controversial. Although considerable caution must be used in interpreting any of these data, the cumulative impact of research on our understanding of human sexuality and of nonsexual differences between men and women is significant. While these findings do not negate the importance of nurture to any individual's behavior or capabilities, they place learning in a much different context than has been the norm. One salient topic that follows from this research on sex differences in the brain is the relationship of brain function to aggressive behavior.

Aggression, Violence, and Related Behaviors

Although emerging knowledge about the human brain challenges widely shared assumptions about behavior and cognition in general, among the most significant political findings are those bearing on aggression (Eisenberger et al. 2007). To date, most attention has focused on hormones, neurotransmitters, the limbic system, and the "fear circuit" in the PFC (Wahlund and Kristiansson 2009; Coccaro et al. 2007). Moreover, aggression is studied under a constellation of affective antecedents and behavioral parallels: fear, anger, hatred, antisociality, impulsivity, criminal behavior, and violence.

Human aggression can be defined as any behavior directed toward another individual that is carried out with the proximate intent to cause harm. Problems of aggression and violence are widespread and affect interpersonal relationships, intergroup interactions, and society at large (Ramírez 2010). Violence is aggression that has extreme harm as its goal. All violence is aggression, but many instances of aggression are not violent. Anderson and Bushman (2002) distinguish two types of aggression. *Reactive* aggression is triggered by negative experiences and is impulsive, often unplanned, driven by anger, and occurs as a reaction to some perceived provocation. This type of aggression is thought to result from a highly responsive threat detection system and a diminished capacity to regulate the intensified emotional responses. It has also been termed *affective, impulsive,* or *hostile aggression.* In contrast, *instrumental* aggression is proactive rather than reactive. It is premeditated, often targeted, goal-directed aggression that is associated with psychopathy and often involves diminished emotional sensitivity, empathy, and remorse. Blair and colleagues (2006) argue that reactive aggression is related to impairments in executive emotional systems, while instrumental aggression is more likely related to impairment in the capacity to form associations between emotional unconditioned stimuli, especially distress cues and conditioned stimuli. Although many neurobiological studies fail to distinguish between reactive (impulsive) and instrumental (premeditated) aggression, in virtually all cases their focus is on the former. Therefore, unless otherwise specified, aggression here will refer to impulsive, not premeditated, actions.

No credible scholar today would argue that the causes of aggressive behavior are either wholly environmental or wholly biological. Although in exceptional cases, such behaviors might be primarily biologically or environmentally based, in a preponderance of cases aggression cannot be attributed to any single factor (Garrett and Chang 2008). A massive literature across many disciplines shows that the violent behavior even of a single person is the result of a combination of overlapping and often reinforcing forces. The most appropriate approaches to the study of aggression, therefore, are those that explicate how biology and environment are related, how a complex set of biological factors interacts with and influences a mixture of environmental factors to produce an aggressive behavior or behavioral pattern (Díaz 2010). As noted by Mehta and Beer (2010), aggression is a complex social behavior that is regulated by multiple social and biological factors that likely work together as part of an integrated system.

It should also be noted that the line between biological and social or environmental factors is blurred because many brain deficits linked to aggressive behavior are themselves the product of environmental insults. The use of alcohol has long been known to provoke aggressive and violent behavior in some people, as have low-cholesterol diets, steroids, and drugs of abuse. Moreover, environmental carcinogens, mutagens, and teratogens are capable of producing tumors of or developmental injuries to the brain. For instance, lead exposure in fetuses and children is extremely risky for neural development. The full impact of workplace and other environmental neurotoxins is far from being recognized (Blank 1999). Also, because of these complexities, it is hazardous to generalize from individuals to groups in looking for biological ties to aggressive behavior. With these caveats in mind, however, it is important to look at the scope of neuronal contributions to aggressive behavior. A consensus statement of the Aspen Neurobehavioral Conference, for instance, concluded that while social and evolutionary factors contribute to violence, brain dysfunction is often critical. "Study of the neurobehavioral aspects of violence, particularly frontal lobe dysfunction, altered serotonin metabolism, and the influence of heredity, promises to lead to a deeper understanding of the causes and solution of this urgent problem" (Filley et al. 2001, 1).

Although specific genetic or neuronal patterns that generally explain aggression in humans have not yet been found, certain anomalies that predispose a person to impulsivity and uncontrollable aggression have (Raine 2008; Glenn and Raine 2009; National Institute on Drug Abuse 2010). Episodic dyscontrol, the result of seizures in the limbic system, for instance, is a well-documented disorder that can lead to abrupt, unexplained, acts of rage. Moreover, traumatic brain injury (TBI) can lead to aggressive behavior (Silver, Hales, and Yudofsky 2008). Because numerous changes in CNS regions, pathways, and neurotransmitters are involved in the expression and modulation of aggression, any insult resulting in local or diffuse cerebral damage may affect this system to some degree, explaining why irritability and aggression are so common in TBI patients (Reeves and Panguluri 2011).

A current theory is that violent impulses originate deep in the limbic system, in the amygdala and hypothalamus, and that it's the job of the higher brain regions, particularly the PFC, to act on or moderate these impulses (Garrett and Chang 2008; Coccaro et al. 2007; Wahlund and Kristiansson 2009). Therefore, lesions in the PFC or disrupted functioning of the neurotransmitter serotonin, which is important for impulse control, can impair inhibitory fear reduction, thus leading to aggression (Mobbs et al. 2007). Because of the damage, the PFC can no longer serve as an inhibitory brake on the fear center (for more details, see chapter 6) or properly regulate social behavior (see, e.g., New et al. 2007 for the contribution of the amygdala-prefrontal disconnection in borderline personality disorder). In their fMRI study of spousal abusers, for instance, Lee, Chan, and Raine (2008) observed that relative to controls, abusers showed more limbic and less frontal activation to aggressive stimuli.

They concluded that the "violent acts committed by batterers against their partners may in part be explained by a functional brain abnormality, that the relative balance of activity between the cortical and the subcortical regions is important in the inhibition of aggressive behavior, and that reduced prefrontal but increased limbic activation could predispose to unbridled, dysregulated aggression" (Lee, Chan, and Raine 2008, 656; Cunningham, Van Bavel, and Johnsen 2008).

Frontal lobe injury also has been correlated with hyperactivity and impulsiveness that can lead to aggressive behavior (Enserink 2000) and is a consistent feature of aggression in schizophrenia (Hoptman and Antonius 2011). The causes of the injuries can be the result of birth damage, childhood illnesses, neurotoxin exposure, or accidents. A study of sixteen California death row inmates found that all had significant impediments, including TBI, poly-substance abuse, or neurological disorders (Freedman and Hemenway 2000). Moreover, all came from families with a history of violence, and fourteen had been brutalized sexually and physically as children. Farrer and Hedges (2011), too, found a significantly higher prevalence of TBI in incarcerated groups compared to the general population.

Furthermore, imaging data have shown notably increased prefrontal activity in subjects exposed to violent images, suggesting that aggressive disorders might be explained by a short-circuiting in this region. Moreover, early developmental damage in this region is associated with impairment of social and moral reasoning, as is common in psychopaths (Schmook et al. 2010). Brain scans of convicted murderers have found radically decreased prefrontal activity, particularly among those who had killed on impulse. Compared to controls, murderers also have increased amygdala activity and decreased activity in the corpus callosum, leading to weakened communication across the hemispheres. The causes of such dysfunctions are not well understood, although again, birth injury, childhood infection, neurotoxin exposure, and brain trauma are likely candidates. Teicher (2002), for example, found that the negative effects of child abuse on the functioning and development of the brain are long term and enduring. The impact of child abuse is not solely behavioral; it actually changes the structure and functioning of the brain or of particular neurotransmitters and their receptor sites.

In addition to the contribution of deficits in regions of the PFC and the anterior cingulate cortex, chemically there is evidence that anterior brain dysfunction in aggressive individuals involves disruptions in serotonergic function (Seo, Patrick, and Kennealy 2008). As a result, contemporary research has focused on the inhibitory role of serotonin, which was first noticed when it was found that the brains of people who became aggressive under the influence of alcohol produced less serotonin than the brains of those who did not become aggressive. Since then, abnormalities of serotonin metabolism (its production, storage, release, or reuptake) have been found disproportionately in impulsively aggressive men, compulsive fire-setters, and the violently suicidal. Moreover, research has shown that individuals prone to aggressive behavior exhibit deficits in electrocortical responses such as P3 wave amplitude. For instance, Venables and colleagues (2011) found a significant negative association between aggressiveness and amplitude of the P3 response to both target and novel stimuli over fronto-central scalp sites. They concluded that while previous studies had emphasized the role of abnormalities in the frontal brain systems that govern affective and behavioral regulation, suggesting that a reduced P3 response in aggressive individuals should be frontally based, aggressive behavior is complexly determined, and reduced P3 amplitude at particular sites is mediated by differing constituent dispositions. At a broad level, for example, angry or reactive aggression is reliably associated with P3 amplitude reduction, whereas instrumental or proactive aggression is not (Patrick 2008).

Aggression often begins when self-control stops. In turn, self-control requires considerable brain food in the form of glucose. Thus, people who have difficulty metabolizing glucose are at greater risk for aggressive and violent behavior. In their four-part study, DeWall and colleagues (2011) found that participants who consumed a glucose beverage behaved less aggressively than did participants who consumed a placebo beverage. They also found an indirect relationship between diabetes (a disorder marked by low glucose levels and poor glucose metabolism) and aggressiveness through low self-control. Moreover, a PET study of the brains of convicted murderers found that they (particularly those who had killed on impulse) had significantly lower glucose metabolism in the PFC, an indication that this region was not functioning as it should (Raine 2002). In their study of impulsive aggressive veterans from a VA trauma clinic, Teten and colleagues (2008) found that deficits in empathy were associated with higher rates of verbal aggression, and alexithymia, an emotional awareness deficit with emotional and language processing characteristics, was uniquely associated with impulsive aggression.

Not surprisingly, numerous studies highlight the importance of the amygdala and other limbic regions in the identification of and response to emotionally salient information (Said, Dotsch, and Todorov 2011;

Salzman and Fusi 2010). As the major site of fear conditioning, the amygdala receives danger signals and relays them in a manner that results in protective responses without conscious awareness (Custers and Aarts 2010). Amygdala activity has been tied to generalized social phobia, a condition characterized by avoidance of situations where scrutiny by others is possible (Stein et al. 2002); borderline personality disorder (New et al. 2007); and numerous neuropsychiatric and neurodevelopmental disorders that heighten aggressive behavior (Piech et al. 2011).

If its threshold is set too low, normally benign aspects of the environment are perceived as dangers, interactions are limited, and anxiety may arise. If set too high, risk taking increases and inappropriate sociality may occur. Given that many neurodevelopmental disorders involve too little or too much anxiety or too little of too much social interaction, it is not surprising that the amygdala has been implicated in many of them. (Shumann, Butler, and Amaral 2011, 745)

In a study of more than 500 subjects, Davidson, Putnam, and Larson (2000) found that violent people express all three factors: increased activity in the amygdala, diminished activity in the PFC, and disrupted functioning of serotonin.

Supporting the interactive nature of environment and brain, hair analyses of serial murderers and violent offenders have demonstrated excessive concentrations of lead and cadmium. Masters and Coplan (2001; Coplan et al. 2007) also found that when lead and manganese interact, the deleterious effects are magnified because they compromise the serotonin, dopamine, and other neurotransmitter systems that are integral to self-control and to setting thresholds for violent behavior. Moreover, EEG research has found relationships between abnormal electrical discharges in the brain and behavioral problems. Such studies demonstrate that abnormalities are identifiable in 15 to 50 percent of violent people as compared to 5 to 20 percent of those persons with no history of violence. Brain laterality, too, has been found to be associated with antisocial behavior, with some evidence of a higher incidence of left-handedness among criminals (Szabo et al. 2001).

Hormonal levels, which are critically regulated by the brain, have also long been linked with aggressive behavior. Despite ongoing controversy over the existence of a direct link between testosterone and criminality, high levels of testosterone have been reported in populations of aggressive individuals such as criminals with personality disorders, alcoholics, and spousal abusers (Siever 2008; Archer 2006; Archer, Graham-Kevan, and Davies 2005). Mehta and Beer (2010) found that testosterone increased aggressive behavior by reducing the ability of the medial OFC to govern self-regulation and impulse control following social provocation, while Derntl and colleagues (2009) found that increased levels of testosterone improved the amygdala's ability to process threat-related stimuli. Testosterone thus increases the propensity toward aggression because of reduced activation of the neural circuitry of impulse control and self-regulation. These findings suggest that when confronted with human facial expressions, testosterone prepares one for action by enforcing more automatic and autonomic processes that lead to attention shifts and a decrease in subconscious fear, thereby facilitating approach behavior (Derntl et al. 2009).

The disparate evolutionary aggression strategies of males and females result in widely varying proactive and reactive responses in level and frequency. Not surprisingly, testosterone also appears to play a decisive role in competition. A study of competing males found that "aggressive behavior and change in testosterone concentrations predicted willingness to reengage in another competitive task" (Carré and McCormick 2008, 408). Furthermore, McDermott and colleagues (2007) found that high-testosterone subjects were much more likely to engage in unprovoked attacks against their opponents than were their lower-testosterone counterparts. Although female participants were as likely as the men to fight back once they were provoked, men were much more likely to initiate a conflict. Moreover, the preferred types of aggression differ for males and females, with males preferring direct aggression and females indirect (Anderson and Bushman 2002, 33). Among males, those with higher levels of testosterone are more likely to be involved in violent acts, although there are significant individual differences. Indeed, males between the ages of twelve and twenty-five are the principal perpetrators and victims of violence, and this may be due to heightened testosterone levels, which escalate in early adolescence (Craig and Halton 2009).

Aggression and Violent Games
In light of the easy availability and increasing variety of violent media, especially graphic video games, the extent to which exposure to media violence increases aggressive behavior has stimulated extensive research on how such exposure affects behavior and cognitive functioning (Kronenberger et al. 2005; Bushman and Huesmann 2006). The general aggression model posits that long-term exposure to violent media content alters internal states (arousal, cognition, and affect), leading to consolidation of aggressive mental schemas (Carnagey, Anderson, and Bartholow

2007). For example, one early fMRI study found that persons exposed to violent images had significantly increased blood oxygen levels in the PFC, indicating that the PFC might serve as a critical filter between the violent images and the decisions made in response to them. It was suggested that future research might reveal that the PFC is somehow short-circuited among people with disorders marked by inappropriately aggressive behavior. In their study of media violence, Kalnin and colleagues (2011) found that adolescents with a history of aggressive-disruptive behavior displayed abnormalities in activation of frontal and limbic regions compared to adolescents with no diagnosis. In particular, youth with high exposure to media violence, especially those with disruptive behavior disorder, had a lower amygdala response to violent stimuli as a result of desensitization (Bartholow, Bushman, and Sestir 2006). Limbic activity was inversely related to frontal and anterior cingulate activity, particularly in individuals with better emotional control, owing to the top-down effects of these regions (Kalnin et al. 2011; Hölzel et al. 2010).

In their MRI study of the impact of violent videos on the brains of normal adolescents, Strenziok and colleagues (2010) found that exposure to the most violent videos inhibited emotional reactions to similar videos over time. Data on brain activation patterns indicated that the area known as the lateral OFC, which is thought to be involved in emotions and emotional responses to events, showed increasing desensitization over time, most markedly for highly violent videos. The authors conclude that continued exposure to violent videos will make an adolescent less sensitive to violence, more accepting of violence, and more likely to commit aggressive acts as the emotional component in the frontal lobe that normally acts as a brake on such behavior is degraded.

Genes, Brain, and Aggression

Given the connections among genes and the brain raised by the interactive model in chapter 1, it is reasonable that this relationship is crucial as it relates to aggression. With its role in human adaptability and survival, it would be remarkable if traits that result from variation in brain function were not influenced partly by genes. According to O'Donovan and Owen (2009), there is now convincing evidence that discrete genes are implicated in aggression. One hypothesis is that gene abnormalities result in structural brain abnormalities, which result in emotional, cognitive, or behavioral abnormalities, which in turn predispose one to antisocial behavior (Raine and Yang 2006). There is increasing evidence for brain impairments in antisocial groups, with particularly strong evidence for impairments in the PFC. Questions then arise as to what specific genes predispose an individual to commit crime, which genes code for the brain impairments found in antisocial groups, how do they change brain processes to give rise to antisocial behavior, and what role the environment plays While these questions may seem inexplicable, recent scientific advances hold the promise of addressing them and transforming our understanding of such behavior. To date, at least seven genes meet the dual criteria of being associated with antisocial/aggressive behavior in humans or animals and influencing brain structure: MAOA, 5HTT, BDNF, NOTCH4, NCAM, tlx, and Pet-1-ETS (Raine 2008).

The monoamine oxidases A and B (MAOA and MAOB) are two closely related enzymes, the products of two abutting X-linked genes, that play an important role in the metabolism of biogenic amines both in the CNS and in the periphery (Craig and Halton 2009). In general, while MAOA preferentially oxidizes biogenic amines such as serotonin, norepinephrine, and epinephrine, MAOB is important in dopamine metabolism and vitiates dietary amines, including phenylethylamine. By far the most investigated gene here is MAOA, a gene that is associated both with antisocial behavior and with a reduction in the volume of the amygdala and OFC. It has been found that the enzyme that this gene codes for degrades dopamine, norepinephrine, and serotonin (Eisenberger et al. 2007). Males with a common variant in the MAOA gene have an 8 percent reduction in the volume of the amygdala, anterior cingulate, and OFC, the brain structures that are involved in emotion and are found to be compromised in antisocial individuals (Meyer-Lindenberg et al. 2006). Abnormalities of MAOA binding are an important neurochemical target for research since lower binding is associated with greater aggression and greater binding is associated with major depressive disorder (Soliman et al. 2010).

Not surprisingly, therefore, in both animal and human populations, aggressive behavior has been linked to a genetic deficiency in MAOA (Eisenberger et al. 2007). In fact, this gene has earned the nickname "warrior gene" because it has consistently been linked to aggression in observational and survey-based studies (McDermott et al. 2008). A recent PET study, for instance, found that reduced PFC MAOA binding is associated with self-reported aggression in males (Alia-Klein et al. 2008). Males who are deficient in MAOA, consequent to an X-chromosome defect at the vicinity of the MAOA gene, demonstrate frequent aggressive behavior (Soliman et al. 2010). Moreover, a genotype associated with low levels of MAOA in an in vitro assay has been associated with greater risk of developing antisocial behavior (Buckholtz and Meyer-Lindenberg 2008). Despite mounting evidence suggesting a relationship between the MAOA-uVNTR polymorphism and aggressive behavior, it is unclear how this genetic polymorphism predisposes individuals to aggressive behavior.

Eisenberger and colleagues (2007) examined the relationship between MAOA and trait forms of aggression and interpersonal hypersensitivity with neural responses in brain areas associated with rejection-related distress. They found that individuals with low-activity MAOA demonstrated higher trait aggression and higher trait interpersonal hypersensitivity than those with high MAOA. Even though aggressive behavior clearly relates to affective processes, few neuroimaging studies have investigated how the MAOA polymorphism relates to the neural activity associated with these affective processes. Instead, neuroimaging studies have focused primarily on how the MAOA polymorphism relates to executive attention or inhibitory control during cognitive tasks, typically observing that the MAOA polymorphism relates to altered activity in neural regions involved in triggering and instantiating cognitive control (Meyer-Lindenberg et al. 2006).

In addressing the question of environmental influence, McDermott and colleagues (2008) concluded that aggression occurs with greater intensity and frequency as provocation is experimentally manipulated upward, especially among low-activity MAOA subjects. There is mixed evidence of a main effect for genotype and for a gene by environment interaction, such that MAOA is less associated with the occurrence of aggression in a low-provocation condition but significantly predicts such behavior in a high-provocation situation. Furthermore, there is evidence a functional polymorphism in the MAOA gene can mediate the impact of traumatic early life events on the propensity to engage in violence as an adult. Specifically, children who had suffered abuse and who had the low-activity form of MAOA were much more likely to develop antisocial problems as adults (McDermott et al. 2008).

Although biological aspects of delinquency, anger, and impulsivity are not fully understood, we know that the dopaminergic system plays an important role in these behaviors and psychological traits. In particular, midbrain dopamine neurons are activated in response to novel rewards, contributing to the learning of cues that predict reward occurrence and modulating novelty seeking and impulse control. Low levels of dopamine are associated with attention deficit/hyperactivity disorder, and the dopaminergic system also contributes to anger and delinquency (both aggressive and nonaggressive). In turn, delinquency, anger and impulsivity, have strong genetic components, and a number of studies have searched for specific associated genes (Ebstein 2006). Among them, the dopamine receptor D4 gene (DRD4) located on chromosome 11 has received special attention for its potential impact on reward motivation, attention, and approach behavior. Recent studies indicate that dopamine is not only released not only in response to such prompters as food, sex, and stimulant drugs but also in response to aggression (Couppis, Kennedy, and Stanwood 2008). By extension, individuals who experience a blunted dopamine response may be motivated to seek experiences that activate midbrain dopamine release.

The findings for the DRD4 gene, however, have been mixed, implying that behavioral expression of the DRD4 gene could be moderated by other factors. Indeed, several studies have found that exposure to a negative social environment exacerbates the risks associated with the DRD4 gene (Dmitrieva et al. 2011). Moreover, as with MAOA, the link between DRD4 and anger has been demonstrated for males but not for females. This gender-specific vulnerability to DRD4 could be attributed to gender differences in biological factors, such as differences in estrogen and progesterone that modulate the dopaminergic activity, differences in sex chromosome genes, differences in DNA methylation, or differences in autosomal genes (Harrison and Tunbridge 2008), or to gender differences in the exposure to contextual risk factors. Dmitrieva and colleagues (2011) conclude that the association between the DRD4 VNTR polymorphism and adolescent delinquency, short temper, and thrill seeking is stronger for males than for females, and that this gender-specific expression of the DRD4 can be explained by gender differences in exposure to violence and parental monitoring of youth activities.

Summary: Brain and Aggression

We are unlikely ever to fully understand, much less to predict or treat, aggression and violence by looking only at genes and the brain, because at base, such behaviors are a product of political and social, not solely biological, problems. As noted elsewhere, the critical question to ask is "how, not whether, biology and environment are related—how a complex of biological factors interacts with a complex of environmental factors producing a single isolated aggressive behavior or a patterned concatenation of aggressive behaviors" (Blank 2006, 15). Depending on its motivating stimuli, aggression can be classified as (1) triggered by fear, and simply the reaction of escape from an adverse situation, (2) reactive aggression, occurring in response to emotions such as anger, frustration, food deprivation, pain, or fatigue, or (3) instrumental aggression, directed toward achieving specific planned goals (Mercadillo and Arias 2010). While a better knowledge of the role of the brain and human behavior is not sufficient to explain human aggression, "one cannot understand violence without a thorough understanding of the human mind" (Pinker 2002, 317). In other words, although it is unlikely that we will ever be able to describe a particular aggressive act or even type of aggression in terms of specific neuronal activity, it is crucial that this neural dimension be an integral part of any respectable paradigm of behavior.

Conclusions

This overview of the linkages between the brain and behavior demonstrates that we can no longer dismiss the role of the brain in explaining addictive behavior, sex differences and sexual orientation, and aggressive and violent behavior. Although the emerging evidence from neuroscience research does not negate the important contribution of environmental influences on behavior, it irrefutably argues for the inclusion of a broader model that includes a strong biological foundation. The next chapter turns attention to the implications of this research for individual responsibility and the criminal justice system, while chapter 6 returns to the role of the limbic system in feelings of fear and empathy and the ramifications of this connection for in-group and out-group relationships that might help us better understand nationalism, terrorism, ethnic conflict, and recurring wars.

5

Individual Responsibility and the Criminal Justice System

Someday neuroscience could well force the legal system to revise its rules for determining culpability and for meting out sentences. It could also shake up society's understanding of what it means to have "free will" and how best to decide when to hold someone accountable for antisocial actions. (Gazzaniga 2011, 54)

This chapter introduces emerging issues facing the criminal justice system that accompany new knowledge about human behavior from neuroscience research. After a short overview of the general issues, it examines the global issues regarding the implications of neuroscience for our notions of individual responsibility and the debate over free will and personal autonomy. The chapter then summarizes current neuroscience research into brain disorders and psychopathy and discusses the impact of this unfolding knowledge on the rules of evidence. This is followed by a more focused analysis of the growing pressures on the courts to countenance evidence based on brain imaging technologies in criminal and civil cases to demonstrate either aggravating or mitigating circumstances. Attention is then turned to the highly contentious issues surrounding its use for lie detection, and finally to the issue of court-ordered medication.

The nineteenth-century Italian criminologist Cesare Lombroso is credited with being the first to propose a theory of criminality positing that criminal behavior was biologically derived and could be predicted by various physiognomic features (Looney 2010). In the early twentieth-century eugenics movement, criminal behavior was linked to feeble-mindedness, insanity, alcoholism, and prostitution and was seen as a degenerate and heritable identity. The search for genetic determinants and biological markers of criminal behavior has continued to this day and has generated passionate debate as the courts face repeated attempts to introduce biological evidence to show diminished responsibility (including the XYY chromosome abnormality in men in the 1960s and 1970s, premenstrual syndrome in women in the 1980s, and CT and PET brain imaging results in the 1980s and 1990s). To date, all of these have largely failed to displace conceptions of responsibility that operate within the practice of the criminal law. The question here is whether the new neuroscience will succeed where these past attempts have failed.

Although it remains to be seen whether the diffusion of neuroscientific knowledge and technologies into the criminal justice system will significantly alter legal principles, there is already extensive analysis of the issues these developments raise for the law. These issues include the basis of criminal responsibility, the impact of neuroscience on normative judgments in law and justice, the potential for neuroscience to clarify the difference between deception and self-deception, and the impact of neuroscience on notions of punishment. Some argue that before long, we will be using neurotechnology to enforce and adjudicate the law (Dunagan 2010). There is no doubt that neuroscience is illuminating our understanding of moral reasoning and judgment, thus precipitating issues of criminal intention and responsibility. Moreover, neuroimaging technologies stand to be a very controversial but pervasive new tool for law enforcement. Correspondingly, advanced methods of brain intervention and behavior-altering drugs raise questions of their appropriate use in enforcing and adjudicating the law. Greely (2005), for instance, suggests that neuroscience might eventually be used in the courtroom to detect lies or to compel truth, to determine bias (on the part of jurors, witnesses, or parties), to elicit or evaluate memory, to determine competency (e.g., competency to stand trial, to be executed, or to make medical decisions), to verify the presence of intractable pain, to prove addiction (or susceptibility thereto), to show a disposition to sexual deviance or predatory impulses (for purposes of involuntary civil commitment), or to show future dangerousness.

Driving the call for the utilization of these new tools is the fact that individuals today have an unprecedented capacity to inflict wide destruction, thus reducing the gap between violent intention and action. The net annual burden of crime in the United States alone has been estimated to exceed $1 trillion, making criminal behavior a costly large-scale social problem and a critical target for scientific investigation (Buckholtz et al. 2010). In attempts to preempt violence at its source, the brain, a new generation of mental surveillance tools will be utilized for interrogation of suspects and lie detection. Although to date, no court in the United

States has allowed brain scan lie detection to be used as evidence, in Mumbai, India, in 2008 a prosecutor successfully used an MRI study of the defendant to show that the defendant had "experiential knowledge" of events only the murderer would know (Erickson and Felthous 2009, 121). A more immediate issue facing U.S. courts is whether to allow evidence of brain damage or disorders that are verified by brain imaging to reduce criminal culpability. Other issues center on the potential for coerced use of drugs or the acceptance of "corrective" surgery for behavioral problems. One question that has already confronted the courts is whether mentally ill defendants can be forced to take medication to stand trial. This dilemma is likely to shift to the use of preemptive mandated interventions to modify the brain chemistry or function—or, more ambitiously, the environment—of those persons diagnosed through neuroimaging tests as likely to behave violently. Of critical importance is the impact of neuroscience technologies and findings on the concepts of free will and individual responsibility.

Free Will and Individual Responsibility

The relationship between mind and matter has long been a matter of great concern to scholars. In historical discussions, the question was often framed either in terms of the existence and nature of an immortal soul or as a metaphysical debate on what sorts of entities the world is made of and how we can know about them (Häyry 2010). As noted earlier, neuroscience is changing our understanding of the brain and its relation to the mind and to human behavior, thereby giving a new inflection to the question of free will and moral responsibility. Although some claim that such findings demonstrate that our traditional assumptions of conscious free will and moral responsibility are outdated, others argue that such research is irrelevant. In any case, neuroscience challenges our conceptions of free will and individual responsibility, and so has repercussions for our understanding of criminal behavior and legal culpability. As noted by Árnason (2010), findings of brain imaging studies suggest that many people whom we now hold responsible for their actions in fact have diminished moral responsibility and legal culpability because of the structure or function of their brain. Erickson argues that "it is inescapable that the novel and powerful technology of brain imaging . . . drives [neuroscientists'] conception of the mind" (2010, 36).

Free will as a concept has intrigued philosophers for centuries and remains a problem.[1] This is not surprising, since this concept, along with responsibility and freedom, plays a central role in the way we view ourselves (Müller and Walter 2010; Kamm 2009). The belief that we act freely and are morally responsible for our actions is at the base of retributive justice systems and, some argue, necessary to justify punishment, blame, or moral condemnation (Häyry 2010). Moreover, the personal stance we take toward each other is based on a belief that humans are capable of moral responsibility and deserve moral consideration. Free will is also associated with human dignity. Humans are to be treated as ends in themselves because they are the originators of their own ends or purposes. Free will and rationality are thus intractable. Free will is the power to originate choices, "the power of agents to be the alternate creators (or originators) and sustainers of their own ends and purposes" (Kane 1996, 3). Thus, it is argued, a belief in free will presumes a special status for humans.

It is important to note for this discussion that our concepts of justice, punishment, and deserts are rooted in our notions of individual responsibility. Some argue that without free will, moral responsibility is empty since at its base it requires that we be truly deserving of praise or blame because it is up to us what we do. Thus, it is an essential feature of life in modern Western society that normal human beings who have reached some level of maturity regard themselves and one another as responsible beings. Although we make exceptions for those who lack free will, we take responsible beings more seriously than nonresponsible ones. We treat them as persons, not objects; we credit them and hold them to blame, and attribute qualities and events to them more deeply than to others.

A foundational premise of cognitive neuroscience is that all aspects of the mind are ultimately reducible to the structure and function of the brain: it is the understanding of the mind *as* brain, and seeks to provide comprehensive explanations of human behavior in purely material terms. To this end, it reflects the dominant approach of modern science, which seeks to understand and explain all observable phenomena as functions of their component parts. According to this disposition, questions of

1. The goal here is not to provide an exhaustive review of the philosophical debate over free will. There are almost as many positions on free will as there are persons who have written about it. Major categories of free will thinkers include libertarians, determinists, compatibilists, and incompatibilists, but the variations are endless.

biology are thought to be reducible to matters of chemistry, which are, by extension, reducible to problems of physics. In principle, this approach ultimately leads to the analysis of all phenomena in terms of the relationships of motion and rest among their most elemental particles. According to Snead (2008), this commitment to using material causation to explain the mind and human behavior carries with it profound implications for perennial concepts such as the existence of the soul, free will, selfhood, and consciousness.

Research on the human genome, cognitive sciences involving artificial intelligence, psychoanalysis, and other theories of unconscious motivation, behavior modification, and ethology concurrently offer formidable support for external determinants of human behavior. While rapid advances in these areas certainly point to more complete explanations of human behavior, the question is whether they necessarily lead to the conclusion that free will and moral responsibility, "as they are viewed in philosophical discourse and everyday life, are not to be counted as candidates among the class of real entities" (Double 1991, 5)? If, as Double argues elsewhere, free will is nothing but our "venting of non-truth-related attitudes" (1996, 3), where does this leave us? While the notion of free will has always been problematic, and will be more so in light of findings from genetics and neuroscience, the debate over free will not end: the idea of free will, though instinctive and subjective, is useful in understanding human behavior.

The notion that individuals have the capacity to make choices free of any deterministic force beyond their control is central to rational models of human behavior and a critical tenet of democratic theory, which assumes that citizens have the capacity to make decisions free of external and internal constraints. Without this assumption, electoral systems are meaningless. Similarly, justice systems depend on some degree of free will to assign responsibility to individuals for their actions and must make specific exemptions in cases of diminished capacity. As discussed later, the evidence of brain and genetic contributions to criminal action is progressively being used as a defense, to suggest a lack of free will on the part of the defendant. The "devil made me do it" defense of the past has become "my genes made me do it" or "my neurons made me do it." This argument is also being used to excuse other antisocial behaviors and unhealthy lifestyles.

Does our expanding knowledge of genetics and neurology mean that individuals are no longer responsible for their actions because they lack conscious control? There are strong reasons to believe the answer to this question is "not really." The fact that we are now aware that all expressions of what we view as the mind are affected by the biochemical and electrical state of the brain should not force us to abandon the notion of individual responsibility, although it does require a refinement of it. Despite evidence of brain-behavior linkages, neural functioning remains a weak predictor of behavior. Even obvious cases of brain damage to the frontal lobes do not consistently lead to behavioral abnormalities or deficits. Furthermore, neuroimaging routinely identifies abnormalities that have had no discernible effect on the person's behavior, and most individuals with antisocial behavior exhibit normal brain functioning as measured by current technologies. Alcoholics in twelve-step programs can and often do refrain from drinking. Even some pedophiles have been known to control their compulsions. Despite the knowledge of neuroscience, then, humans do retain the capacity to make conscious decisions—this is what continues to separate us from other mammals. The notion of individual responsibility is still functional, although the traditional notion based on tabula rasa has long been outdated.

Compatibilists argue that determinism does not matter; what matters is that individuals' wills are the result of their own desires and are not overridden by some external force. For the compatibilist, voluntary actions are simply those that are not coerced. Moreover, a compatibilist could distinguish between levels and claim that determinism at one level need not imply determinism at all levels. For instance, one could argue that determinism at the neurobiological level does not necessarily entail determinism at the level of cognition (Buller 2010). It should come as no surprise that we are not fully rational, entirely conscious creatures whose actions are determined solely by logic and reason, like Mr. Spock in *Star Trek*. Humans are constrained by brains that have evolved from primitive times, when emotions of fear and aggression were crucial to survival. Although free will in an absolute sense is, and most likely always was, a philosophical artifact with little grounding in reality, it remains relevant, though qualified. With few exceptions, individuals ultimately must bear responsibility for their actions.

Broome, Bortolotti, and Mameli (2010) provide an interesting take on what they frame as the threat to the "commonsense" ways of attributing moral responsibility. The commonsense conception tells us that there are circumstances in which the agent is entirely and fully responsible for an action that results in undesirable outcomes and circumstances in which the agent is not at all responsible for the bad outcome. The commonsense conception also tells us that between these two situations there is a whole range of in-between cases where the agent is responsible for the bad outcome to some extent but not fully, and where the punishment should be proportionate to the extent to which the agent is responsible. Moreover, for both the commonsense and the current legal conception, the responsibility status of the agent is determined by the causal process that led to the outcome and, more precisely, by the specific nature of the psychological process that led to the particular action. Broome and colleagues (2010) contend it is premature to say what global impact the neurosciences will have on our intuitions about responsibility, and far too early to say whether the commonsense notion of responsibility will survive this impact, and if so, in what form.

The concept of individual responsibility faces pressures, however, not only from emerging knowledge in genetics and neuroscience but also from changes in social values that would isolate individuals from personal responsibility for their decisions. Although genetic and neurological findings are likely to contribute to this process, they themselves must be placed in the broader perspective of changes in social values, which have a momentum of their own. The biggest danger is that genetic and neurological arguments will be used to reinforce the view that free will is no longer relevant and that moral responsibility is therefore impossible.

Individual Responsibility and Diminished Capacity

The decision to impose criminal responsibility rests on an assumption about the defendant's decision to engage in proscribed conduct. We punish only those who we believe had the capacity to make a choice. A new generation of neuroimaging technologies, however, provides insights into structural and functional abnormalities in the brain that may limit the autonomy of some dangerous offenders and unravel the fabric of the criminal justice system (Shuman and Gold 2008). How will the results of these technologies be received by the courts? Are they relevant to existing formulations of the prima facie case, the insanity defense, or mitigation of sentence; and will changes in the science or the law be required to accommodate this knowledge?

In the legal context, there are both global challenges and more specific challenges to responsibility and culpability that rely on neuroscience (Árnason 2010). The global sort of challenge argues that neuroscience is revolutionizing our understanding of free will and responsibility, discrediting any justification of punishment based on retributive justice (Greene and Cohen 2004; Burns and Bechara 2007). According to this view, a scientifically enlightened approach to punishment must be consequentialist, that is, one that determines punishment according to its expected effects on future behavior, rather than being concerned with restoring a balance of justice (Greene and Cohen 2004). Buller (2010), however, challenges this conclusion and maintains that neuroscience will not necessitate any substantial change in the current legal intuitions. The idea that we are just machines whose thoughts and actions are simply and entirely physical processes is not a new one. While evidence shows us what the brain is doing, it cannot tell us what the mind is doing. To say that the law assumes that people are agents and have individual responsibility is not thereby to make any philosophical commitment to the genuineness or reality of free will since the law is not interested in free will in any deep metaphysical sense. Rather, the law assumes that we have free choice and that, in our day-to-day lives, our actions are intentional and voluntary (Buller 2010, 197).

Roskies (2010) contends that neuroscience has affected our conception of volition very little: the numerous reports of the death of human freedom have been greatly exaggerated, and notions of intention, choice, and the experience of agency have been maintained. Where neuroscience has affected our conception, it has typically challenged conventional views of the relationship between consciousness and action—for example, more aspects of behavior than previously imagined are governed by unconscious processes. Furthermore, neuroscience demonstrates that free choice is not a unitary faculty but rather a collection of largely separable processes that together make possible flexible, intelligent action. In fact, neuroscience might affect our notion of volition in the future by elucidating the neural systems and computations underlying different aspects of it and in practice could result in a better way of categorizing its components (Brass and Haggard 2008; Pacherie 2008). To date, the most significant contribution neuroscience has made is to compel us to formulate novel questions about the nature of voluntary behavior and provide new ways of addressing them (Roskies 2010).

Müller and Walter (2010) argue that a sharp line between full autonomy, on the one hand, and complete lack of autonomy on the other has never existed. Although there might be a few cases on either side, the gray area between them is sizable. The concept of autonomy is not a categorical property (present or absent) but rather a measured and changeable one that depends on certain biological and social prerequisites. Although at some level autonomy must be conditioned by proper functioning of the brain, it can be influenced temporarily or long term by many factors, including brain disorders or damage, medication, drugs, or physical interventions in the brain. An associated concept is the voluntariness of a person's action. Generally what a human being does counts as an act only if it is of a kind that the person can do or refrain from doing at will. Thus, a key distinction for the courts is between those acts that are voluntary and those that are not voluntary. To act voluntarily is to exercise a two-way capacity to do or refrain from doing the act, knowing that it is within one's choice (Bennett and Hacker 2011).

Likewise, individual responsibility will always be found on a continuum and depends on contextual factors as well as on internal brain-based factors. Moral responsibility and legal culpability, too, are not either/or attributes but span a range from full responsibility to no responsibility, with most though not all criminal acts falling somewhere in between. Even if neuroscience ultimately shows that we lack free will in a full sense, this does not mean we lack autonomy or are absolved of responsibility. Nevertheless, the more we learn about the neurological bases of defective decision making, aggressive behavior, or lack of self-control, the more apparent will be those cases where one is seen as not (fully) responsible for one's actions (Hughes 2010). Neuroimaging can play a role in showing whether a defendant in a criminal case has brain abnormalities that affect conditions for moral agency, such as self-control, rationality, or the ability to act on reason. As a result, neuroimaging may be helpful in avoiding undeserved punishment (Bennett and Hacker 2011; Árnason 2010; Vincent 2009).

In the United States, the insanity defense is available to provide a legal excuse for those whose criminal acts are due to serious mental illness. Its evolution is integrally related to the advancing conceptions of psychopathic disorders and their relevance to this defense. As legal and mental health professionals discuss the ideal insanity defense and whether psychopathic disorders should be qualifying conditions, any practical outcome of such deliberations must take into account the politics of this strategy. Although society is far from reaching a consensus as to whether neurological explanations should lead to exculpation, some argue that the insanity defense itself, even for serious, psychotic mental disorders, is "withering under relentless attack" and must be advanced and defended without diluting its salience by including psychopathic disorders (Felthous 2010).

According to Baskin, Edersheim, and Price (2007), even if we were to connect the theoretical dots between neuropathology and behavior, this correlation may not be legally relevant. These complicated questions of responsibility and agency are far from being resolved. To date, no brain image can identify thoughts or ascribe motive. Neuroscience cannot distinguish thought from deed and has little, if any, predictive power (Baskin, Edersheim, and Price 2007, 267). Ultimately, we encounter the same moral and methodological problems predicting violence with neuroimaging as we do trying to predict violence with other clinical instruments. In most instances violence is a complex, multifactorial, and socially driven behavior that cannot be reduced to a unitary brain function or region (Baskin, Edersheim, and Price 2007).

Even when neuroscience evidence is deemed inadmissible during the guilt phase of a trial, or is admitted but does not result in acquittal, it still could be relevant and admissible in sentencing, which by its very nature demands consideration of a broader range of information (Shuman and Gold 2008, 732). Therefore, the use of neuroimaging evidence has generally been more successful in showing diminished culpability at the sentencing stage than in disproving criminal intent (Snead 2007, 2008; Mobbs et al. 2007). The federal sentencing guidelines make it clear that impulse control disorders can justify a downward departure from the recommended sentences and allow a lower threshold for reduced mental capacity if the defendant is unable either to absorb information normally or to exercise the power of reason, or if the person knows what he is doing and that it is wrong, but cannot control his compulsion. When a criminal court establishes that a defendant did commit the act he is accused of, it still must determine criminal intent (*mens rea*) and, if so, whether there are any mitigating circumstances that make the defendant less blameworthy (Árnason 2010).

Neuroimaging has been used in courts as evidence both to show that a defendant lacked the mental capacity to form criminal intent and to show diminished culpability (Bennett and Hacker 2011). One well-publicized use of fMRI to show diminished responsibility was in the U.S. Supreme Court case of *Roper v. Simmons* (125 S. Ct. 1183 [2005]). In this case, a man was found guilty of having murdered a woman when he was seventeen years old. At the sentencing stage, amicus briefs submitted by the American Medical Association and American Psychiatric Association utilized neuroimaging evidence to argue that he did not deserve capital punishment since adolescents are less blameworthy than adults. Both briefs pointed to neuroimaging studies of brain development and function in adolescents showing that they have a disposition to be risk seeking, to be more aggressive, and to have poorer decision-making capacity than adults. This has been linked to evidence that the limbic system of the brain develops earlier than the prefrontal cortex (PFC). In other words, the parts of the brain most involved in decision making and self-control are not fully developed until after twenty years of age, whereas the regions of the brain linked to impulsive behavior are more active than later in life. Thus, it was argued that it is inappropriate to hold juveniles to be as responsible for their actions as adults. Simmons was not sentenced to death, though it was unclear to what extent the neuroimaging evidence contributed to the Court's decision (Árnason 2010).

The Court in *Graham v. Florida* (560.S. Ct.[2010]) expanded that limitation and ruled that for crimes other than homicide, a sentence of life without the possibility of parole for a person under the age of eighteen violated the Constitution's prohibition of cruel and unusual punishment. In the future, justice might require that convicted juveniles undergo neuroimaging to determine whether their white matter structure is adult-like. The results of such a test could then be used to provide guidance to the court on sentencing (Gazzaniga 2011).

Brain Damage, Psychopathy, and the Law

The general theory of crime proposes that it can be explained by the combination of situational opportunity and lack of self-control (Mathias, Marsh-Richard, and Dougherty 2008). Impulsivity is an important component of self-control and is increasingly being utilized to assess forensic populations. Neuroscience confirms that some forms of abnormal behavior are due to reduced abilities of self-restraint and self-control consequent on loss of gray matter, likely the result of loss of synapses in a particular part of the brain. Impulsive behavior is characterized by acting without thought, that is, on compulsion. Delay aversion behavior is behavior indicative of difficulty in resisting the temptation of a small immediate satisfaction for the sake of a large delayed reward. Impulsiveness and delay aversion—acting without prior thought as to the consequences of one's actions—are associated with abnormal activity in the orbitofrontal cortex (OFC) (Bennett and Hacker 2011). In turn, impulsiveness is a major characteristic of borderline personality disorder, in which a person exhibits paranoid ideas, difficulties with relationships, angry outbursts, and violent behavior that often leads to criminality.

As discussed in chapter 4, the relationship between the top-down control of the reflective PFC and the bottom-up drive of the impulsive limbic system appears to play a significant role in aggression, although the details of these mechanisms are not clear (Siever 2008). Therefore, when the limbic system becomes hyperactive, it can overwhelm the control mechanisms of the PFC, resulting in impulsive behavior (Rangel, Camerer, and Montague 2008). Substance abuse subjects, for example, show a greater response to reward and a lower response to punishment than normal subjects (Burns and Bechara 2007). This difference in response can have an adverse effect on decision-making capabilities because rewards are overvalued and unpleasant consequences are undervalued. When damaged by trauma, tumor, or stroke, frontal lobes can cause impulsive actions, in effect removing the segment of the brain that says, "Don't do that! It's not appropriate! There are going to be consequences!" The question is whether the courts will allow claims from the accused that they acted out of overwhelming impulsiveness to be verified by allowing MRI studies showing substantial loss of gray matter in their OFC. Present research indicates that intermediate levels of the capacity to exert control over impulsive actions are likely to be associated with intermediate levels of gray matter loss. This being so, however, neither the gray matter loss nor the behavioral traits can indicate that under the particular circumstances in which the accused acted, he was able to exert self-control or not (Bennett and Hacker 2011).

Psychopathy
Psychopathy is a clinical construct characterized by a pattern of interpersonal, affective, and behavioral characteristics, including egocentricity, deception, manipulation, irresponsibility, impulsivity, stimulation seeking, poor behavioral control, shallow affect, lack of empathy, guilt, or remorse, and a range of unethical and antisocial behaviors. Although it is often described as a constellation of traits, there is growing evidence that psychopathy represents a unified personality disorder (Glenn, Kurzban, and Raine 2011). These characteristics appear to be heritable, manifest early in childhood, and are relatively stable throughout adolescence and into adulthood (Hare and Neumann 2008). Epidemiological studies demonstrate an increased risk for psychopathy in individuals who have encountered traumatic or stressful situations, including childhood abuse, violent crime, divorce, unemployment, and medical illness (Gao et al. 2010; Hyde et al. 2011; Caspi et al. 2010).

Individuals with psychopathy are at increased risk for both instrumental and reactive aggression (Glenn, Kurzban, and Raine 2011; Anonymous 2009), and, although their behavior is not inexorably criminal, psychopathy is a strong predictor of relapse in violent offenses (Müller 2010). Considerable research suggests that psychopaths have deficits in emotional processing and inhibitory control, engage in morally inappropriate behavior, and generally fail to distinguish moral from conventional violations (De Oliveira-Souza et al. 2008). Although these observations have led to the supposition that psychopaths lack an understanding of moral right and wrong, Cima, Tonnaer, and Hauser (2010) conclude that psychopaths do understand the distinction between right and wrong but do not care about such knowledge or about the consequences of their inappropriate behavior. A recent study of the relationship between psychopathy and five domains of morality suggested that psychopathy is associated with reduced empathic concern for others that affects their concerns about harming others and fairness (Glenn and Raine 2009). Raine and Yang (2006) note that while psychopaths may not be insane in any strict legal sense, since they are cognitively capable of distinguishing right from wrong, there is a question whether they can be considered fully responsible for their criminal behavior since, as a result of neurobiological impairments beyond their control, they lack the capacity for the feeling of what is right.

Currently, psychopathy does not exist as a DSM-IV diagnosis but is commonly assessed using the Psychopathy Checklist Revised (PCL-R), developed by Robert Hare and C. S. Neumann (2008). The PCL-R consists of two parts: factor 1 measures personality traits such as superficial charm, grandiose sense of self-worth, shallow affect, lack of remorse, and lack of empathy, while factor 2 scores behavior and lifestyle patterns such as impulsivity, poor behavioral control, parasitic lifestyle, juvenile delinquency, and criminal versatility. The PCL-R has been criticized for mixing personality traits with lifestyle and behavior, and some authors subdivide psychopaths into primary and secondary psychopaths (Müller 2010). Primary psychopaths have more pronounced psychopathic personality traits, scoring high on factor 1, while secondary psychopaths display antisocial behavior and lifestyle, scoring higher on factor 2 (Skeem et al. 2007). For Wahlund and Kristiansson (2009), primary psychopaths are more calculated and disciplined and often use predatory violence, while secondary psychopaths are more impulsive and often commit affective violence. Hare and Neumann (2008) estimate that 1 percent of the population is psychopathic, even if many have never committed a crime.

Imaging studies have found that structural and functional brain changes are related to the degree of psychopathic traits (Harenski, Kim, and Hamann2009; Müller 2010). Therefore, a psychopath as commonly perceived by the public most likely represents the extreme end of a continuum of symptom severity (Glenn, Kurzban, and Raine 2011). Furthermore, while psychopathy is usually described as pathology, from an evolutionary perspective, while it might be a disorder, psychopathy could also be construed as an adaptive strategy (Glenn, Kurzban, and Raine 2011). Certain psychopathic features, then, might not necessarily be a bad thing for society, and in some professions, such police officers or politicians, they might even be beneficial (Abbott 2007).

One distinctive characteristic of psychopathy is a lack of empathy. Blair et al. (2006) found that psychopaths lack affective empathy (see also Blair 2005, 2007a). A prevailing hypothesis is that the prefrontal amygdala connections are disrupted in psychopaths, leading to deficits in contextual fear conditioning, regret, guilt, and affect regulation (Mobbs et al. 2007). Dysfunction in this circuitry may be the result of core impairments that arise from atypical stimuli—reinforcement learning and the representation of outcome information (Crowe and Blair 2008). Such impairments could then interfere with moral socialization and decision making. Moreover, many of the studies reviewed by Shirtcliff and colleauges (2009) signaled reduced amygdala activation in individuals scoring high on the PCL-R, suggesting a link to the neurobiology of callousness and unemotional trait (CU). Marsh and colleagues (2008) found that in addition to reduced amygdala responsiveness, youth with CU traits had reduced connectivity with regulatory brain areas. In addition, studies have shown that individuals with psychopathy fail to react to threatening stimuli (Birbaumer et al. 2005), which could be due to a dysfunctional or slow autonomic nervous system or an inhibitory mechanism stemming from other areas of the brain.

Psychopaths also have reduced peripheral autonomous tonus, low heart rates, and changes in skin conductance. Interestingly, a low resting heart rate measured during childhood and indicating a low peripheral responsivity has been shown to be the best biological marker for antisocial behavior in child and adolescent samples (Raine 2002). Despite impairment of executive function, frontal lobe volume was not reduced in psychopathy, although Müller and colleagues (2008) reported volume reduction in the PFC and Dolan and colleagues (2002) revealed a 20 percent volume reduction of the temporal cortex in impulsively aggressive psychopaths. Kiehl and colleagues (2004) found that psychopaths performed worse while processing abstract words. They hypothesized that psychopaths might be impaired in the recognition of complex social emotions such as love, empathy, or guilt and argued that, because of temporal brain changes, psychopaths are impaired in integrating words that lack a concrete meaning. Subjects with high scores for psychopathy were also less likely to form a cooperative and reciprocal style of interaction (Rilling et al. 2007).

According to Buckholtz and colleagues (2010), alterations in the function of the brain's reward system may contribute to a latent psychopathic trait. They found that individuals who scored high on traits like egocentricity, manipulation of others, and risk taking had a hypersensitive dopamine response system. The amount of dopamine released was up to four times higher than in those who scored lower on the personality profile. Since psychopathic individuals are at increased risk for developing substance use problems, Buckholtz and colleagues (2010) investigated possible links between the brain's reward system (activated by abused substances and natural reward) and the behavioral trait (impulsivity/antisociality) characteristic of psychopathy. Using both PET and fMRI, they found that linking traits that suggest impulsivity and the potential for antisocial behavior to an overreactive dopamine system could help explain why aggression may be as rewarding for some people as drugs are for others.

Because psychopathy is associated with two extremes, Hecht (2011) hypotheses an interhemispheric imbalance in the psychopath's brain. On the one hand, psychopathic individuals lack any concern for social norms, have difficulty maintaining healthy social relationships, and have a seriously compromised capacity to feel certain emotions, such as empathy, guilt, or fear. In addition to these deficits, some behaviors and tendencies, such as impulsivity, stimulation seeking, aggression, and risk taking, are exaggerated in psychopathy. There is evidence that social behavior in general, as well as feelings of empathy, guilt, and fear, is mediated predominantly by regions in the right hemisphere (RH), whereas impulsivity, stimulation seeking, aggression, and risk taking are linked primarily to left hemispheric (LH) activity. Thus, from a neurobiological perspective, it appears that psychopathy may be associated with an altered and imbalanced interhemispheric dynamics in the form of a relatively hyperfunctioning LH or a hypofunctioning RH, or both (Hecht 2011, 3).

Psychopathy is of particular interest for the law because it often involves aggression that is "premeditated, emotionless, and instrumental in nature" (Glenn and Raine 2009). Unlike the primarily reactive aggression observed in other disorders, psychopaths engage in aggressive acts for the purpose of benefiting themselves. This is noteworthy in light of the view that psychopathy represents an alternative life history strategy that is evolutionarily adaptive because it suggests that aggression, risk taking, manipulation, and promiscuous sexual behavior may be the means through which psychopaths gain advantage over others. Recent neurobiological research supports the idea that abnormalities in brain regions vital to emotion and morality may allow psychopaths to pursue such a strategy without facing the social emotions, such as empathy, guilt, and remorse that typically discourage instrumentally aggressive acts, and may even experience pleasure while committing these acts.

Psychopathy is not presently a recognized basis for an insanity defense. In fact, psychopaths are often deemed more dangerous than offenders without the pathology and tend to receive longer or harsher sentences (Gazzaniga 2011). In his fMRI study of 2,000 inmate volunteers, Kiehl found that compared to the average offender, 60 percent of psychopaths reoffend within 200 days. For maximum-security juveniles, 68 percent of those at high risk for psychopathy reoffended. "Psychopathy is currently considered the single best predictor of future behavior. If you have a diagnosis of psychopathy and you're going for parole or something, they view that as a risk factor" (quoted in Mooney 2012). Therefore, a neuroimaging tool that could reliably identify psychopaths could be useful at innumerable stages in the legal process.

A related area where neurological findings might be on a collision course with recent policy initiatives has to do with crimes motivated by hate toward members of particular groups. The assumption is that the perpetrators of such acts are conscious of the reasons why they hate and are capable of controlling their loathing and fears. In a word, they are responsible for their hatred. However, if fear is organized deep within subcortical memory circuits in the amygdala where it arouses strong emotional passions at levels inaccessible to consciousness or willed deliberation, or if the individual suffers from deficiencies in the capacity of the PFC to control those emotions, can we justifiably blame him or her for acting on these deep-seated emotions? While such knowledge of the brain does not excuse hatred and fear, it insinuates that we are not always dealing with rational, conscious action.

Neuroscience as Evidence in the Courts

Although brain scans and other types of neurological evidence have seldom been a major factor in trials to date, eventually they could transform judicial views of personal credibility and responsibility (Gazzaniga 2011). The ability to link patterns of brain activity with mental states could overturn traditional rules for deciding whether a defendant had control over his or her actions and gauging to what extent he or she should be punished. Increasingly, attorneys are asking judges to admit neuroscience data into evidence to demonstrate that a defendant is not guilty by reason of insanity or that a witness is telling the truth. The U.S. Supreme Court in *Daubert v. Merrell Dow Pharmaceuticals* (509 U.S. 579 [1993]) required parties attempting to introduce evidence in federal court based on a scientific technique to provide evidence of the technique's reliability and validity, peer review of data generated by its use, documentation of a known error rate, and general acceptance by experts in the field. Under *Daubert,* judges have the responsibility to assess the impact of allowing scientific evidence in their court. If they believe the jury will consider the scans as one piece of data supporting an attorney's or a witness's assertion, or if they think that seeing the images will give jurors a better understanding of some relevant issue, they might approve the request. Contrarily, if they conclude that the scans will be too persuasive for the wrong reasons or will be given too much weight simply because they look impressive scientifically, they will reject the request. In legal terms, judges need to decide whether the use of the scans will be "probative" (tending to support a proposition) or "prejudicial" (tending to favor preconceived ideas) and likely to confuse or mislead the jury. To date, judges typically have decided that brain scans will unfairly prejudice juries and provide modest or no probative value, or on the grounds that the science does not support their use as evidence of any condition other than physical brain injury (Gazzaniga 2011). "For neuroimaging to meet these legal and medical standards with scientific integrity, scientists must convincingly correlate the dynamic images in a person's brain with the way the person is thinking or acting at that moment" (Baskin, Edersheim, and Price 2007, 239).

Stoller and Wolpe (2007) see four areas of brain imaging applications that could be useful to the courts. At the most basic level, a party to a lawsuit might use brain images to show the structure of the brain. The images might show a brain that does not have the shape or size or component portions of a "normal" human brain or to demonstrate damage to a previously normal brain. They see this as least problematic. Litigants, however, might attempt to use brain images to go beyond brain structure or condition to explain a particular loss of function such as paralysis or inability to speak. This type of use of brain images requires linking the brain, and perhaps certain areas of the brain, to specific human capacities and activities. While these first two applications are straightforward, more challenging in light of current imaging capabilities is their use regarding behavior, for instance presenting brain images in arguing for or against a person's responsibility for his or her actions or for making predictions about future behavior. A litigant might present brain images to establish that a person is dangerous—that he or she is likely to commit violent acts in the future. A litigant might even attempt to use brain images to establish a wide range of propensities or personality traits such as recklessness, racism, or unhappiness. And finally, neuroscience can be used to determine whether a person is telling the truth or lying. Extensive use of brain images to detect conscious and unconscious lies could potentially transform current methods of fact finding in legal disputes.

Almost half a century ago, H. L. A. Hart stated that certain mental conditions have an "excusing" effect on punishment, whereby "the individual is not liable to punishment if at the time of his doing what would otherwise be a punishable act he is . . . the victim of certain types of mental disease" (1968, 28). The Supreme Court, however, upheld a state statute allowing for increased "punishment" in *Kansas v. Hendricks* (521 U.S. 346, 346 [1997]), in the form of civil commitment, for sexually violent predators who suffer from a "mental abnormality." Thus, American jurisprudence is faced with a paradoxical role of mental conditions that can serve to either aggravate or mitigate sentences. This dual purpose of entering mental conditions as evidence becomes a pressing legal issue as those conditions associated with brain structure or function become more detectable through neuroimaging.

While neuroimaging technology has advanced at an amazing pace over the past fifteen years, finding its way into the courtroom at an accelerating pace, neuroscience cannot yet predict neurological, psychiatric, or behavioral deficits. As a result, for Martell (2009), there is currently little room for neuroscience in the courtroom. Although this might change in the future, for now the relevance of neuroscientific evidence is limited by the state of the science. For instance, current scientifically supported uses for structural brain imaging are limited to providing evidence of brain lesions resulting from head injury, stroke, or neurological disease. Similarly, current Society for Nuclear Medicine guidelines indicate that the only appropriate uses for SPECT include the detection and evaluation of cerebrovascular disease, dementia, epileptic foci, and suspected brain trauma (Juni et al. 2002).

In contrast, Martell (2009) suggests that neuropsychological testing can identify specific behaviors that are associated with brain function in a valid and reliable way and serves as a natural bridge between the structural and functional neurophysiological parameters measured by neuroimaging and the actual cognitive and behavioral strengths and weaknesses of the individual that may be relevant in court. Scientifically unsupported uses of neuroscience evidence, however, include the extrapolation from structural or functional imaging to violent or criminal behavior and the use of brain scans to diagnose or confirm any psychiatric disorder. Gazzaniga (2011) suggests that in future cases, neuroscience evidence might be admissible if it has the more limited goal of simply assessing the character and overall honesty of defendants and in this way supplementing or replacing character witnesses. Federal Rule 608(b) provides that once the character of a witness has been attacked, counsel can introduce as evidence opinions about the witness's "character for truthfulness or untruthfulness." Gazzaniga wonders whether machine-produced evidence that someone tends toward dishonesty might be more prejudicial than evidence from traditional sources.

Although the neuroscience-centered "event-feature-emotion complexes" (EFEC) framework of Moll and colleagues (2005) allows the prediction of changes in moral behavior by including both cognitive and emotional dysfunction related to moral emotions and judgment, questions are raised as to how this affects the philosophy of law. For example, Bloch and Mcmunigal (2005) suggest there are four classic justifications for punishment: (1) *retribution* focuses on a "get tough on crime perspective," whereby criminals are punished for their wrong deeds; (2) *deterrence* suggests that punishment should be designed to prevent future criminal behavior; (3) *incapacitation* implies that those who commit crimes should be taken off the streets to prevent future crimes from occurring; and (4) *rehabilitation* proposes that punishment is administered to rehabilitate criminals. According to the authors, each of these justifications for punishment suggests different views of how the EFEC framework will have an impact on the law. Retribution and deterrence both seem incompatible with a more deterministic neuroscience view of criminal behavior, while incapacitation and rehabilitation both speak to a more deterministic philosophy of moral behavior (Knabb et al. 2009).

One issue unique to the United States is whether a mentally disabled death row defendant can be executed. The Supreme Court in *Panetti v. Quarterman* (551 U.S. 930 [2007]) ruled that a seriously mentally disabled death row defendant has a constitutional right to make a showing that his mental illness "obstruct[ed] a *rational understanding* of the State's reason for his execution." Scholars have begun to consider the impact of neuroimaging evidence on capital trials, raising the question of whether reliance on such testimony can actually make "sentencing more rational and humane" (Perlin 2010a). Predictably, a review of the literature on neuroimaging reveals a broad array of positions, promises, and prophecies. Snead (2008) contends that the ambition of cognitive neuroscientists is to use the new powers conferred by neuroimaging to overthrow retributive justice as a legitimate justification for criminal sanctions. Eagleman adds that there is a new potential to use "detailed combinations of behavioral tests and neuroimaging to better predict recidivism" (2008, 38), while Sapolsky (2004) predicts that a more subtle approach to offenders whose behavior can be linked to trauma in certain neural systems appears to be inevitable and, if so, could turn the criminal justice system and the justification for punishment on its head.

Functional Imaging and the Law

There is much imagination, hype, and even science fiction in "neurolaw," according to Annas (2007). Other scholars charge that "researchers, clinicians, and lawyers are seduced into becoming true believers in the merits of [brain imaging] for understanding the relationship between brain and behavior" (Tancredi and Brodie 2007, 289; also see Fischbach and Fishbach 2005; Simpson 2008). Currently, as discussed in chapter 2, the main difficulties of brain measurement and imaging are their unreliability and the openness of the results to conflicting interpretations (Häyry 2010; Poldrack 2011). Tovino agrees that fMRI offers only "an illusory accuracy and objectivity" (2007). Similarly, Baskin, Edersheim, and Price (2007) conclude that neuroscience is not currently helpful in determining structural correlates of predatory violence or cognitive constructs such as psychopathy. For them, "The use of this kind of speculative neuroimaging in the courtroom presents methodological and ethical problems that currently outweigh any probative value" (267). Arrigo (2007) goes even further, arguing that the law's use of fMRI technology for purposes of interrogating those accused of or responsible for criminal transgressions is an inherently flawed application: "Through the deployment of fMRI brain scanning, the subject is commodified, technologized, and reontologized. In the realm of medicolegal science, virtual realities and pseudoforms of neuronal activity are reproduced and disseminated as their meanings become more authentic and true than the originals from which they are derived" (479).

Although these emerging technologies challenge the courts, this is nothing new. As Khoshbin, and Khoshbin (2007) illustrate, in previous generations people looked to inheritance (genetics), anatomical features (phrenology), a history of emotional trauma or unresolved psychic conflict (psychoanalysis), and socioeconomic deprivation (sociology and economics) to explain why some commit crimes and others do not. Tomorrow, it seems that people will look increasingly to neuroscience. Tovino (2007) notes that earlier methods of body imaging and brain mapping raised similar issues, and recommends using our experiences with these technologies to guide emerging neuroimaging policy. Even if functional neuroimaging is not ready for routine courtroom use, serious reflection and imaginative speculation on what new brain imaging technologies can and cannot tell us and of what legal use they might be are now essential.

Judicial opinions involving phrenology, radiography, PET, and SPECT reveal several themes, including the general duty of the law to keep up with advances in medicine and science and the more specific duty of the law to adopt technologies that will assist the jury in seeking the truth. Contrasting themes, however, reflect an uneasiness about the illusory objectivity of body imaging and brain mapping and the difficulty of balancing advances in science and medicine against the risks associated with junk science and charlatans. Khoshbin and Khoshbin, (2007) argue that functional brain images should not be admitted for the purpose of establishing responsibility for, motivation for, or propensity to commit a particular behavior, or to show an inability to control a particular behavior. Indeed, given the current state of scientific knowledge about the brain, the courtroom is an inadequate forum for determining the "truth" of such evidence.

Until recently, determining the mental state of a defendant fell largely on the shoulders of court psychologists and experts in psychiatry for qualitative assessments related to insanity pleas or mitigation at sentencing. Thus, neuroimaging might provide additional, pertinent biological evidence as to whether an organically based mental defect exists (Batts

2009). With increasing frequency, criminal defense attorneys are integrating neuroimaging data into hearings related to determinations of guilt and sentencing mitigation or to provide insight for parole hearings. For instance, if a tumor or lesion that allegedly altered behavior is removed, one might make a case that the chance of reoffending is low (Batts 2009). One possible scenario is a legal system inundated with criminal defense attorneys attempting to rationalize away any crime as the result of neurological disease, and juries overwhelmed by the vividness and dramatic nature of such data.

Recent research suggests that presenting neuroimaging evidence at trial may sway juries' beliefs about a defendant's culpability. In one study, mock jurors acknowledged that they gave credence to neuroimaging visual aids in mitigating the sentence. Jurors who stated that the neuroimaging data had influenced them were nearly six times as likely to render a not guilty verdict (Gurley and Marcus 2008). Given this, Batts (2009) asks how the average jury, comprised of laypeople not well versed in neurological deficits, is supposed to distinguish a normal, functioning brain from a damaged one when experts are unable to testify with reasonable certainty that a brain lesion caused the defendant's behavior at the time of the crime. So, although neuroimaging evidence might pose unfair defense bias in some criminal trials, since a large percentage of defendants pleading insanity do in fact have identifiable brain damage, such evidence is certainly relevant to many cases and cannot be excluded wholesale (Batts 2009). Mediating how and when to introduce such evidence will require that judges and juries become educated about the relevant neurological disorders that may have an impact on responsibility. It is not difficult to foresee a future where law schools teach relevant areas of neuroscience, creating a new type of "neurolaw" expert capable of bridging this gap.

One of the recurring themes in the use of imaging evidence in the courts is that it is particularly vulnerable to misuse because of its visual attractiveness. This visual power can easily result in misunderstanding over what the images show and mean (Khoshbin and Khoshbin 2007). For instance, Erickson notes the "gloss of intrigue and seduction" inherent in neuroimaging testimony (2010, 36). Others have argued that the visual "allure" of such testimony can "dazzle" and "seduce" jurors in ways that are inappropriately persuasive (Feigenson 2006). Moreno (2009) says, flatly, "brain research is *sexy*."

Data from fMRI, SPECT, and PET scans can be referenced and presented in dazzling multimedia displays that may inflate the scientific credibility of the information presented. Imaging, available in brilliant colors, with its apparent simplicity and vividness, accompanied by exotic names for brain regions, can prove irresistible to many defense attorneys, judges, and jurors. Sophisticated clinicians may present oversimplified, contradictory testimony that may be met with naive acceptance or considerable skepticism by judges, juries, and society in general. (Basken, Edersheim, and Price 2007, 267)

Therefore, it is possible that the vibrancy of this testimony may have a distortive impact on jurors, thus raising the question as to whether judges are less susceptible than jurors to the radiant heuristic in this setting (Perlin 2010a). Is there a danger that courts will interpret a brain image as having more evidentiary value than it actually possesses? For instance, if a scan had a reliability rate of 0.72, might a juror, seeing a "picture" in vivid color, ignore the 0.28 chance of error and assume absolute accuracy? And although such risks of misinterpretation are genuine, are they any greater than the risks associated with viewing any color picture or any multicolor diagram, images often introduced in trials? "Once we comprehend the range of standards the legal system now uses, the scalar and not binary character of reliability and validity, and the legal system's venerable reliance on techniques for identifying deception that are worse than even the most modest claims for neural lie detection, the case against fMRI becomes less compelling" (Schauer 2009, 102).

Salerno and Bottoms (2009) argue that neuroimaging research can help the legal system better understand how jurors decide. In particular, neuroscience research addressing how emotion affects the moral reasoning process has the potential to inform the debate about whether there should be more regulation of the admission of highly emotional evidence such as autopsy photographs and victim impact statements. Although victims' rights advocates support the admission of such evidence in courts, others are concerned that jurors' emotional responses to this evidence might unfairly prejudice them against a defendant. Of course, both the defense and the prosecution regularly attempt to sway jurors' emotions and, in turn, their verdicts. It is important to note that emotional evidence on either side has the potential to affect jurors' judgments through legally desired probative channels or legally untenable prejudicial channels. Again, it is up to the judge to decide whether the proposed evidence fits the former or the latter category, but neuroscience can provide insights into the process (Salerno and Bottoms 2009; Markowitsch and Staniloiu 2011b).

Superficially, the status of fMRI under the Self Incrimination Clause hinges on whether evidence gathered through it is more like spoken testimony against oneself or like DNA samples, fingerprints and blood tests. If like the former, then such evidence would not be admissible. If like the latter, such evidence would be. It is highly likely that fMRI's legal fate will be determined by how apt judges regard these respective analogies. (Thompson 2007, 344)

Another issue facing the courts is the huge expansion of experts. According to Khoshbin and Khoshbin (2007), the explosion of imaging technologies has, in a relatively short time, resulted in a need for highly specialized training in producing and interpreting images. Under *Daubert,* which provides the current standard for application of FRE 702 for all federal and most state courts, a witness must be "qualified as an expert by knowledge, skills, experience, training, or education" to testify to "scientific, technical, or other specialized knowledge" if that knowledge "will assist the trier of fact to understand the evidence or to determine a fact in issue." In 2000, an amendment to FRE 702 added the following language regarding acceptance of testimony by the courts: "if (1) the testimony is based upon sufficient facts or data, (2) the testimony is the product of reliable principles and methods, and (3) the witness has applied the principles and methods reliably to the facts of the case."

Khoshbin and Khoshbin (2007) suggest that medicine and neuroscience could provide guidance on the clinical purposes for which imaging techniques have been validated, the state of research, and the processes underlying the technologies for judges who make evidentiary determinations in the cases before them. They also argue that professional medical societies should provide written guidance concerning the validity of brain imaging techniques and the professionally accepted clinical purposes for which those techniques may be used, as well as specific guidance for evaluating the training and professional qualifications of expert witnesses. They also suggest that the Institute of Medicine should work with the relevant scientific and medical professional societies and that the President's Council on Bioethics should continue to serve in an advisory and educational role to the public with respect to policy and ethical issues that may be raised by the use of neuroimaging technologies.

Meanwhile, the number of courts allowing defendants to present neuroimaging evidence is growing. In insanity defense cases, defendants have a right to access the assistance of a psychiatrist, and in at least one case, this right was extended to neuroimaging tests. One high-profile case involving brain imaging is *People v. Weinstein* (1992), in which the defense petitioned to admit PET scans and an MRI of the defendant's brain in the guilt phase of his murder trial over the issue of whether he should or should not be held criminally responsible for strangling his wife, owing to the presence of an arachnoid cyst on the surface of his brain. The defense evidence included testimony that the cyst may have caused diminished function in the tissue surrounding it. Following a pretrial hearing on this evidence, the prosecution accepted Weinstein's guilty plea to the reduced charge of manslaughter, possibly because it believed that the jury would be unduly persuaded by the scans (Khoshbin and Khoshbin 2007). Similarly, in *McNamara v. Borg* (923 F.2d 862) (9th Cir. 1991)., the defendant was allowed to introduce PET scan in evidence in the sentencing phase of his murder trial. The defendant testified that he was suffering from schizophrenia at the time he committed the crimes. The jury sentenced him to life in prison rather than to death, with some jurors later admitting that the PET evidence persuaded them to grant leniency.

Neuroscience and Lie Detection

The most salient issue surrounding neuroimaging evidence to date has centered on its use in the determination of truth or lying. Despite the long history of debate over lie detection, the pursuit of a reliable method for detecting deception continues to be an enduring goal in the neurobehavioral sciences (Illes and Bird 2006). Extensive use of brain images to detect conscious and unconscious lies could potentially transform current methods of fact finding in legal disputes (Stoller and Wolpe 2007). Proponents of brain imaging as a method of lie detection believe that their methods are, or will soon be, far superior to the polygraph and that they are immune to efforts to "fool the test." Researchers are developing methods that purportedly go beyond the detection of conscious lies and show what information is in a person's brain, even in the absence of any verbalizing by the person. Pettit (2007) suggests that "lie detection" is an inappropriate term for these new methods; perhaps "truth detection" or "truth investigation" would be more accurate. However, given the history of denying polygraph evidence by the courts, it seems fair to say that judges will be extremely skeptical of the claims made by proponents of truth detection and will demand compelling scientific evidence.

Various technical forms of lie detection have been proposed that make use of EEG, PET, functional near-infrared spectroscopy (fNIRS), transcranial magnetic stimulation (TMS), fMRI, and many combinations of these studies. Although each technique has some advantages, according to Shamoo (2010) most have failed to provide consistently reliable data to inspire much confidence. For instance, EEGs use electrodes placed on the skin to measures changes in the neuronal potentials termed event-related potentials. Nevertheless, a measurement of a single event is very difficult, and thus a summation and averaging waveform must occur, raising many questions as to its suitability. PET, on the other hand, cannot provide a measurement of a single transient event. As well, the need to inject an individual with a radioactive substance raises many ethical and legal issues. Although fNIRS is noninvasive, it does not penetrate a thickness of more than 2 centimeters (Heckman and Happel 2006). Ironically, Wolpe, Foster, and Langleben (2005) contend that the very advantage of simplicity fNIRS has over other techniques raises added concerns regarding its potential abuse.

The initial indication from TMS research is that, under certain circumstances, magnetic stimulation can slow deceptive responses, distinguishing them from truthful ones. In the TMS study by Luber and colleagues (2009), the slowing effect was found in the group mean data. Although this sort of effect is useful to research exploring brain mechanisms of deceptive behavior, it is far from observable in an individual in the courtroom. However, given advances in the existing technology that improve spatial and temporal targeting with EEG- and fMRI-guided TMS, it is possible that TMS or other noninvasive brain stimulation technologies may be employed as lie detectors in the future. However, while TMS is noninvasive and portable, it has been linked to seizures in a small percentage of those tested (Heckman and Happel 2006).

Not surprisingly, fMRI is one of the most promising technologies to detect lying and truth telling (Shamoo 2010; Keckler 2006). As discussed in chapter 2, fMRI measures changes in the nuclear spins of protons. The nuclear spin properties of oxygenated hemoglobin are different from those of the deoxygenated hemoglobin. In fMRI, they measure the changes between the two states of oxygenation in hemoglobin to correlate regions and function of the brain to a particular stimulus or behavior. These blood-oxygen-level-dependent (BOLD) signals are then correlated with MRI. Since lying requires more metabolism and thus more oxygenated blood than the relaxed normal function of telling the truth, this will show up on the scan. Langleben and colleagues (2002) have used fMRI with sophisticated computer software, termed the "machine learning method," and found that the anterior cingulate cortex and the superior frontal gyrus became active when subjects lied. The investigators contend that, under strict laboratory conditions, they are able to differentiate brain activity associated with lies in 88 percent (for untrained subjects) to 99 percent (for trained subjects) of cases. Similarly, Kozel and colleagues (2004, 2008) reported over 90 percent accuracy in identifying neural patterns in individuals that they attribute to lying. Another recent addition is the machine learning classifier analysis of fMRI data (Pereira et al. 2010). This highly advanced software analyzes the data for "pattern localization" in the brain image and "pattern characterization" to interpret these patterns.

A very controversial technique, developed and marketed by Lawrence Farwell, is "brain fingerprinting," which attempts to discern whether a person has a record of a particular event or an image, such as a crime scene, stored in his brain. It is based on the assumption that each bit of information in the brain is stored in neurons and that when a salient piece of information is recognized, specific neurons fire in response. Brain fingerprinting uses EEG sensors to record electrical brain signals emitted when subjects encounter various stimuli. It analyzes multiple EEG responses to each stimulus and claims to be able to determine whether the brain is giving an "information present" or "information absent" response (Stoller and Wolpe 2007). It has been found that when the neurons fire, a P300 wave is generated. Because the subject has been told to pay particular attention to the target, this stimulus will be salient to the subject and will create a baseline P300 response, or memory- and encoding-related multifaceted electroencephalographic response (MERMER). If the subject has knowledge, the probes will be noteworthy to the subject and, like the targets, will elicit a MERMER. If the suspect lacks such knowledge, the probes do not elicit a MERMER. By comparing the brainwaves elicited by probes, targets, and irrelevants, Farwell claims his system can determine whether the probes represent information that is known to the subject, even if the subject claims no knowledge of or familiarity with it.

Although Farwell claims that test results have found have 100 percent accuracy in determining whether subjects did or did not recognize the probe stimuli, none of these results have been independently verified. Additionally, brain fingerprinting faces various obstacles that could render it unreliable when used in a real-life situation. For one, since the P300 wave measures recognition, the suspect must remember the particular detail being shown to him. Conversely, the stimulus cannot be salient to the suspect for any reason other than its association with the crime, including media coverage of the crime. There are also questions concerning whether a memory of a crime might change over time, or in response to drugs or stress. Last, evidence suggests that, as with fMRI, there might be effective countermeasures to brain fingerprinting that allow a subject to "defeat" the test. Still, it is well established that the P300 event is a response to a salient stimulus, even if the cognitive processes underlying it are unclear (Stoller and Wolpe 2007, 363).

Most of the research on these methods has occurred under controlled conditions and with a small number of test subjects, but in the field individuals undergoing testing for lying or deception are subject to many variables that cast doubt on the outcome (Buller 2005). A critical consideration in the real world is that all of the above technologies are susceptible to countermeasures, including drugs, alcohol, relaxation techniques, and backward counting, as well as physical countermeasures such as inducing pain (Wolpe, Foster, and Langleben 2005). For instance, because fMRI tracks the levels of oxygen in blood, differences in blood flow can lead to divergent results. Countermeasures include thinking of emotionally charged scenes during the baseline readings or simply moving one's toe (Green 2005, 54). It has long been recognized that countermeasures against polygraphy are effective, and similar measures could be used by subjects who were familiar with the test procedure for these new technologies. For example, if subjects knew that reaction time is an important outcome measure in these studies, they could intentionally "jitter" their responses to create statistical uncertainty and obscure the difference between lie and truth responses (Luber et al. 2009). Furthermore, some of these tests, such as TMS, have high spatial resolution, and so even subtle movements of the head could interfere with experimental results. Techniques to block the countermeasures, perhaps based on field testing, would be necessary to address these concerns.

Even if fMRI can be useful in detecting some types of lies, its success rate is not consistently higher than the results obtained using more traditional lie detection methods (Vrij 2008). Many confounding conditions, variables, and behaviors can lead to automatic responses and render polygraph results inaccurate. Because the polygraph measures increased arousal, it can be useful in detecting guilt and fear. People can experience inner turmoil while lying; however, they can also experience psychological stress because of the fear of being interrogated. Guilt and fear, then, can occur for many reasons other than deception. In contrast, if a liar feels no guilt or fear, the deception may go undetected.

Predictably, the prospect of using neuroscience to detect deception in legal settings has generated widespread resistance, with many insisting the research is flawed science, containing weaknesses of reliability (the degree of accuracy), external validity (do laboratory results predict real-world outcomes?), and construct validity (do studies test what they purport to test?). In light of these reservations, White (2010) concludes that it is unlikely that fMRI scans will gain general admissibility as courtroom evidence. Tancredi and Brodie (2007) agree that the use of fMRI for lie detection is fraught with problems that call its use in the legal system into serious question (Miller 2010; Meegan 2008). However, Thompson (2007) warns against couching objections to using fMRI merely in terms of technological inadequacy. "But limiting ourselves to the technological elements implies that should the technology prove feasible, there would be no other objections . . . we should err on the side of caution and restrict the use of fMRI on the basis of its unique ability to compromise involuntarily an individual's mental privacy, rather than merely on the grounds of lack of reliability given the state of the art" (343–344).

Admitting that these technical shortcomings are of concern, Schauer (2009) contends that the critics are mistaken in believing that scientific standards alone should determine when these methods are ready for legal use. He points out that the courts now routinely admit many types of evidence, such as witness accounts, that are far more dubious than the lie detection science. There are always concerns about credibility based on the misperception, hedging, fudging, and slanting that is rampant in a system dominated by testimony from self-interested parties. In law as in science, "compared to what?" is an important question. Thus, for Schauer (2009), evaluating fMRI lie detection requires knowing what it would supplement or replace. Currently, the jury or a judge determines whether a witness is telling the truth. When cross-examination provides little clarification, which it seldom does, juries are instructed to evaluate the deportment of a witness to determine veracity. In doing this, however, they rely on numerous myths, urban legends, and pop psychology with little reliability. They distrust witnesses who perspire, fidget, or fail to make eye contact and trust those who speak confidently while looking directly at them while in their Sunday best. Research shows that ordinary people's ability to distinguish truth from lies rarely rises above random, and juries are unlikely to do better. Therefore, the question is whether neuroimaging should be rejected in favor of keeping traditional methods that have been at the heart of the jury system for centuries, that are extremely unlikely to be changed substantially, and that are demonstrably unreliable. The law has its own standards for determining admissibility, and those standards are more lenient than scientific standards. Schauer (2009) argues that jurors should be allowed to consider the result of a lie detection test that has a 60 percent accuracy rate because it could provide reasonable doubt as to guilt or innocence.

Issues surrounding the use of neuroscience technologies for the purpose of lie detection are compounded by companies marketing these services. According to Stoller and Wolpe (2007), the escalating use of these new technologies challenges our existing system of legal protection against invasions of privacy and coercion to produce evidence. As a result, the efforts to use new lie detection technologies in the legal setting will trigger a range of constitutional challenges potentially under the First, Fourth, Fifth, Sixth, Seventh, and Fourteenth Amendments (Greely and Illes 2007; Luber et al. 2009; Boire 2005). Given the violative nature of entering the suspect's mind, it is likely that the Fifth Amendment right to avoid self-incrimination would preclude forced use of these methods in U.S. courts (White 2010). In addition to the Fifth Amendment protections against self-incrimination, a witness might also claim Fourth Amendment protection against unreasonable search and seizure. Although doing so to protect the contents of one's mind, so to speak, is unprecedented, it might be progressively relevant as the ability to detect veracity progresses. Also, First Amendment claims could be raised as interference with free responses to questions could be viewed as a violation of the right to free mentation and speech. This would be the first time that the limits of "cognitive liberty" would be defined and tested; thus there is no clear legal precedent (Wolpe, Foster, and Langleben 2005).

Wolpe, Foster, and Langleben (2005) warn of the dangers to cognitive liberty that these technologies present and claim that for the first time, we would need to "define the parameters of a person's right to cognitive liberty, the limits of the state's right to peer into an individual's thought process with or without his or her consent, and the proper use of such information in civil, forensic, and security settings" (39). In general, concerns about cognitive liberty break down into two categories: the fear that fMRI scans are mind reading and the worry that such scans amount to an unauthorized search. However, White (2010) contests this, arguing that these concerns are unwarranted. Voluntary markets in fMRI lie detection scans do not pose any more of a threat to cognitive liberty than commercial polygraph examinations, and fMRI is not mind reading. This detection process hardly amounts to reading thoughts.

In contrast, however, the potential benefits to society of such a technology, if used appropriately, could be considerable. Moreover, Tancredi and Brodie (2007, 291) suggest that future developments in imaging, such as the BIRN initiatives, which standardize the hardware of fMRI and improve reliability and reproducibility tests; efforts to combine fMRI with other imaging techniques, task-functioning tests, and clinical evaluations; and the increased use of 3-tesla MRI and higher-strength MRI should lead to more consistent and detailed findings, thus making them more useful to the legal system. Lie detectors using fMRI will also likely be refined, thereby allowing precise and direct imaging that far outstrips what can be achieved using existing detectors. Moreover, the problems of confounding factors, such as subjects experiencing anxiety from unrelated thoughts, will likely be worked out. In addition, TMS could be used to inhibit or activate parts of the brain that create confusion during lie detection and aid in sorting out confounding factors. As a result, fMRI lie detectors could become an important adjunct for law enforcement entities, including the FBI, the military, and Homeland Security, which are already exploring the utility of these technologies (see chapter 7).

Use of Lie Detection Techniques in the Courts

The courts have begun to struggle with admissibility of incipient deception-detecting technologies. Given a history of rejection of lie detection spanning over eighty years since *Frye v. United States* (293 F. 1013, D.C. Cir. [1923]), which denied the admissibility of a polygraph test as evidence, expert testimony about lie detection faces obstacles not faced by other expert testimony. Although one might have thought that *Frye* would change as methods of lie detection improved, to this day courts overwhelmingly reject attempts by parties to introduce results of polygraph tests into evidence. According to Pettit (2007), one explanation is that developers of polygraph testing have been unable to make sufficient improvements in reliability to justify admissibility under the applicable legal standards. If this were the only rationale for judicial hostility to the use of lie detector tests, the proponents of lie detection through brain imaging might have reason for optimism. However, it is clear that the courts have other reasons besides concerns about reliability for exclusion. For instance, some judges have argued that lie detector evidence invades the province of the jury. What role would remain for the jury if courts routinely admitted lie detector evidence and a truly reliable scientific method of determining the truth and accuracy of witness testimony as well? Would not jurors expect lie detector results for every witness and disbelieve witnesses who did not come with proof that they had passed the test? Indeed, would there be any need for trials at all (Pettit 2007)?

One of the first cases to confront the use of brain-scanning technology for lie detection ended in a federal district court in Tennessee. In *United States v. Semrau* (No. 07–10074 M1/p), a magistrate excluded evidence offered by the commercial fMRI lie detection company CEPHOS and detailed why he found that the unfair prejudicial influence of the technology substantially outweighed its probative value. The magistrate's main objection was that the defense expert conducting the lie detection test could not tell the court whether the answer to any particular question was true or false, and thus the test lacked sufficient scientific reliability, under FRE 702. In fact, the expert testified that he could tell only whether the defendant was answering the set of questions about the case truthfully overall (Gazzaniga 2011).

The use of brain fingerprinting has also arisen in recent cases. The first time that brain fingerprinting testimony was admitted was at a hearing on a postconviction claim in Iowa in 2000, during which Dr. Farwell testified that his testing of a convicted murderer showed that his brain did not contain information about the murder but did contain information consistent with his alibi. The district court judge denied relief on the basis that his claim was time barred. On appeal, the Supreme Court of Iowa held that the claim was not time barred and granted a new trial on the basis of the Brady violation, but it specifically stated that the brain fingerprinting evidence was not necessary to the result and that the Court gave it "no further consideration" (*Harrington v. State*, 659 N.W.2d 509, 516 n.6 [Iowa, 2003]). In another case involving a convicted murderer's claim for postconviction relief (*Lebron v. Sanders*, 2005 U.S. Dist. Lexus 35588 [S.D.N.Y. Dec. 25, 2005]), the defendant sought brain fingerprinting testing of himself and several other persons to determine their extent of involvement in the crime. In this case the judge refused to order these tests because the defendant had failed to show how his claims (arrest without probable cause and coercion of his guilty plea) "will be assisted by any of the tests he is seeking and, particularly why physical or mental tests taken eleven years after his guilty plea would be indicative of the physical or mental state of the persons at the time of his arrest and conviction" (Id. p. 1).

In still another claim for postconviction relief by a man convicted of murder, the defendant submitted an affidavit from Dr. Farwell stating that he had conducted brain fingerprinting testing on him, asking him about "salient details of the crime scene," and that Slaughter's brain response showed that these details were absent from his knowledge

(*Slaughter v. State*, 105 P.3d 832 [Okla. Crim. App. 2005]). The Court of Criminal Appeals of Oklahoma expressed several reasons for concluding that this evidence did not meet the standards for granting postconviction relief. In part, the court found that there was no support for Farwell's bare affidavit that referred to a comprehensive report that was never presented to the court. Furthermore, there was "no real evidence" that the *Daubert* standards had been met, and insufficient evidence that the data offered would "survive a *Daubert* analysis" (Id.). Finally, the court gave credence to the state's argument that the salient details of the crime were presented at Slaughter's trial, thus casting doubt on the brain fingerprinting conclusion that these details were absent from Slaughter's brain. Although Pettit (2007) questions what the court would have done if Farwell and Slaughter's lawyers had done a better job in presenting and supporting the brain fingerprinting evidence, the court expressed strong skepticism that would appear quite difficult to overcome.

Coerced Medication or Treatment

At the minimum, contemporary research on the brain demands a wider debate of the potential coerced use of neurotechnologies and drugs. Furthermore, it will require reevaluation of the conceivable court-ordered use of highly selective anti-aggression drugs or implant chips in the PFC to compensate for reduced brain function (Enserink 2000). A question already facing the courts is whether to order medication for mentally ill defendants so that they can stand trial (Perlin 2010b), an issue that has already made it to the U.S. Supreme Court, which found that the

Constitution permits the government involuntarily to administer antipsychotic drugs to a mentally ill defendant facing serious criminal charges in order to render that defendant competent to stand trial, but only if the treatment is medically appropriate, is substantially unlikely to have side effects that may undermine the fairness of the trial, and, taking account of less intrusive alternatives, is necessary significantly to further important governmental trial-related interests. (*Sell v. United States*, 539 U.S. 166 [2003], at 179)

Moreover, the Court's ruling in *Jones v. United States* (463 U.S. 354 [1983]) makes it probable that defendants who invoke this right to refuse will remain institutionalized longer. Additionally, in *Riggins v. Nevada* (504 U.S. 127 [1992]), the Court held that a defendant competent to stand trial had a right to refuse medication at his trial when he was proffering an insanity defense. To some extent, the case law creates for counsel an intolerable Hobson's choice: if a client is in great psychic pain (with ruinous hallucinations and delusions), and the attorney suggests that the client take medication, that decision could have an eventual serious (even deadly) impact on a capital crime client (Perlin 2010a).[2]

The Court has made it clear that the most important issue to be considered in a forced-medication case is the patient's litigational status (Perlin 2010a): Is he currently incompetent to stand trial? Is he proffering an insanity defense? Is he currently incompetent to be executed? If a defendant who is currently incompetent to stand trial seeks to refuse the imposition of medication while awaiting trial, and assuming he meets the standards articulated in *Sell*, his decision to refuse medication may result in institutionalization in a maximum-security forensic facility for far longer than if he had acquiesced. Although to date, this debate has focused on court-mandated medication, in the future it is likely to shift to the use of brain implants and other devices and to the potential use of preemptive mandated interventions to modify the brain chemistry of those persons diagnosed through brain imaging tests as at high risk for violent behavior. Paradoxically, new knowledge about the interaction of the brain and the childhood environment also intensifies the need to develop preventive policies in response to early maladaptive behaviors, particularly in males.

An extension of this issue is whether the courts will allow individuals to be coerced, explicitly or implicitly, in the workplace or in schools to take cognitive-enhancing drugs. Chandler (2011) discusses three doctrines—the doctrine of mitigation, the standard of care in negligence, and child custody determinations in family law—to show how the law may pressure people to consent to treatment by offering a choice between accepting medical treatment and suffering a legal disadvantage. He suggests there is a greater range of social pressures that might compel the uptake of novel neurotherapies than one might presume. Once treatments are developed and offered with therapeutic benefits in mind, their very existence gives rise to unintended legal consequences, thus raising caution for physicians and for legal policy makers. How should physi-

---

2. The question of a defendant's right to refuse medication that would make him competent to be executed has been considered, with conflicting results, by state and federal courts: compare, for example, *State v. Perry*, 610 So. 2d 746 (La. 1992) (right to refuse) and *Singleton v. State*, 437 S.E.2d 53, 60–62 (S.C. 1993) (same) with *Singleton v. Norris*, 992 S.W.2d 768 (Ark. 1999) (no such right). Moreover, the Supreme Court in *Buchanan v. Kentucky* (483 U.S. 402 [1987]) more than hinted at this possible dilemma (Perlin 2010b, 481).

cians, who are required by medical ethical principles to obtain valid consent to treatment, react to a patient's reluctant consent that is driven by legal pressure? From the legal policy perspective, are our legal doctrines satisfactory, or should they be changed because, for example, they unduly promote the collective interest over individual freedom to reject medical treatment, or because they channel us toward economically efficient treatments to the detriment of more costly but potentially superior approaches of dealing with behavioral problems (Chandler 2011)?

As noted above, another issue that courts will face is whether to order preemptive treatment for individuals identified as having serious brain abnormalities that might cause them to be violent. For instance, the evidence linking gray matter loss in particular parts of the brain to the loss of different restraint capacities points to the possibility of designing interventions that return gray matter to normal, and hence restore the normal capacity for restraint (Bennett and Hacker 2011). The power of certain drugs such as clozapine and lithium to promote the growth of new synapses has been extensively studied in animal models (Kim and Thayer 2009). More recently, it has been shown that olanzapine can arrest the loss of gray matter in certain parts of the brain of young adolescents diagnosed with schizophrenia (Mattai et al. 2010). Other potential coerced treatments could include deep brain stimulation and the physical interventions discussed in chapter 2.

Conclusions: Neuroscience and the Courts

For better or worse, forensic neuroscience is already bringing evidence from the study of brain structure and function, and the new applications that follow it, into the courtroom. Least controversial is its application in civil litigation, where brain imaging can provide useful evidence of brain lesions and neuronal disease that might be relevant to establishing causation and assessment of damages. Therefore, neuroimaging findings have been used as demonstrative evidence of brain damage, such as head injury or exposure to neurotoxins, in support of tort claims, as well as to sustain findings of dementia among those involved in competency hearings or undue influence cases.

However, the entrance of forensic neuroscience into criminal law has caused heated controversy. Despite this robust and vocal opposition, as neuroscience advances, there will be increased pressure to allow its findings to be introduced as evidence to establish diminished competency to proceed or participate at various points in the criminal process, as well as permanent incompetence to stand trial or vulnerability to interrogative suggestibility (Martell 2009, 127). Furthermore, the issue of coerced medication or treatment is bound to escalate in light of advances in neuroscience. Chapter 6 turns attention to the social and cultural implications of the brain sciences, while chapter 7 returns to the commercial aspects and military uses of many of the same technologies described here.

6

Implications for Social Behavior, Racism, and Conflict

This chapter discusses how emerging knowledge of the brain might help us better understand social relations, racism, ethnic conflict, and recurring wars. For example, there has been an increased interest in the conflict resolution literature regarding the impact of neuroscience findings on rational choice theory, and thus on the foundations of foreign policy. Likewise, there is considerable research on the neural foundations of bias and its ties to intergroup conflict. Imaging studies have found that even among seemingly unprejudiced people, racial category labels prime stereotypes and out-group cues such as faces activate negative evaluative terms. Moreover, research on the limbic system suggests that dependence on rational incentives is unlikely to resolve the fears and hatreds among conflicting groups and that even the leaders of the opposing sides are limited in their capacity to convince their followers to accept peace without ameliorating emotional biases deep within the brain.

As policy makers deal with a more unpredictable and threatening world setting, how can they best assimilate the emerging knowledge about the brain's contribution to aggression, bias, and intergroup conflict into foreign policy? Are there more effective ways than force or traditional mediation with handpicked leaders to resolve such problems? What good does it do to remove a ruthless dictator if his followers hold similar views of outsiders? Moreover, what dangers are there when leaders characterize certain easily identifiable ethnic groups as the enemy? Or when they profile based on physical characteristics? What deep-seated evolutionary fear circuits do such policies trigger? In contrast, to what extent do leaders exploit the fear system to build support for their policies, and what are the ramifications of such fear system engagement in the long run?

Social Neuroscience

What determines how humans interact socially? Why do we sometimes unite but at other times refuse to cooperate? Why do some people take risks to achieve their goals, whereas others prefer to stay on the safe side? In the past few decades, social scientists have made considerable inroads in understanding the mechanisms of human social interaction and cooperation (Kosfeld 2007). Under the Standard Social Science Model, the prevailing focus has been on the external social, environmental, and institutional determinants of human behavior, while the internal biological mechanisms that originate and regulate individual decision making have been more or less ignored or dismissed as irrelevant—the biological being viewed as an empty organism or a black box. With the emergence of such fields as social neuroscience and neuroeconomics, however, neuroscientists have begun to look inside the black box and offer a more inclusive and detailed view of it. They have merged methods from the social sciences and cognitive neuroscience in an effort to understand what role the human brain plays in generating decisions. Game theory, in conjunction with a variety of scientific techniques, including functional imaging, transcranial magnetic stimulation (TMS), and intranasal administration of neuropeptides, is now being used to analyze how human brain function and structure shape behavior.

Humans are fundamentally social creatures whose mental and physical health depends on their social position and relationships. Accordingly, the human brain has evolved to promote social coordination, communication, interaction, and collective enterprises (Cacioppo 2006). Living in a social group is a major adaptive advantage, but it presents its own challenges. To get along and get ahead, it is necessary to learn who is friend and who is foe. Also, it can be advantageous to form alliances with certain group members in one context but to outmaneuver them in another. The social brain hypothesis suggests that, from an evolutionary standpoint, living in bigger, more complex social groups selected for larger brain regions with a greater capacity for performing relevant computations. Interspecies comparisons in nonhuman primates have demonstrated that amygdala volume is associated with troop size, suggesting that this brain region supports skills necessary for a complex social life and allows humans to build a richer social world (Weaver 2010). Moreover, species characterized by interaction in larger social groups have a larger corticobasolateral complex within the amygdala that conjointly expanded with the evolutionarily more recent cortex and the lateral geniculate nucleus. Social behavior, at the base of political behavior, then, refers to a range of activities that affect and reflect how individuals react to their environments, with a specific emphasis on reactions to other people and to human institutions (Lupia 2011).

Social neuroscience investigates the biological mechanisms that underlie social structures, processes, and behavior and the interactions among social and neural structures and processes. Such an endeavor is challenging because it necessitates the integration of multiple levels. Mapping across systems and levels (from the genome to social groups and cultures) requires interdisciplinary expertise, comparative studies, innovative methods, and integrative conceptual analysis (Cacioppo and Decety 2011). Some branches of social neuroscience, particularly those that rely on neuroimaging, have focused on the neural mechanisms involved in thinking about the traits, mental states, and behavioral predispositions of others and have identified a network of brain regions consistently active during such mentalizing, including the medial prefrontal cortex (PFC) and the posterior part of the superior temporal sulcus (Cacioppo and Decety 2011, 163). Moreover, most of the complex functional constructs used in social neuroscience, such as self-regulation, prejudice, attachment, empathy, and trust, cannot be seen as "natural kinds" situated in one location in the brain or mapping to an underlying biological mechanism in a one-to-one fashion but instead must be broken down into subcomponents. Research on the "self" is a case in point. The self is a complex construct and has included assorted dimensions, including the ecological self, the present self, the distant self, the experiential self, the prereflexive self, the mental self, core self, the minimal self, the spatial self, the emotional self, the autobiographical self, and the narrative self. The question of the representation of the self in the brain has been the target of recent neuroscientific investigations, which have uncovered the involvement of numerous brain regions, including the medial PFC, ventro- and dorsolateral PFC, lateral parietal cortex, bilateral temporal poles, and insula (Cacioppo and Decety 2011, 167).

For instance, Bickart and colleagues (2011) confirmed that amygdala volume correlates with the size and complexity of social networks in adult humans. Study participants who had vaster and more complex social networks had larger amygdala volumes, an effect that was not dependent on the age of the subjects or their own perceived social support or life satisfaction (Weaver 2010). Although a question remains as to how the amygdala contributes to social networks, the decision to develop and maintain relationships is likely linked to its role in response to faces, emotions, and emotional memories. While it is probable that social behavior relies on a much broader set of brain regions, an exploratory analysis of subcortical structures did not find strong evidence for similar relationships with any other brain structure except associations between social network variables and thickness in three cortical areas, two of them with amygdala connectivity. On this basis, Bickart and colleagues (2010) concluded, "Humans are inherently social animals. We play, work, eat and fight with one another. A larger amygdala might enable us to more effectively identify, learn about and recognize socioemotional cues in conspecifics, allowing us to develop complex strategies to cooperate and compete" (164).

One such area is the concept of attractiveness. Employment hiring recommendations, starting salaries, performance ratings, earnings, and labor force participation have all been found to be influenced by physical attractiveness. Recent work in political science has also shown a role for physical attractiveness, especially a very strong role for "facial dominance," in voter decisions (Spezio et al. 2008; Tingley 2007). In this regard, a recent fMRI study (Tsukiura and Cabeza 2011) found that the neural mechanisms for judging facial attractiveness and moral goodness overlap. Activity in the medial orbitofrontal cortex increased whereas activity in the insular cortex decreased as a function of both attractiveness and goodness ratings. Within each of these regions, the activations elicited by attractiveness and goodness judgments were strongly correlated with each other, supporting the idea of similar contributions of each region to both judgments. Moreover, activations in orbitofrontal and insular cortices were negatively correlated with each other, suggesting an opposing relationship between these regions during attractiveness and goodness judgments. These findings have implications for understanding the neural mechanisms of the beauty-is-good stereotype, and also take one into the cultural realm.

Cultural Neuroscience

One of the most rancorous ideological battles continues over the relative importance of the cultural and biological foundations of human behavior. As we saw in chapter 4, those who deny a potent biological base will have a difficult time in light of findings from genetics and neuroscience. Drawing on both the growing body of evidence on cultural variation in psychological processes and the recent development of social and cognitive neuroscience, the emerging interdisciplinary field of cultural neuroscience attempts to understand how culture as an amalgam of values, meanings, conventions, and artifacts that constitute daily social realities interacts with the mind and its underlying brain pathways (Kitayama and Park 2010). According to Chiao (2010), the use of neuroimaging and event-related potential techniques to study the effects of culture on the spatial and temporal features of the neural response to perceptual, cognitive, social, and emotional tasks is central to cultural neuroscience. This research has also led to neuroanthropology, a humanistic science for the study of the culture-brain nexus, which Dominguez Duque and colleagues (2010) contend can make a significant and distinctive contribution to the study of the relationship between culture and the brain (also neuroaesthetics; see Chatterjee 2010).

Until recently, the relevance of culture for brain development, structure, and function and vice versa was unrecognized, but this has changed with the integration of cross-cultural psychology, cognitive neuroscience, and molecular biology to study how neural development, structure, and function vary from one cultural group to the next (Seligman and Brown 2010; Han and Northoff 2008). The first generation of cultural neuroscience studies has yielded a number of important insights. First, culture seems to have a pervasive effect on all levels and dimensions of neural activity, from low-level perceptual and attention processes to high-level cognitive, affective, and social functions, including activity associated with language, music, mental calculation, emotions, mentation, and self-knowledge and awareness (Seligman and Brown 2010). In a word, differences in cultural systems are reflected in differences in brain systems.

Second, cultural differences appear to be reflected either in different arrangements of neural circuits or in different levels of activation. Current findings also indicate that cultural experiences not only modulate but also govern preexisting patterns of neural activity and are thus constitutive of that experience (Han and Northoff 2008). The constitutive character of cultural experience is particularly salient when one considers preliminary evidence that cultural differences can be expressed in actual brain structural changes, such as gray matter volume (Han and Northoff 2008; Turner and Whitehead 2008). For example, early perceptual processing regions in the ventral visual cortex reveal cultural variation in response to faces and complex visual scenes. Gutchess and colleagues (2010) report that during categorization, East Asians recruit regions in the frontoparietal network, whereas Westerners recruit temporal lobe and cingulate cortex to a greater extent. Similarly, in an earlier study, Gutchess and colleagues (2006) found that the left middle temporal cortex, implicated in the processing of isolated objects, was activated more in Westerners than in East Asians when subjects were shown identical pictures of objects with background scenes. In a comparison of Caucasian American and Korean subjects, Cheon and colleagues (2011) found that cultural variation in preference for social hierarchy led to cultural variation in in-group preferences in empathy resulting from an increased engagement of brain regions associated with representing and inferring the mental states of others. Still other studies have confirmed the importance of cultural and religious values to neural processing of the self within the medial PFC, a region associated with self-relevant processing and social cognition. Additionally, the human amygdala shows heightened response during own-culture fear perception but not to facial qualities that predict voting behavior, suggesting both universality and cultural specificity in the amygdala's response to social-emotional stimuli (Chiao 2010).

While fruitful work has focused on how culture influences the contents of cognition, culture can also exercise a profound effect on the *how* of cognition—the mechanisms by which cognitive tasks get done. Many fundamental processes of daily cognitive activity involve the operation of cognitive tools that are not genetically determined but rather invented and culturally transmitted. Further, these cognitive creations become firmware, thereby constituting a reengineering of the individual's cognitive architecture by his or her cultural context, resulting in a variety of disparate cognitive phenotypes (Wilson 2010). Wilson claims that cognitive retooling in humans is ubiquitous in everyday cognition, results in reorganization of the neural system, is founded in embodied representations, and is made possible only by the evolution of an unprecedented degree of voluntary control over the body. Furthermore, there is emerging evidence that humans possess a unique, cooperatively based social cognition that allows a particularly rich form of social learning and cultural transmission (Warneken and Tomasello 2009). While cultural accumulation cannot explain human cognitive universals, it is nevertheless a highly important and neglected source for determining the adult cognitive phenotype. The consequence of this is that cognitive phenotypes should be expected to differ across cultures, subcultures, and even local groupings such as families (Wilson 2010; Han and Northoff 2008). While brain regions include strong predispositions to perform certain cognitive functions, there is a considerable "fringe of variability" (Dehaene et al. 2005) that allows each individual's experiences and learned cognitive habits to rewire the system.

Losin, Dapretto, and Iacoboni (2010) conclude that the combination of culture and neuroscience is additive and synergistic and that cultural neuroscience studies have the potential to transform existing methodologies and produce unique findings. They argue that one of the most productive areas of cross-cultural comparison is social cognition. A theoretical framework that has dominated this field is the classification of cultures as either individualistic or collectivistic. Individualism is characterized by the assumption that individuals are separate in identity and responsibility from one another and is typically associated with Western cultures, particularly in the United States. Collectivism is characterized by the idea that individuals are defined by and obligated to their social groups and is typically associated with Asian cultures. While this framework has been used to explain cross-cultural differences in visual perception, causal attribution, motivation, and emotion, research on individualism versus collectivism has been criticized because these concepts are applied broadly and often are not treated as independent constructs but rather as two ends of the same continuum. Losin and colleagues (2010) conclude that future investigations in cultural neuroscience will allow us to investigate the neural correlates of individualistic and collectivistic orientations.

Herding in Humans

One concept for understanding group behavior is herding, which is defined as the alignment of the thoughts or behaviors of individuals in a group through local interaction without centralized coordination. Herd mentality and behavior have been a ubiquitous part of human behavior since individuals began to construct tribes and migrate collectively (Raafat, Chater, and Firth 2009). Although most social science interest in herding has focused on stock markets, herding theory has clear relevance for many areas of politics, from voting behavior to political protests to mob violence. Herding is an important and well-documented feature of human behavior across a number of domains and is reflected in such psychological concepts as mass hysteria, the bandwagon effect, and group-think. Several studies, for instance, have shown that happiness and obesity tend to spread through social networks in a manner analogous to a contagious disease (Fowler and Christakis 2009; Christakis and Fowler 2007).

According to Raafat, Chater, and Firth (2009), social neuroscience is ideally positioned to connect these levels because, while social structures may be emergent organizations beyond the individual, they require biological systems in the individual to create them. Although there are as yet few studies investigating the neurobiological correlates of herding mentality, conformity, and emotional contagion, there are many areas in which cognitive neuroscience can be applied to herding. Novel neuroimaging techniques (scanning many individuals at a time) could be used to capture brain-based correlates of herding, whereas the emerging field of neuroeconomics (Fehr and Camerer 2007; Sanfey 2009) offers the possibility of characterizing and building biological models of herding (Coricelli and Nagel 2009). One potential research area on susceptibility to herding concerns the impact of social networking and its increased capacity for sharing information immediately to ever-proliferating points of contact for an individual (Pentland 2007; Raafat, Chater, and Firth 2009).

The Brain and Trust
At the core of social relationships is the concept of social trust. Trust pervades human societies. It is indispensable in friendship, love, families, and organizations, and it plays a key role in economic exchange and politics (Kosfeld et al. 2005). In the absence of trust in a country's institutions and leaders, political legitimacy breaks down. While trust can be a risky decision because the trusting person risks being exploited, the hope is that trust will be rewarded by reciprocal behavior, thereby increasing the efficacy of the social interaction (Kosfeld 2007). When people trust and trust is rewarded, all are typically better off compared with a situation in which no one trusts and people act in an untrustworthy manner. Thus, the economist Kenneth Arrow described trust as "an important lubricant of a social system" (Arrow 1974). However, since the risk involved in the trusting decision can never be resolved completely, to achieve benefits from the social interaction the trusting person must always overcome his or her aversion to the risk of being exploited.

Under recently, little was known about the biological basis of trust. Although social research has identified many conditions that affect the level of trust, such as repeated interaction, the possibility of sanctions, and the activation of social norms, a question remains as to what is it in the brain that makes humans trust each other despite the risks. In their study, Koscik and Tranel (2011) hypothesize that damage to the amygdala can interfere with normal patterns of interpersonal trust and reciprocity and result in increases in reciprocity despite betrayals of trust. They conclude that "the amygdala is necessary for adaptive and appropriate social functioning at the intersection of evaluation and social interaction. Amygdala damage may result in maladaptive social behavior due to a lack of appropriate and accurate evaluation of social situations and social partners" (2011, 610).

In addition, researchers have targeted oxytocin, a neuropeptide produced in the hypothalamus and stored in the posterior pituitary, as potentially contributing to increased levels of trust. Oxytocin is naturally released into the bloodstream and widely distributed throughout the CNS, but it can also pass through the blood-brain barrier after intranasal administration. Oxytocin reduces anxiety and the neuroendocrine responsiveness to social stressors and acts as a central regulator of stress-protective social behavior (Heinrichs et al. 2003). Evidence concerning oxytocin function has led to the supposition that it enhances the willingness of humans to show trust toward others and accept social risks arising from interpersonal interactions (Baumgartner et al. 2008). In a recent study of trust by Kosfeld (2007), subjects received either a single intranasal dose of oxytocin or a placebo. Oxytocin increased subjects' willingness to trust their trustee, although it did not make them more optimistic about the latter's trustworthiness (also see De Dreu et al. 2010). The results indicate that oxytocin somehow helps humans overcome their natural aversion to uncertainty regarding the behavior of others, as well as raising their willingness to trust their trustee.

The discovery that oxytocin increases trust in humans is likely to have important clinical implications for patients with mental disorders associated with social dysfunctions such as social phobia or autism, where sufferers are often unable to evince even basic forms of trust toward others. Therefore, the administration of oxytocin in combination with behavioral therapy might yield positive consequences for the treatment of these patients. At the same time, however, the results of these experiments raise potential fears of abuse, ranging from unscrupulous employers or insurance companies using oxytocin to induce trusting behavior in their employees or clients to fraudulent salesmen spraying customers. "Governments might use chemical means to enhance citizens' trust in its policies and actions; one can only imagine how this might be used to sway public opinion or stifle dissent. Political candidates could use it to reap the benefits of unearned trust at the polls" (Penney and McGee 2005, 1). According to Kosfeld (2007), however, these fears are groundless because the surreptitious administration of a substantial dose of oxytocin, for example through air conditioning, food, or drink, is technically impossible, and forced nasal administration is likely to raise the recipient's level of distrust, therefore negating the intended effect.

Like trust, forgiveness is an important cognitive process for the survival of human beings in a society because it enables us to avoid unnecessary conflicts (Farrow and Woodruff 2005). Thus, when faced with violations of social rules, people make moral judgments as to whether the violation is forgivable. From a cognitive neuroscience perspective, forgivability judgments require an understanding of others' actions and intentions, as well as an evaluation of the outcomes of the actions. These judgments, therefore, are likely to activate multiple cognitive processes that are supported by several brain regions. To study this phenomenon, Hayashi and colleagues (2010) used PET to investigate the neural mechanisms underlying the willingness to forgive another person's moral transgression involving deception. Behavioral data showed that a perpetrator's dishonesty and the seriousness of the scenario decreased a subject's willingness to forgive the moral transgression. Relative to honest responses, a perpetrator's dishonest responses were associated with right ventromedial PFC (VMPFC) activity in the subject, possibly reflecting the subject's recognition of the perpetrator's deception. Further analysis revealed that the left VMPFC tempers judgments regarding the forgivability of moral transgressions.

Moral Judgment

In light of the central role that morality plays in the constitution of human nature, debates over the moral nature of humans have been at the core of discussions among theologians and philosophers for millennia. Why do people jeopardize material resources or physical integrity to help or punish perfect strangers, out of a sense of fairness, concern for others, or observance of cultural norms? Similarly, what makes individuals engage in often self-sacrificing behaviors to support abstract causes, beliefs, and ideologies or to violently protest against perceived acts of injustice? To understand these complex human behaviors, from altruism to patriotism, investigation of the neural and cognitive mechanisms underlying the moral mind is of paramount importance (Moll and de Oliveira-Souza 2007). Moll, de Oliveira-Souza, and Zahn (2008) suggest that while moral cognition draws on general social cognitive and motivational abilities, its most distinctive feature is the ability to altruistically motivate social behavior. In turn, these motivations depend on the representation of complex moral sentiments and values.

Evidence suggests that this moral sensitivity emerges from a sophisticated integration of cognitive, emotional, and motivational mechanisms that are internalized through an active process of cultural learning during sensitive periods of individual development (Moll, de Oliveira-Souza, and Zahn 2008). Therefore, moral sensitivity relies on an array of complex motivations that enable social cohesion and reorganization (Spezio and Adolphs 2007). The identification of the neural components and their relationships to psychological processes underlying moral cognition offers indispensable knowledge for our understanding of the strengths and weaknesses of our moral nature. Moll and colleagues (2008) argue "that deeply entrenched neurohumoral mechanisms, which are largely shared with other social species, provide the motivational force underlying human moral sentiments and are uniquely combined with cognitive abilities only recently developed in our species, such as conceptual abstraction and elaborate representation of future consequences of actions" (175). These motivational systems are not merely evolutionary remnants that must be controlled by the rational cortical systems. Instead, social attachment and aversion, deeply influenced both by social learning and by individual biological differences, are crucial motivators of moral actions and are inextricably linked to moral evaluations and judgments. Churchland (2011), for instance, argues that human morality is formed on a neurobiological base that originally evolved to promote care of self and offspring and was later adapted to foster more sophisticated cooperative interactions. It is thus the natural result of a complex, evolved, and learned set of capacities. Moreover, fMRI studies have been useful in showing that there is a stable network of brain regions involved in moral cognition, irrespective of task constraints.

Ever since early reports of social behavioral changes after brain injury, neuroscience has worked to uncover crucial evidence bridging brain and morality (Moll and de Oliveira-Souza 2007). Functional neuroimaging and brain lesion analysis have promoted sophisticated cognitive models and tools that are fueling rapid advances in our understanding of human morality, which relies on partially overlapping abilities, such as the capacity to make moral judgments and experience moral emotions and to behave according to moral standards. In this vein, a study by Koenigs and colleagues (2007) provides crucial evidence that bilateral damage to the VMPFC increases "utilitarian" choices in moral dilemmas (i.e., judgments favoring the aggregate welfare over the welfare of individuals), strongly supporting the notion that normal moral judgment springs from a complex interaction of cognitive and emotional mechanisms relying on specific neural structures.

Moretto and colleagues (2010) provide further evidence that the VMPFC constitutes the neural foundation for judgments about personal moral dilemmas by mediating the anticipation of the emotional consequences of personal moral violations. Ciaramelli and colleagues (2007) add that the VMPFC is necessary to oppose personal moral violations, possibly by mediating anticipatory, self-focused emotional reactions that may exert strong influence on moral choice and behavior. Moreover, it has been found that VMPFC-lesioned patients, relative to healthy individuals and patients with brain damage in other cerebral regions, are more likely to endorse personal moral violations in order to maximize favorable consequences. According to one study, this intensified utilitarian pattern of moral judgment results from impaired affective and intuitive processes that normally oppose deviations from moral values and rules shared by a social group (Greene 2007). Moretto and colleagues (2010) conclude that this deficit might prevent these patients from anticipating the negative emotional consequences of moral violations and, as a result, to conform their behavior to moral norms and values shared by their social group.

Similarly, Mendez (2009) found that our innate moral sense is based in a network centered in the VMPFC and its connections that shapes moral emotions and moral drives, such as the avoidance of harm to others and the need for fairness and punishment of violators. The neurobiological evidence indicates the existence of automatic pro-social mechanisms for identification with others that are part of the moral brain. Patients with disorders involving this moral network have attenuated emotional reactions to the possibility of harming others and may perform sociopathic acts. Results from fMRI and lesion studies, then, indicate that the PFC is essential for successful navigation through the complex social world and for implicit and explicit social cognitive and moral judgment processing. Forbes and Grafman (2010) agree, but also found evidence of considerable overlap among regions active when individuals engage in social cognition or assess moral appropriateness of behaviors, a finding that underscores the similarity between social cognitive and moral judgment processes in general.

A recent fMRI study investigated conflict situations in which a moral standard clashed with a personal desire and necessitated making a choice between a morally guided or a hedonistic behavior (Sommer et al. 2010). When compared to neutral conflicts, moral conflicts elicited higher activity in a broad network that included the medial PFC, the temporal and parietal cortex, and the posterior cingulate. Further analyses revealed that, in contrast to morally guided decisions, self-indulgent decisions were associated with significantly higher rates of uncertainty and unpleasant emotions and induced substantial activation of the amygdala/parahippocampal region. "Therefore, the activation of the amygdala region during the processing of moral conflicts resulting in an immoral or personal desire-oriented behavior may indicate the subject's unpleasant emotions associated with what is colloquially referred to as 'bad conscience'" (Sommer et al. 2010, 2025). In their study investigating lawyers and other academics during moral and legal decision making, Schleim and colleagues (2011) found activation of brain areas comprising the moral brain in both contexts. However, there was stronger activation in the left dorsolateral PFC and middle temporal gyrus when subjects made legal decisions, suggesting that such decisions were made according to more explicit rules and demanded more complex semantic processing.

These studies reflect a near consensus about the brain regions involved in moral decision making, including the involvement of both cortical and subcortical-limbic structures (for a very comprehensive review, see Raine and Yang 2006). While it is therefore reasonable to assume that both emotion and cognition play important roles in moral judgment, disagreement remains over how they interact to produce moral thought and choices (Leben 2010). One view is that, although emotion and cognition operate together to produce the decision, they are dependent on largely separable neural systems. This is evident in the context of difficult moral decisions associated with response conflict that lead to competition between emotional (limbic) and cognitive (especially the PFC) brain regions, where automatic emotional responses must be suppressed by "rational" top-down processes so that better decisions leading to overall "greater benefit" can be made.

A competing view presumes that emotion and cognition are nondissociable elements underlying moral motivations and that such motivations are represented within corticolimbic neural assemblies (Moll, de Oliveira-Souza, and Zahn 2008). Conflicting moral decisions, therefore, do not entail a clash between emotion and cognition but among two or more choices that rely on corticolimbic assemblies encoding distinct motivationally salient goals. As such, a cognitive process that is devoid of motivational salience would never be able to overcome a motivationally laden choice.

Greene and colleagues (2004) conducted influential studies supporting the dual-track thesis in which they scaned subjects as they made

Box 6.1
Moral Dilemmas

Dilemmas that evoke conflicts between moral intuitions have generated a large literature. One important class of cases is the so-called trolley problem. When people are faced with the hypothetical choice of switching a runaway trolley away from a track containing five people onto a track containing only one, most choose to throw the switch, killing one to save five. When faced with the choice of pushing a bystander in front of the same trolley, also saving five people (the footbridge problem), most people refuse. Why the difference? One natural explanation is that, in the footbridge problem, there is a conflict between reason and feeling. Reason declares that one must die to save many. Our sympathy for the one, therefore, conflicts with this reasoned judgment; when we contemplate doing the deed in an up-close and personal way, that sympathy can be strong enough to override the judgment of reason (Klein 2011, 143).

judgments about dilemmas such as the trolley problem (see box 6.1). They distinguished between personal (where the subject has to imagine intentionally creating serious harm directly to another person) and impersonal dilemmas and concluded that personal dilemmas are more emotionally engaging. While personal moral dilemmas activate the VMPFC and other regions that have been associated with emotional processing, impersonal dilemmas, by contrast, activate the inferior parietal lobes and the middle frontal gyrus, areas previously associated with working memory, a presumably more cognitive process. Furthermore, since moral dilemmas activate distinct cognitive and emotional tracks, these two processes can issue conflicting commands.

This dual-track theory, however, has provoked extensive debate. For instance, Klein (2011) has argued that rather than two distinct networks, experiments on moral deliberation reveal that portions of a single unified network are involved. Even if the same network is involved in all social cognition, it is exceedingly unlikely that every area will be involved to the same degree on every task. Some tasks will require more imagination, others more integration, others greater sensitivity to others' mental states, and so on. Moral decisions, then, are bound up with thinking about ourselves as part of a moral community. To judge our own actions as good or bad is in part to judge whether others have reason to praise or blame us. To judge others is in part to determine their attitudes toward other members of the moral community: contempt for others deserves blame, while sincere intentions might not. To judge our own action as wrong is partly to judge ourselves as having made a breach in our relationships with others in the moral community and seek to repair it (Klein 2011, 12).

Similarly, studies on moral decision making have looked at differences between moral judgments by contrasting utilitarian (well-being maximizing) and deontological (duty-based) content. In this approach, the trolley problem (box 6.1) pits the utilitarian stance of sacrificing one life to save five (the view that we should maximize aggregate well-being regardless of the means employed) against the deontological ethical view, which holds that we must obey certain duties even when this leads to a worse outcome. Thus, many deontologists consider it wrong to kill a stranger under any circumstances (Kamm 2009). Utilitarian judgments were found by Greene and colleagues (2004; see also Greene et al. 2008) to be associated with longer response times and increased activation in areas implicated in deliberative processing. Contrarily, deontological judgments were associated with greater activation in areas related to affective processing, such as the VMPFC and the amygdala. These differences in neural activation have been interpreted to reflect distinct neural subsystems that underlie utilitarian and deontological moral judgments (Greene et al. 2008). However, by using a wider range of dilemmas and controlling for the distinct contribution of content and intuitiveness, Kahane and colleagues (2011) found that behavioral and neural differences in responses to such dilemmas are the result of differences in intuitiveness, not differences between utilitarian and deontological judgment. Despite content, counterintuitive moral judgments were associated with greater difficulty and with activation of neural areas that involve emotional conflict. By contrast, intuitive judgments were linked to activation in the visual and premotor cortex. Kahane and colleagues (2011) thus reject theories that have associated utilitarian and deontological judgments with distinct neural systems.

Human behavior is motivated not only by materialistic rewards but also by abstract social rewards, such as the approval of others and notions of fairness (Güroğlu et al. 2010). When choosing an action in social situations, the brain must convert different types of reward into a common scale in order to evaluate each option. Izuma, Saito, and Safdato (2010) investigated the neural correlates of such valuation computations when individuals decided whether to donate to real charities or keep the money for themselves, in the presence or absence of observers. Exceptionally high striatal activations were observed when a high social reward was expected (donation in public) and when there was the potential for monetary gain without social cost (no donation in the absence of observers). According to the authors, these findings highlight the importance of the ventral striatum in representing both social and monetary rewards as a "decision utility," and add to the understanding of how the brain makes a choice using a "common neural currency" in social situations (629).

Empathy

Highly related to trust and moral judgment for understanding both individual and group behavior is the concept of empathy, which entails the capacity to share the affective experiences of others. Empathy enables us to communicate and interact with each other effectively and to predict the actions, intentions, and feelings of others. This ability to share others' feelings results in a better understanding of the mental states and actions of the people around us and can promote pro-social behavior (Singer and Lamm 2009). Philosophers and psychologists have long conjectured about empathy, but recent research has begun to provide a better grasp of why, when, and how we experience empathy and whether we can use that knowledge to increase pro-social behavior and reduce conflict. As with trust and altruism, there is strong evidence that empathy has deep evolutionary, neurological, and biochemical underpinnings. Even more advanced forms of empathy in humans are built on basic forms and remain connected to core mechanisms associated with affective communication, social attachment, parental care, and motivation to help others (Cacioppo and Decety 2011, 164).

In recent years, social neuroscience has made considerable progress in revealing the mechanisms that enable a person to feel what another is feeling. Empathy, which implies a shared interpersonal experience, is implicated in many aspects of social cognition, notably pro-social behavior, morality, and the management of aggression (Decety 2010). Sharing the emotions of others is associated with the activation of neural structures that are also active during the firsthand experience of that emotion. In general, empathy is viewed as a first, necessary step in a chain that begins with affect sharing, followed by understanding the other person's feelings, which in turn motivates other-related concern and leads to the engagement of helping behavior. Empathy and pro-social behavior are thus closely linked on a conceptual level. Recent studies show that empathy is a highly flexible phenomenon and that vicarious responses are malleable with respect to a number of factors, such as contextual appraisal, the interpersonal relationship, and the perspective adopted during observation of the other (Singer and Lamm 2009). Of note, as with fear, part of the neural activation shared between self- and other-related experiences is automatically activated.

Most studies of empathy utilize the observation of pain in others to evoke empathic responses (Lamm, Meltzoff, and Decety 2010; Singer and Lamm 2009). An accumulating number of fMRI studies have demonstrated striking similarities in the neural circuits involved in the processing of both the firsthand experience of pain and the second-hand experience of observing it in others. One common finding is that perceiving pain in others activates the same neural networks that are activated when we experience pain ourselves. Such accounts hold that the capacity to project ourselves imaginatively into another person's position by simulating their mental activity using our own mental apparatus lies at the root of our mature mind-reading abilities.

According to Decety (2010), human empathy involves several components, affective arousal, emotion understanding, and emotion regulation, each with unique developmental trajectories. These components are implemented by a complex network of interacting neural regions, including the insula, medial and orbitofrontal cortices, amygdala, and anterior cingulate cortex, as well as by autonomic and neuroendocrine processes implicated in social behaviors and emotional states. According to Singer and colleagues (2004), empathy works by tapping into a brain mechanism that already exists for our own pain, thus making us believe we are feeling pain emotionally even when we are not feeling it physically. This emotional activity has been found to be stronger in some people than in others, particularly in those who have a stronger emotional bond to the sufferer. Thus, empathy can help us understand what someone else is feeling, be it pain, joy, or anger, and in that way may help us ward off danger.

In their fMRI study, Decety, Echols, and Correll (2010) explored whether feelings of empathy are modulated by the target's stigmatized status and whether the target bears responsibility for that stigma. Subjects were exposed to a series of video clips featuring individuals experiencing pain who were either (1) similar to the participant (healthy), or (2) stigmatized but not responsible for their condition (infected with AIDS from an infected blood transfusion), or (3) stigmatized and responsible for their condition (infected with AIDS from intravenous drug use). Results showed that participants were significantly more sensitive to the pain of AIDS transfusion targets than to the pain of healthy or

AIDS drug targets, as evidenced by higher pain and empathy ratings and greater hemodynamic activity in areas associated with pain processing. Furthermore, behavioral differences in responses to healthy and AIDS drug targets were moderated by the extent to which participants blamed AIDS drug individuals for their condition. The more participants blamed the target individuals, the less pain they attributed to them compared with healthy controls, thus revealing that empathic resonance is moderated early in information processing by a priori attitudes toward the target group.

Similarly, many studies have found that females have on average a stronger drive to empathize, while males on average have a stronger drive to systemize (Baron-Cohen 2010). Singer and colleagues (2006) concluded that when it comes to deriving pleasure from someone else's misfortune, men seem to enjoy it more than women. Using fMRI, they compared how men and women reacted when watching other people suffer. If the sufferer was someone the subjects liked, areas of the brain linked to empathy and pain were activated in both sexes, but only women had a similar response if they disliked the sufferer, while men instead exhibited a surge in the reward circuits. Thus, women had a diminished but still active empathic response if they disliked the target, whereas for men it was completely absent. Empathic responses to other people, particularly in men, are not automatic but depend on the emotional link to the suffering person. Men also admitted to having a much higher desire for revenge, and derived satisfaction from seeing the unfair person being punished.

Similarly, Ma and colleagues (2011) found more empathy toward friends than strangers, although the response toward friends was salient only when subjects were not involved directly in their loss, and concluded that both familiarity and self-engagement are factors that influence empathy. Similarly, Lamm, Meltzoff, and Decety (2010) concluded that it is more challenging to empathize with dissimilar others in a situation that is aversive for the observer, although it is possible. "This flexibility is a cornerstone of our ability to empathize with diverse others—from animals to anthropomorphized objects such as pets to people from different cultural backgrounds" (374). In another fMRI application, Zahn and colleagues (2009) found that subjects with higher empathic concern had increased activity in the subgenual cingulate cortex when experiencing feelings of guilt for acting counter to social values, but not with altruistic donations to societal causes.

In their study, Cheon and colleagues (2011) compared the effect of cultural variation in preference for social hierarchy on the neural basis of intergroup empathy by imaging Korean and American subjects while they observed scenes of racial in-group and out-group members in emotional pain. They found that the Korean participants reported experiencing greater empathy and that viewing subjects in pain elicited stronger activity in the left temporoparietal junction, a region previously linked to mental state inference for in-group as compared to out-group members. Furthermore, preferential reactivity in this region to the pain of in-group members was associated with greater preference for social hierarchy and in-group biases in empathy. This suggests that cultural variation in preference for social hierarchy leads to cultural variation for in-group preferences in empathy because of the increased engagement of brain regions associated with representing and inferring the mental states of others.

Shirtcliff and colleagues (2009), however, propose that neurobiological impairments in individuals who display little empathy are not necessarily to the result of a reduced ability to understand the emotions of others but instead are connected to decreased physiological arousal to their own distress. One manifestation of this reduced stress reactivity may be a dysfunction in empathy leading to callousness. Callousness and unemotional (CU) traits are related to maladaptive social information processing (Frick and White 2008). While empathy can promote affiliation and pro-social behavior, CU traits are associated with antisocial behavior and are particularly useful in predicting which antisocial and violent youth will continue offending into adulthood (Shirtcliff et al. 2009). Deficits in emotionality, disruptions in empathy and the failure to respond to the distress cues of others (i.e., callousness) are at the core of the impaired decision-making capabilities (Kimonis et al. 2008). It is notable that individuals with high CU traits often show reduced amygdala activation (Shirtcliff et al. 2009). While healthy individuals experience heightened amygdala activation in response to the distress of others and find this experience unsettling, individuals with reduced amygdala activation in response to their own distress have difficulty processing others' distress. In contrast, healthy individuals learn to avoid the distress of others either by performing actions that reduce their own distress (i.e., empathic or pro-social behaviors) or by learning to avoid actions associated with their distress (i.e., not engaging in anti-social behaviors).

In his book, *Zero Degrees of Empathy*, Baron-Cohen (2011) sees "evil" as the product of a failure to empathize. He terms psychopathic personality disorder, borderline personality disorder, and narcissistic personality disorder "zero negative" because affected subjects' lack of empathy is considered to be solely detrimental, and autism spectrum disorders "zero positive" because, while affected persons lack empathy, they also have systematizing natures that are valued because their focus on detail and patterns is a key component of advanced processes. He attributes these traits to dysfunction in the brain's empathy circuit, genetic abnormalities, and poor developmental conditions that preclude the forging of an emotionally resilient psychology that is normally developed through secure attachment. Baron-Cohen contends this helps explain human cruelty; for instance, humiliation can be used to establish status, to signal collaboration with a dominant person, or to respond to one's own perceived oppression. Additionally, dehumanization may arise when perpetrators have to sidestep their intact empathy mechanisms in order to permit murder via indirect methods, such as by mercenaries, through the use of "impersonal" technologies such as drones, or by forcing victims to kill one another. According to Preston (2011), an interdisciplinary framework that combines neuroscientific knowledge with findings from social and political science may allow us to capture the "richness of the human context" in such inhumane circumstances.

A major question in drawing connections between this research and group conflict is how empathic brain responses and individual differences in empathy are linked to pro-social behavior. What is the role of personal distress versus sympathy in predicting helping and other forms of pro-social behavior? For example, is the absence of such behavior a result of deficits in affective sharing, insufficient regulation of high personal distress, or a combination of the two? Furthermore, very little is known about the malleability of the empathic and emotional brain. Can we train people to become more empathic, and if so, which processing level, bottom-up or top-down, should be utilized for such a training to be most effective and persistent?

Fear Conditioning

In contrast to empathy, which facilitates sharing the affective experiences of others, fear and related emotions are at the base of in-group versus out-group tensions. As noted in chapter 4, along with frontal lobe abnormalities and hormonal and neurotransmitter dysfunctions, aggressive behavior has been found to be highly correlated with abnormalities in the limbic system, particularly the amygdala, which is responsible for the detection of threat and the orchestration of stress responses in the brain and the body (Rodrigues, LeDoux, and Sapolsky 2009). It is important to recognize that the fear process initiated in the amygdala is largely beyond conscious control. When the amygdala detects a threat, its outputs lead to the activation of a variety of target areas that control both behavioral and physiological responses designed to address the threat (Lang and Davis 2006). In addition to the expression of defensive behaviors such as freezing, amygdala activation leads to responses in the brain and the body that support the fear reaction. Monoaminergic systems are activated, resulting in the release of neurotransmitters such as norepinephrine, acetylcholine, serotonin, and dopamine through-out the brain (Rodrigues, LeDoux, and Sapolsky 2009). This leads to an increase in arousal and vigilance and heightened processing of external cues, as well as endocrine system effects (LeDoux 2007; Ramos and Arnsten 2007).

Moreover, connections from the amygdala to the brain stem lead to the activation of the sympathetic nervous system, which involves the release of the neurotransmitters epinephrine and norepinephrine from the adrenal medulla and norepinephrine from the terminals of sym-pathetic nerves throughout the body. Adrenal medullary hormones and sympathetic nerves produce an array of effects, including increasing blood pressure and heart rate, diverting stored energy to the muscles and inhibiting digestion (Rodrigues, LeDoux, and Sapolsky 2009). Fear arousal initiated by a perceived threat, therefore, leads to activation of the stress response, a state of alarm that precipitates an array of auto-nomic and endocrine changes designed to aid self-preservation.

If the perceived harm passes, messages are sent to the amygdala to confirm that the threat has passed. As noted above, this entire process often occurs without conscious awareness of the stimulus (Armony and Dolan 2002). The advances in imaging techniques discussed in chapter 2 have been crucial in uncovering the specific brain structures and mechanisms involved in this complicated amygdala-based fear circuit.

Beyond signaling fear itself, the amygdala plays a central role in evalu-ating stimuli as unpleasant. MRI studies show regionalized increases in brain activity in the midbrain dopamine system. In uncertain situations, the evaluation process may also call for heightened involvement of the PFC for decision making and planning of an appropriate behavioral response. Zaretsky and colleagues (2010) suggest that this brain network is involved in generating subjective assessment of social affective cues. The amygdala is activated most strongly when lacking sufficient data regarding a potential threat in the surroundings (Hsu et al. 2005). Especially in ambiguous situations, the amygdala may act to decrease the threshold of relevant sensory systems in order to gather additional information for resolving the ambiguity of the signal and increasing its predictability (Zaretsky et al. 2010, 2272). If this information resolves the uncertainty, amygdala activity subsides. In this way, the amygdala plays a crucial role in formulating a predictive decision under uncertainty.

Neuroimaging studies have reported greater activation of the amygdala in response to emotional facial expressions, especially fear, indicating that it is capable of rapidly processing fearful and sad facial expressions (Fenker et al. 2010; Kim et al. 2010). Pictures of frightening faces initiate a quick rise and fall of activity in the amygdala (Sato et al. 2011). It has been shown to respond more to facial expressions than to neutral faces and more to positively valenced and negatively valenced faces than to faces in the middle of the continuum (Said, Dotsch, and Todorov 2010; Todorov and Engell 2008). Williams and colleagues (2006) found that even subliminal facial expressions of fear stimulate the amygdala in normal persons. Moreover, Olsson, Nearing, and Phelps (2007) found that fear acquired indirectly through social observation, with no personal experience of the aversive event, engages similar neural mechanisms as fear conditioning, thus validating the "extension of cross-species models of fear conditioning to learning in a human sociocultural context" (3; Fischer et al. 2003). Researchers are also deciphering the fear process, hoping to determine the effects of age, gender, and illness on the system. For example, Todd and colleagues (2011) found that happy faces are more salient for children than angry faces and that both children and adults show preferential activation to mothers' over strangers' faces.

Studies examining genetic links in the fear process have concluded that subjects who inherit certain variations in the gene for the serotonin transporter molecule experience more amygdala activity, and thus more fear (Vastag 2002). Other studies suggest that stress may intensify this response (Price and Drevets 2010). The reactivity of the amygdala to threat-related cues has been linked to trait anxiety, anxiety disorders, and depression, and therefore represents a risk factor for negative psychological outcomes (Hyde et al. 2011; Dickie and Armony 2008; Fakra et al. 2009). Variability in anxiety, an individual's behavioral sensitivity to threat and stress, reflects both the underlying reactivity of these neural circuitries and a risk for psychopathology associated with stress and adversity (Hariri 2009). Not surprisingly, bilateral damage to the amygdaloid complex is associated with decreased appreciation of danger and recognition of fear. Such patients judge faces rated as untrustworthy and unapproachable by healthy subjects, as approachable and trustworthy (Tranel et al. 2006). In contrast, exaggerated amygdala volume is related to a heightened susceptibility to internalizing disorders commonly related to a fearful temperament, such as major depressive and anxiety disorders (Van der Plas et al. 2010). Stein and colleagues (2002), for instance, found a key role of the amygdala in social phobia, a condition characterized by fear and the avoidance of situations where people might be scrutinized by others.

Although most attention has focused on the amygdala's role regarding fear conditioning and negative events, Baxter and Murray (2002) contend there is considerable evidence that it also has a role in the processing of positive emotions, particularly in specific kinds of stimulus-reward learning. However, in one study where there was more amygdala response to movies of facial expressions than to pattern motion, Van der Gaag, Minderaa, and Keysers (2007) concluded that under more ecologically valid conditions, the contribution of the amygdala to the detection of fearful facial expressions must be less direct than previously assumed. Piech and colleagues (2011) investigated whether a face stimulus could be conditioned by using a voice that had negative emotional valence and was collected from a real-life environment, and found there was greater amygdala activation in response to faces that had been paired with the voice than to those that had not. The finding that an ecologically valid stimulus elicited the conditioning and amygdala response suggests that our brain automatically processes unpleasant stimuli in daily life (Iadaka et al. 2010).

This emphasis on the role of the amygdala does not diminish the interrelated role of a broad network of brain regions, including the medial prefrontal and anterior cingulate cortices, in mediating limbic-initiated fear and generating emotional responses (Etkin, Egner, and Kalisch 2010). Davidson, Putnam, and Larson (2000) posit that impulsive fear and conflict arise as a consequence of faulty regulation of emotion, not the initiation itself. They suggest that while normal individuals have the capacity to voluntarily regulate their negative emotions, aided by restraint-producing cues such as facial signs of anger, persons predisposed to act aggressively have an abnormality in the central circuitry responsible for these adaptive strategies. They conclude that the propensity for reactive (impulsive) aggression is associated with a low threshold for activating negative affect and a failure to respond appropriately to the anticipated negative consequences of their aggressive behavior. Practically speaking, such findings have clear implications for how we react to others in a social setting, particularly regarding aggressive behaviors. The neurobiology of hate or ideology remains unsettled, but "as surely as there are neurobiological mediators of aggression, there are neurobiological mediators of ideology" (Perry 2002).

Group Bias and Racism

It has long been observed that humans tend to quickly categorize other people on the basis of social distinctions such as race, gender, and age and that these rapid, often automatic responses guide much human cognition and behavior. As noted in chapter 1, the brain is parsimonious and designed to utilize adaptive shortcuts that efficiently use limited mental resources to deal with a complex, dynamic sensory and social world. In the social context, these category-based responses are often interpreted as prejudice or stereotyping because they exclude the effortful process of getting to know an individual in detail and represent a simplistic framework within which to conduct social interaction (Derks, Inzlicht, and Kang 2008). Social categorization saves time and mental energy, but quick heuristics to assess a person can harm both the perceiver and the target. Therefore, controversy surrounds the inevitability or controllability of these rapid category-based responses by the brain, although there is clear evidence that individual differences, stimulus context, and social goals can modulate seemingly automatic biases (Wheeler and Fiske 2006). At their worst, however, these reflex actions can result in racism and produce conflict between in- and out-groups, leading to social discontent and possibly physical confrontation.

Although behavioral analyses are a logical choice for understanding the wide-ranging consequences of stereotyping and prejudice, studies using neuroimaging and electrophysiological research can clarify the neural mechanisms that underlie their activation and application (Ito and Bartholow 2008). One of the most consistent findings is that the medial PFC is activated when people think about other people; thus, it is intimately involved in social processing and notions of equality. However, in *Envy Up, Scorn Down,* Fiske (2011) found no activation of that region when subjects viewed images of homeless or drug-addicted individuals, implying that these groups are recognized by the brain as less typically human, that is, as social out-groups that elicit disgust (Tracy 2011). Fiske (2011) concludes that, while we are all wired for comparison to know how we measure up against others and where we stand in the world, some individuals are more vulnerable to these impulses than others.

Two emotions in particular, envy (the result of comparisons revealing someone else's superiority and our own inferiority) and scorn (the result of the reverse), lie at the heart of a vast number of interpersonal, societal, and international problems, according to Fiske (2011). Envy and scorn in turn are tied to two broad classes of motives, competence and warmth. Harris and Fiske (2006) employ the so-called stereotype content model, which predicts that perceived trait warmth and competence interact to differentiate affective reactions to members of distinct social groups. In a two-dimensional space of four quadrants, social groups perceived as high in both competence and warmth evoke the emotion of pride, groups perceived as high in competence but low in warmth receive the ambivalent emotion of envy, and groups perceived as high in warmth but low in competence receive the paternalistic emotion of pity. Finally, out-groups perceived as low in both warmth and competence elicit extreme disgust. Perceiving out-groups as disgusting suggests that people see them as so strikingly different that they do not evoke an exclusively social emotion. Infrahumanization theory (Leyens et al. 2003) postulates that people see some groups as less human than others and judge out-groups as not experiencing complex emotions to the same extent that the in-group does. Given the increasing number of ethnic conflicts in the post–Cold War security environment, some of which have resulted in ethnic cleansing or genocide, there does indeed appear to be varying degrees of dehumanization during intergroup conflict (Harris and Fiske 2006; Friend and Thayer 2011).

Frequently, this categorization can be initiated on the basis of face perception (Ito and Bartholow 2008). Individuals are able to recognize and remember countless numbers of faces, which facilitates subsequent interpersonal communication. This knowledge of face processing in the brain has been used to investigate the neural basis of a pervasive social and cognitive phenomenon known as the own-race effect, according to which individuals demonstrate greater difficulty in recognizing other-race faces than own-race faces. This social-cognitive phenomenon has been noted as an impediment to interracial social communication, as well as in eyewitness identification (Wright et al. 2003).

The neural circuitry involved in the categorization by race-of-face has been extensively examined by fMRI studies (for a review, see Eberhardt 2005). Because of its role in social cognition, reduced activity in the medial PFC suggests that people belonging to extreme out-groups may be perceived differently, maybe even as less human (Harris and Fiske 2006). Moreover, activation in the middle fusiform gyrus has been found to be modulated by race, being larger for own-race faces. Similarly, activation in the dorsolateral PFC, involved in executive control, is tempered by racial bias in response to other-race faces. Although a complex network of multiple neural systems has been implicated, substantial evidence suggests that amygdala activity reflects arousal triggered by precipitous unconscious assessment of potential threat signified by sensory, social, and emotional stimuli.

Activation of the amygdala has been observed in response to other-race faces during social categorization tasks (Wheeler and Fiske 2006), even when presented subliminally (Cunningham, Raye, and Johnson 2004). Moreover, amygdala activation to other-race faces correlates with measures of implicit social bias (Phelps et al. 2000), supporting the involvement of the amygdala in explicit and implicit race evaluation. Although Hart and colleagues (2000) argued that their research was aimed not at uncovering any race differences but rather at assessing fMRI responses to out-group versus in-group faces across subjects of both races, the results demonstrated that black and white participants responded in opposite ways to identical stimuli. Blacks exhibited greater amygdala response habituation to black faces than to white faces, while whites exhibited greater amygdala response habituation to white faces. Furthermore, it has been found that activation in the amygdala familiarizes faster to own-race as opposed to other-race faces (see Walker et al. 2008).

Overall, imaging studies have confirmed differential amygdala activation to face stimuli of different races, with more amygdala activation in response to out-group faces (Phelps 2001). This pattern typically occurs when participants engage in perceptual encoding of faces or make social category judgments (Ito and Bartholow 2008). As stated by Phelps, Cannistraci, and Cunningham (2003), "In spite of a decline over the past several decades of prejudicial attitudes toward black and white social groups as measured by explicit self-report, there is robust evidence of racial bias using indirect assessments that bypass awareness and conscious control . . . the correlation between amygdala activation and indirect measures of race bias suggests that the amygdala is engaging in the automatic processing of social group information from facial stimuli" (204).

Walker and colleagues (2008) found that the race of face has significant effects on face processing starting from early perceptual stages of structural encoding, but also that social factors may play an important role in mediating these effects. In particular, social experience has been shown to mold the way in which individuals recognize familiar faces, as well as own-race versus other-race faces (Walker et al. 2008). Behavioral research suggests that increased experience with faces of one's own race as compared to other races can shape the initial way in which faces are encoded (Walker and Tanaka 2003). In one study, this tendency toward greater amygdala activity in response to racial out-group than to in-group faces was eliminated when participants made other nonsocial judgments about the individuals (Wheeler and Fiske 2006). It also appears that the features of the target individuals can moderate amygdala activity in response to race. Ronquillo and colleagues (2007) found that dark-skinned targets elicited more amygdala activity than light-skinned targets. Another study found greater amygdala activity in response to black than white faces among white participants when the targets were looking at the perceivers, but not when the targets' eyes were averted or closed (Richeson et al. 2008). The authors argued that averted or closed eyes signal a low potential for threat, and attenuate racial differences in amygdala activity.

Masten, Telzer, and Eisenberger (2011) examined the neural responses that occur during perceived racial discrimination from the standpoint of the targets of discrimination. To do this, they scanned black participants while they were excluded by whites and then measured distress levels and race-based attributions for exclusion. They found that black participants who were more distressed from social exclusion showed greater social pain–related neural activity and reduced emotion-regulatory neural activity. Additionally, those who attributed the exclusion to racial discrimination displayed less social pain–related and more emotion-regulatory neural activity. Their findings highlight the distress associated with negative social treatment in general and suggest that individuals who face negative social treatment more frequently experience this distress more often. Given that experiences of negative social treatment have been linked to increased physiological stress responding and negative mental health outcomes (Slavich et al. 2009), it is likely that these negative events take a cumulative toll. Of note, Masten, Telzer, and Eisenberger (2011) found that those who attributed exclusion to racial discrimination actually had less activity in distress-related regions and more activity in regions previously linked with emotion/pain regulation, thus supporting the notion that making discriminatory attributions (i.e., seeing it as racism) can buffer against negative emotions accompanying the perceived discrimination. When individuals make discriminatory attributions in situations of negative social treatment, blame for the negative treatment is attributed externally (discrimination is the result of others' prejudice or racism), thus reducing negative affect and self-blame.

The "contact hypothesis" (Brown and Hewstone 2005) offers a potential explanation for the disparate effects of race on the early stages of face processing. It proposes that conditions fostering intergroup contact, including contact between ethnic and racial groups, can ultimately reduce prejudice and improve intergroup relations (Pettigrew and Tropp 2006). One interesting possibility is that frequent contact may influence processing of other-race stimuli, such as faces, from early perceptual stages. For instance, research by Walker and colleagues (2008) has demonstrated that individuals with more interracial experience show a diminished own-race effect because they are more likely to encode other-race faces in a similar manner to own-race faces. For example, Walker and Tanaka (2003) found that East Asian subjects were better able to discriminate East Asian faces, while Caucasians were better able to discriminate Caucasian faces, thus indicating that an own-race advantage occurs at the encoding stage of face processing. Conversely, individuals with less other-race contact may allocate their perceptual resources to face categorization, therefore, making other-race face differentiation more difficult (Walker et al. 2008).

A recent study of social cognition in people from different cultural backgrounds found increased activation in the precuneus among Israeli and Arab participants as they read pro-out-group versus pro-in-group statements (Bruneau and Saxe 2010). Activation in this region is believed to signify emotional reasoning during difficult moral judgments of harmful behavior, thus suggesting a strong in-group bias in evaluating the reasonableness of partisan statements about the Middle East. Although these findings support previous research on neuronal and evolutionary reasons for prejudice and stereotypes in intergroup relations, the particularly high level of in-group bias among both Israelis and Arabs is undoubtedly exacerbated by the politically hostile context in which those participants interact on a daily basis, which in turn can make extremely aggressive conflict between such groups significantly more likely to occur and harder to mitigate (Friend and Thayer 2011).

Bias in intergroup relations is an automatic response across populations because group conflict and prejudice are evolutionary traits that improved overall fitness by enabling members of a group to gain access to competitive reproductive enhancing resources and detect coalitions and alliances (Tooby and Cosmides 2010). Despite this, some researchers speculate that race bias can be erased by altering the social context. Kurzban, Tooby, and Cosmides (2001), for example, reported experiments showing that categorizing individuals by race is not inevitable. Instead, they hypothesize that encoding by race is a reversible by-product of cognitive machinery that evolved to detect coalitional alliances. When cues of coalitional affiliation no longer track or correspond to race, subjects markedly reduce the extent to which they categorize others by race. In their study, Harris and Fiske (2007) found that the manipulated social goal of the perceiver can spark medial PFC activity in response to the social group members who elicit disgust. This suggests that other processes, such as direct instructions to individuate or instructions to form a more accurate impression of the target, might also activate this region in these actors. As such, these results provide corroboration of the critical role of social goals in changing perceptions of social groups toward which people feel extreme prejudice.

These studies, however, fail to explain why, despite all the social programing and legislation to reduce racism in recent decades, a strong residue of automatic racial bias is found even in people who consider themselves unbiased. A more fundamental question is what triggers the amygdala to preset these distinctions in the first place, presumably at a very early age. Although there is a growing catalogue of brain imaging research supporting the roles of other brain regions in fear-driven behavior, this summary shows that the limbic system and the PFC coexist in a delicate balance. A better understanding of this relationship is crucial for explaining human conflicts. In turn, this knowledge has considerable implications not only for domestic policy and racial harmony but also for studying international conflict; an example of this application is how our brains identify those who are a threat, or an enemy. The next section is an overview of the potential impact of these research findings on intergroup conflict and foreign policy making and argues for a more in-depth integration of cognitive neuroscience into the traditional study of these areas.

Ethnic Conflict and Recurring Wars

There is, then, a mounting volume of research on the neural foundations of bias and its links to intergroup conflict. Imaging studies have found that even among seemingly unprejudiced people, racial category labels prime stereotypes (Fiske 2002). Out-group cues such as faces or names can activate negative evaluative terms. These biases appear to be automatic, unconscious, and unintentional, thus creating an environment subtly hostile to out-groups that eludes rational decision making. Although a habitually low level of bias can attenuate the activation of stereotypes, category activation can be moderated by short-term motivations. For instance, these fear circuits could be reinforced by external forces, such as the September 11 attacks, which acted to trigger American fears of those who looked or sounded as though they might have Middle Eastern origins, or conversely, the impact of the war in Iraq on the Arab world's view of Americans as the imperialist enemy. Although these predispositions do not eliminate the possibility of changing the social context to minimize amygdala-based fears, policies that ignore their reality are unlikely to succeed in the long run. That the Iraqis "saved" by the coalition forces did not greet them with open arms is not surprising in light of the research findings presented here. Nor should the continued elusiveness of a Middle East peace plan based on seemingly rational negotiations be unexpected.

Fiske (2002) notes that whether bias is conditionally or unconditionally automatic, less prejudiced perceivers can compensate for their automatic predispositions through a conscious effort to not act on them. Among moderate people, bias often manifests not in open hostility to but in withholding positive emotions—basic liking and respect—from out-groups. Such people tend to withhold rewards from the out-group and to attribute its perceived failings to its essence, that is, to innate, enduring attributes, often biological or cultural differences. Thus, Fiske argues that bias among moderates is ambivalent and mixed: "Although various out-groups are all classified as 'them,' they form clusters. Some elicit less respect than others, and some elicit less liking than others" (125).

In contrast to the moderates, extremists openly dislike out-groups, whether religious, cultural, or sexual, and reject any possibility of closeness to them. Moreover, extremists biased against one out-group tend to be biased against others. Strong forms of bias correlate with support of racist movements. Hate crime perpetrators and participants in ethnic violence not surprisingly endorse prejudices and stereotypes that fit extreme forms of discrimination (Fiske 2002). Just as events might be able mitigate bias among moderates, such biases also can be reinforced by immediate events, ranging from a suicide bombing to an unintentional snub, thus radicalizing moderates by playing on their fears. Moreover, Cairns and colleagues (2006) in a study of Northern Ireland found a general in-group bias—for example, feeling thermometer ratings (affect) for the in-group exceeded those for the out-group—but the bias was moderated by participants' in-group identification and religious group.

Hewstone, Rubin, and Willis (2002) suggest that while most research on intergroup bias has focused on its relatively mild forms, events such as ethnic cleansing and genocide demonstrate the potency of social categorization in these extreme forms of intergroup bias. They conclude that is a mistake to consider ethnic and religious mass murder as a simple extension of group bias, even if based on deep-seated hatred for the out-group. Such extreme actions often entail extended external duress and possibly fear for one's own life, and are therefore more complex than intergroup bias per se. "Real-world intergroup relations owe as much of their character to intergroup history, economics, politics, and ideology as they do to social-psychological variables such as self-esteem, in-group identification, group size, and group threat" (Hewstone, Rubin, and Willis 2002, 594).

The question remains as to what extent these factors reflect the actions of the brain, in particular the limbic system. As Pinker (2002, 335) notes, diplomats often try to bring opposing sides together by facilitating trade, cultural exchanges, and people-to-people exchanges. They also attempt to blunt long-standing animosities by fashioning compromises that allow each side to save face and bring in third parties as guarantors, but are often frustrated when at the end of the day the two sides hate each other as much as they did before the intervention. Research on the role of the amygdala and related midbrain regions suggests that dependence on rational incentives is unlikely to resolve the fears and hatreds, and that even the leaders of the opposing sides are limited in their ability to convince their followers to end conflict in the absence of ameliorating emotional biases deep within the brain.

One might use various approaches to try to defuse the amygdala and transform extremists to moderates, but it is unlikely that predispositions wired early in life can be purged without preventing their development in the recesses of the brain in the first place. Studies, for instance, have found that the impact of an abusive childhood can be longlasting (Teicher 2002). If so, ethnic hatreds, hardened by growing up in violent areas such as the Middle East, are most likely enduring and highly resistant to change. Although very early socialization of children to mitigate these emotions might be effective, uncontrolled hatred and fear of groups by adults seems nearly impervious to change by persuasion or compromise. These brain-based mechanisms shaped by fearful environments might explain why long-standing problems in Northern Ireland, Somalia, Cambodia, the Balkans, and the Middle East appear irresolvable.

Many international conflicts are recurrent, and some of these are characterized by periods of violence, including war, that are hard to describe as planned products of rational decision making. Such conflicts account for 40 to 50 percent of all state-against-state wars and are arguably the dominant form of warfare today. Similarly, internal civil wars, fueled by deep enduring animosities, continue to spill over their borders, often embroiling, directly or indirectly, major powers in system-defining confrontations. Not surprisingly, there have been many theories as to what mechanisms sustain a conflict and predispose to periodic violence. According to Long and Brecke (2003), however, none of the prevailing structural and behavioral models of such conflicts suitably includes the emotive dimension.

According to a rationalist approach to decision making, for example, such wars are the product of cost-benefit calculations, that is, if the benefit achieved from attacking outweighs the cost, an aggressive strategy is more likely to be taken. But as noted by Friend and Thayer (2011), it would seem more rational and cost-effective to avoid conflict and make mutually beneficial settlements than to wage war. The fact that this is not the case suggests that additional causal mechanisms are responsible for coalitional aggression. To the extent it is noted at all, emotion is either dismissed as a pathologically irrational factor or viewed as a potentially important variable that is inestimable. For instance, Bueno de Mesquita (1981) argues that "the selection of war or peace is a choice that is initiated by individual leaders who must accept responsibility for their decisions. . . . Their choices depend on their estimation of costs and benefits" (5). Similarly, Howard (2001) concludes that "conflicts between states which have usually led to war have normally arisen, not from any irrational and emotive drives, but from an almost superabundance of analytic rationality . . . [that] enables them to assess the implications that any event taking place anywhere in the world, however remote, may have for their own capacity, immediately to exert influence, ultimately perhaps to survive" (33). Thus, deciding whether to go to war or not is a rational choice for decision makers under certain conditions, while emotion is treated as irrational and therefore exogenous to decision making and action. This clearly contradicts the findings of neuroscience.

In an effort to integrate emotion and reasoning to explain the causes of recurrent conflict, Long and Brecke (2003) present three emotive models of recurrent conflict. The first model is "frustration-anger-aggression" and is derived from the observation that individuals often become aggressive when they are unable to reach a desired goal. In essence, this model argues that the thwarting of an objective interest, in particular an interest in territory, will lead to an instrumental form of aggression designed to reassert a claim or goal, for self-enhancement or self-defense. Although it does not fit the dominant rationalistic framework, the frustration-aggression hypothesis also can be viewed as less instrumental and more emotional in its motivation. The work of Sell, Tooby, and Cosmides (2009), which found that males with enhanced abilities to inflict costs or to confer benefits are more prone to anger, prevail more in conflicts of interest, and consider themselves entitled to better treatment, is also relevant here. Of note, stronger men have had a greater history of fighting than weaker men and more strongly endorse the efficacy of force to resolve both interpersonal and international conflicts.

Their second model, "unacknowledged shame and rage," asserts that certain events trigger an emotionally driven set of decisions and actions that perpetuate war. Specifically, unexpected military defeat, loss of territory, or hostile occupation of territory, followed by the psychological absorption of such a calamity by a collective body, can trigger a shame-rage-aggression sequence. The result is an almost unquenchable thirst for vengeance, the compelling desire to get even, right a wrong, or avenge an injury (Harkavy 2000). Conflict begins after one or both parties suffer a "narcissistic" injury to its (their) group identity, such as the loss of territory, and collective shame is evoked. Shame is a very powerful emotion that involves a negative evaluation of the global self, and anger provides temporary relief by mobilizing the self and avoiding feelings of condemnation. When shame goes unacknowledged, however, another person or group becomes identified with the hurt, thereby leading to further withdrawal, recrimination, and threats. Rage and vengeful violence are the ultimate forms of protection from shame. Furthermore, the shame-anger-aggression emotional sequence is recursive, as one subsequently feels ashamed of one's anger, which generates resentment and hatred because the shame driving the loop is outside awareness. Nations locked in this spiral have great difficulty deescalating or resolving conflicts of interest not because of the content of the issue but because of the emotional milieu in which conflicts of interests occur. In such instances, violence can become compulsive (Long and Brecke 2003).

Long and Brecke's third model is "fear and perceived threat." As demonstrated here, fear is a very powerful survival emotion designed to facilitate or inhibit certain behavior patterns, often without conscious awareness on the part of an individual or a collective. Whether powerful or weak, fear functions as a signal indicating the possibility of danger arising from power relations in which the party is involved. Fear has environmental triggers, some preset during evolutionary adaptation, but many others learned. The change in the environment that triggers fear need not be a hostile act or threat per se but only a novel event that one has learned to associate with threat. Contrary to our usual supposition, we do not fear because we are threatened, we feel threatened because we fear. The object of fear is not retrospective (the change in power balance); it is prospective, such as the imagined possibility of harm or injury resulting from a changing power balance. Fear forces attention on this environmental change, among innumerable other environmental changes, that stirs appraisal and inferences of threat and that informs preference formation and action, including the possibility of aggression (Long and Brecke 2003, 30).

However, Mercadillo and Arias (2010) argue that cooperation and compassion are two forms of behavior with similar developmental, cognitive, and cerebral regulatory bases to the mechanisms activated in violence, even though they result in radically different forms of behavior. They contend that violence and compassion lead to moral action dependent on whether sociocultural contexts are adverse or favorable to human well-being. Both forms of behavior involve an adaptable regulatory mechanism that takes account of learned social behavior codes. They conclude that the neurocognitive system is a flexible and adaptable mechanism that regulates behavior directly, according to the sociocultural context in which individuals live, and that cognitive neuroscience might provide tools for creating and changing mental concepts that could eventually enable human beings to live together in peace.

Implications for International Leadership

The findings here raise two broad issues for political leadership. The first is how leaders can best integrate this emerging knowledge of the brain's influence on aggression, bias, and group conflict into domestic and foreign policy. Are there more effective strategies to deal with such problems than force or traditional mediation with handpicked leaders? What purpose does it serve to remove a ruthless dictator if his followers hold similar views of outsiders? Moreover, what dangers are there when leaders select certain easily identifiable ethnic groups as the enemy, or when they profile based on physical characteristics, and what deep-seated evolutionary fear circuits do such policies trigger? Finally, what are the long-term consequences when leaders exploit the amygdala-based fear system to build support for their policies?

The second issue centers on the leaders themselves, and on the assumption of rational choice that underlies democratic decision making. To what extent do the unconscious fears, biases, and hatreds found in the primitive brain regions influence decisions of even well-meaning, moderate leaders? Should aspirants for leadership positions undergo brain imaging to ensure their emotion regulatory circuitry is functioning adequately? More critically for the world, how can we determine the extent to which damage to the PFC or other critical brain circuitry of a leader is responsible for his or her extreme levels of fear, hatred, and aggression toward out-groups, and often even toward other members of his or her own group? As noted in an editorial on violence and mental illness in the *Archives of General Psychiatry*, "Much of the criticism of violence research arises from the naïve belief that there will be a sudden breakthrough, which, through misinterpretation for a narrow political agenda, will overturn social policy. This is unlikely, as the causes of violence, if they are ever to be found, will undoubtedly be complex and multifactorial" (*Archives of General Psychiatry* 1996, 484). Despite this, the evidence from social neuroscience confirming the importance of emotion and the basic drives of the limbic system in understanding ethnic conflict and recurring wars has soared in the last decade and cannot be ignored. This discussion of politics and neuroscience will be revisited in chapter 9.

7

The Media, Commercial and Military Applications, and Public Policy

This chapter shifts attention from the contributions of neuroscience to our understanding of social behavior, particularly as it relates to cultural differences and in- and out-groups, to the policy arena. As introduced in chapter 3, neuroscience raises a myriad of policy issues at many different levels. Some focus on research priorities and safety and efficacy concerns, some on individual applications, and others address the broader social implications. These latter issues are often the most difficult to engage because they take place in a complex societal environment with many countervailing forces. This chapter examines several of the major factors outside the political institutions themselves that influence neuroscience in society. It first examines the role of the mass media, particularly the controversy surrounding the way in which the media report emerging neuroscience findings. It then offers an overview of the commercial applications of neuroscience, including neuromarketing and the commercialization of brain-based products and services. This is followed by a discussion of the personal, commercial, and policy issues that neuroenhancement raises. The chapter ends with a discussion of the key role of the military and security sectors in brain-centered research and development. It also summarizes current and projected applications of brain intervention in these areas, many of which draw on enhancement and neuroimaging technologies.

The Media and Neuroscience

Increasingly, the popular press and the Internet are rife with often highly optimistic and oversimplified coverage of unfolding research and interventions in the brain. Additionally, health-oriented magazines and television shows extol the virtues of medical innovations and the quick fixes or breakthroughs they promise. By and large, then, media coverage solidifies the public's reliance on a technological fix and stimulates its appetite for new, often expensive, high-technology procedures. Participants at a recent National Science Foundation neuroscience workshop, for instance, expressed dismay at the many fMRI-based behavioral inferences that have found their way into the popular press and are circulated broadly (Lupia 2011). Moreover, at least some of the media outlets seem to relish uncovering and sensationalizing cases in which access to treatment was denied. Not surprisingly, there is considerable debate over the lapses in procedural transparency and rigor that often accompany such media claims (Abbott 2009b).

Gilbert and Ovadia (2011) document many instances of hype and exaggerated coverage of deep brain stimulation (DBS) in scientific journals and the popular press. For instance, even *Nature* presented DBS findings in an overly optimistic article titled "Brain Electrodes Can Improve Learning," which received highly enthusiastic coverage in television news and the press, one such overly hopeful discussion headlined "Deep Stimulation 'Boosts Memory'" (Gilbert and Ovadia 2011, 3). Similarly, Tovino (2009 is concerned that the intense media hype surrounding fMRI technology may cause employers, insurers, criminal justice officials, governmental agencies, and other third parties to believe that investigators using fMRI are actually capable of reading minds and detecting lies.

Distorted reporting, commercial pressures, and other factors can result in the misuse or misapplication of neuroimaging (O'Connell et al. 2011). This pattern of extolling medical and scientific innovation while giving little if any attention to ethical issues and technical constraints risks turning ethical neglect into de facto ethical approval, thereby promoting public acceptance of these technologies without full knowledge of their risks and limitations (Clausen 2011; Racine et al. 2010). In an informative study, Racine, Bar-Ilan, and Illes (2006) reviewed 235 articles on neurostimulation techniques published in the print news media in the UK and United States and found that 51 percent were enthusiastic depictions, while only 4 percent underscored the risks. Moreover, among the articles reviewed, 29 percent contained a "personal twist," including first-person narratives and descriptions of "miracle stories of patients cured of Parkinson's disease, dystonia, and Tourette's syndrome" (Racine et al. 2010).

Another major study (O'Connell et al. 2011) found that the frequency of media articles on neuroimaging increased threefold in two years. As in earlier reports, they found that the viewpoints adopted were predominantly positive, especially those on neuroimaging applications for marketing (77 percent), biosecurity (57 percent), and employment (60 percent). The most prevalent finding was a tendency for positive articles to take an uncritical stance on proposed applications and to avoid ethical issues (68 percent), suggesting an influence of commercial interests on the reporting of research. Moreover, 41 percent of media articles surveyed gave little or no technical detail of the neuroimaging methods, a practice especially notable in articles on marketing and diagnostic applications. Insofar as neuroimaging is far from being validated as a reliable marketing or diagnostic tool (Ariely and Berns 2010), the avoidance of technical issues favors acceptance. Similarly, it could be argued that the lack of ethical discussion in the reporting of neuroimaging applications in the marketing, employment, and biosecurity sectors could be the result of financial pressures to report positively and omit negative details (O'Connell et al. 2011).

Meanwhile, there is mounting evidence that patients educate themselves and build their hopes from sources such as television and the Internet (Schneiderman 2005). From the viewpoint of the layperson or potential psychiatric patient, a highly optimistic media depiction of a technique can be far more influential than the often austere and subtle explanation found in specialized journals. Ford (2009) agrees that such reports generate an "educational vulnerability" for patients. He affirms that very often when patients consider neurosurgical techniques, they have already been preconditioned by buoyant media portrayals of novel brain interventions. Furthermore, passionate, one-sided media accounts can convey to the general public and potential patients unrealistic hope and act as a baseless promotion of the technologies (Schlaepfer and Fins 2010; Bell, Mathieu, and Racine 2009). Thus, each new story raises patients' hopes and promises an answer to a question that their suffering prompts them to ask with more and more desperation as time goes by (Gillett 2011).

Although we expect the media to play an informative and investigative role, with regard to neuroscience media sources seldom provide critical analysis or question the assumptions or limitations of the original study. "Creative" headlines are framed to attract public attention and are commonly reproduced on Internet search sites and numerous blogs dedicated to the subject. The need for more responsible reporting, from popular media as well as from neuroscientists and neurosurgeons themselves, demands not only closer monitoring but also better mechanisms for communication among all parties (Racine et al. 2010; Ford 2009).

Because of the public's limited knowledge and inability to distinguish fact from opinion, distorted reports lacking technical or ethical details provoke social and ethical concerns. To reduce such misconceptions, it has been argued that neuroscientists must be more effective in conveying their discoveries to the public (Racine et al. 2010; O'Connell et al. 2011).

During the course of their comprehensive analysis of media coverage of neuroscience, O'Connell and colleagues (2011) uncovered several other patterns in the general media, including a failure to report how factors such as sample size can affect both the interpretation of results and the ability to generalize the findings. There was also evidence that some journalists deliberately distorted the research to make the story more impressive, and likewise, that some researchers misrepresented or overstated the importance of their research to attract media attention. Moreover, it was found that individuals affiliated with commercial neuroimaging interests receive a disproportionate amount of media attention as a result of calculated marketing to increase revenue (O'Connell et al. 2011). This in turn can lead to a polarized, positive slant, such that the "news" becomes a public relations exercise on behalf of the companies selling products.

Although expectations about scientific discoveries need to be carefully framed within scientifically acceptable limits, frequently marketing and direct-to-consumer neuroimaging services provide little or no detail of technical shortcomings (Ariely and Berns 2010). Commercial neuroimaging companies rarely concede that the reliability of their methods is undermined by insubstantial evidence bases and an absence of replication of their findings by other investigators (Spence 2008). Despite this, leading proponents of commercial neuroimaging services regularly use media outlets to promote their work through solicited op-ed features, personal websites, and merchandising. In contrast, many academic neuroscientists are not media savvy and are unable to communicate their more impartial versions to the public. According to O'Connell and colleagues (2011), as the interaction between neuroscience and the media increases, it will become necessary to establish guidelines for the professional conduct of neuroscientists participating in a dialogue with the media and industry.

Commercial Uses of the Brain Sciences

An interconnected area of apprehension surrounds the commercial uses of neuroscience. Even relatively innocuous applications of brain imaging, such as neuromarketing, strike many observers as problematic (Fisher et al. 2010; Singer 2004). As discussed in earlier chapters, neuroscience raises many legal, ethical, and political issues. Although these issues challenge traditional values and law, the introduction of the commercial dimension complicates matters and in some ways simultaneously "normalizes" neuroscience as just another set of products to be marketed and sold. This section looks at several aspects of this dynamic phenomenon of the move of neuroscience to the marketplace.

Neuromarketing

The primary goal of neuromarketing is to obtain objective information about the inner workings of consumers' brains without depending on subjective verbal reports (Renvoisé and Morin 2007). Marketers traditionally have used a range of market research techniques, from focus groups and individual surveys to actual market tests, to guide the design and presentation of products and to maximize sales by controlling the menu of offerings, pricing, advertising, and promotions. According to Ariely and Berns (2010), the anticipation is that neuroimaging will streamline marketing processes, provide a more efficient trade-off between costs and benefits, and reveal information about consumer preferences that is not obtainable through conventional methods. The assumption is that people cannot fully articulate their preferences when asked to verbally express them and that their brains contain hidden information about their true preferences that can be exploited in influencing their buying behavior.

Much of the work to date has focused on brain responses to visual representations of products, such as pictures of or advertisements for the product (Rossiter et al. 2001). In the future, however, it is expected to provide an accurate tool for determining prospectively what potential products fit underlying consumer preferences, thus allowing more efficient resource allocation by companies in developing the most promising products (Ariely and Berns 2010). Neuromarketing could also be helpful in identifying differences in consumer reactions to diverse types of inputs. For instance, in a study of neural responses to wine, medial orbitofrontal cortex (OFC) responses were higher when subjects were told that the wine was very expensive (Plassmann et al. 2008). Activity in this region also correlated with self-report ratings of how much participants liked the wine, even though all the wines sampled were actually the same. These results suggest that the instantaneous experience of pleasure from a product is influenced by pricing, and that this effect may be mediated by the medial OFC (Plassmann, O'Doherty, and Rangel 2007). Similarly, this information could be used to help target marketing efforts to specific persons or groups or to time pricing moves to capitalize on individual weaknesses that are known to coincide with particular biological states (Ariely and Berns 2010).

Although neuroimaging is often touted as an exciting new tool for advertisers, its effectiveness remains poorly understood, and it is unknown whether it can prospectively reveal the effectiveness of a particular marketing tactic (Lee, Broderick, and Chamberlain 2007). There is evidence that the amount a person is willing to pay for a product or service is correlated with activity levels in the medial OFC and prefrontal cortex (Plassmann, O'Doherty, and Rangel 2007). Rather than demonstrating the existence of a "buy button" in the brain, current evidence suggests that purchase decisions depend on a combination of neurobiological processes, with no single brain region responsible for a consumer choice. However, most attention has focused on dopamine-rich brain regions, with the OFC and striatum consistently implicated in goal-directed action such as valuation (Plassmann et al. 2008; Hare et al. 2008; Yin, Ostlund, and Balleine 2008).

While it is difficult to get an accurate estimation of its true use, since many companies do not publicize such activities, neuromarketing research in academic institutions and consulting startups around the world is rising (Dunagan 2010). By 2010, more than ninety marketing companies were offering services using neuroimaging (O'Connell et al. 2011). The greatest potential currently seems to lie in the design stage of a product, where fMRI is used to gauge the attractiveness of the image, taste, or smell of a new product (Ariely and Berns 2010), but promoters also support its use for identifying subconscious motives for buying products (Häyry 2010). Similarly, there is an emerging market in neuro-education that incorporates insights from neuroscience into education to maximize the brain's capacity to learn (Dunagan 2010). Although the stimulation of a student's brain with high-powered magnets to increase motor skill acquisition seems farfetched today, there are strong pressures to enhance learning in an increasingly competitive educational and workplace environment, and the potential market appears huge (Luber et al. 2009).

Although critics see neuromarketing as dangerous because marketers could bypass neural control systems, the real world of direct-to-brain marketing is more complicated, and the expression of desire in brain readings is unlikely to translate directly to predictable purchasing behavior. Dunagan (2010), for instance, argues that neuroimaging is no more terrifying than participating in a good focus group. While public fears center on invasion of private inner thoughts by neuroimaging, experts are more concerned about commercial pressures on research practices, such as the patenting of methods or the trading of transparency for the marketability of results (Eaton and Illes 2007).

Neuromarketing raises a range of ethical issues, which fall into two major categories: the protection of various parties who may be harmed or exploited by its research, marketing, or deployment and the protection of consumer autonomy (Murphy, Illes, and Reiner 2008). While issues in the former category are straightforward, issues of consumer autonomy may or may not be problematic, depending on whether the technology can ever realistically manipulate consumer behavior such that consumers are unaware of the subversion. Murphy, Illes, and Reiner(2008) call this phenomenon "stealth neuromarketing" and argue that academics and companies using these techniques should adopt a code of ethics to ensure beneficent and nonharmful use of the technology. The prospect of big corporations or political lobbyists enlisting neuroscience to manipulate consumer and voter behavior has inevitably raised apprehension in some quarters. For instance, a watchdog agency founded by consumer advocate Ralph Nader asked the U.S. government to investigate neuromarketing companies on public health grounds (Tingley 2006).

Political Marketing

These issues become even more problematic when neuromarketing techniques are used to sell candidates to the public and to design political campaigns. Given the high political stakes and the vast amount of money involved in U.S. elections, however, it is only a matter of time before such practices become commonplace. Already studies have shown a complex pattern of activation in response to statements about candidates that have been interpreted as evidence that motivated reasoning involves activation in the ventromedial prefrontal cortex (PFC), the anterior cingulate cortex, the posterior cingulate cortex, and the insula (Westen et al. 2006). Other studies have suggested that activation of the medial PFC might be associated with maintaining a subject's preference for a candidate in response to advertisements, whereas activity in the lateral PFC might be associated with changing candidates (Kato et al. 2009).

Although politics in many ways involves marketing, just as any other product area does, it is distinctive in that it often takes a much more negative tone, thus triggering research about the impact of negative advertising. In their study of partisanship and turnout, Stevens and colleagues (2008) found some circumstances under which turnout can be increased by negative appeals, but suggest that voters are "motivated processors" of advertising claims and evaluate the fairness of an ad according to their partisan predispositions. When partisans perceive the criticisms of their own party's candidate to be fair, they are less likely to vote, thus indicating that negative advertising can affect not only the overall turnout but also the composition of the electorate. Similarly, Sides, Lipsitz, and Grossmann (2010) found that citizens' distaste for negativity affects how they view the candidates, politicians, and politics in general, which leads to a demobilizing effect as their low evaluations of the campaign's tone "spill over" and affect their assessments of other aspects of the campaign. In the near future there is bound to be considerably more research on what messages and approaches best resonate with voters and political marketing aimed at how best to package a candidate's appearance, trustworthiness, and message content so as to maximize voter support.

In addition to using neuroscience technologies to market an existing candidate, hypothetically they could be used to design a model candidate (Ariely and Berns 2010). For instance, substantial neuroimaging work has been done on the perception of human faces and features such as facial symmetry, skin color, and attractiveness. Already, key brain structures in visual processing include the fusiform face area for basic face processing and the superior temporal sulcus for gaze direction and intention (Kanwisher and Yovel 2006). Moreover, a study on the effect of political candidates' appearance found that insula activation in response to seeing a picture of a candidate was associated with a greater likelihood of that candidate losing the election (Spezio et al. 2008). With appropriate modeling and data from key constituencies, one could choose the "perfect" candidate in terms of style and appearance, as well as the most marketable message content, by maximizing striatal and OFC responses to platform statements (Ariely and Berns 2010).

Commercial Products of Neurotechnology

As presented in chapter 5, many provocative commercial applications of neuroscience, such as lie detection, are proliferating, and there likely will be pressures to utilize brain scans to screen job and school applicants in the future (White 2010). Recently, fMRI has been commercialized for nonmedical use, and at least two companies have marketed this method of lie detection. The companies, Cephos Corporation and No Lie MRI, offer fMRI scans to clients for a sizable fee. Potential clients for commercial lie detection vary. One company offering fMRI services to the public asserts on its website that fMRI is the "first and only direct measurement of truth verification and lie detection in human history." Another declares that its fMRI tests are "fully automated" and "observer independent (objective)," while a third states that its fMRI testing is "nonsubjective" in that humans do not ask the questions or examine the scans (Tovino 2007). It is, perhaps, telling that current commercial applications of this technology have found the largest market among spouses suspicious of their partners (Huang and Kosal 2008).

The commercialization of fMRI has led to a frenzy of warnings and inspired numerous critical articles and special reports. As noted in chapter 5, there are many technical, legal, and philosophical questions about the appropriateness of using neuroimaging technologies to this end. Many authors have cautioned against the use of fMRI for lie detection and have concluded that premature application of this technology could result in potentially disastrous outcomes (Wolpe, Foster, and Langleben 2005). Phillips (2004) is particularly concerned with the potential use of brain scans to diagnose or make predictions about a behavior or illness: "who might be a paedophile [*sic*], who might be violent, who might develop a mental illness or prion disease. . . ?" (38). There are also criticisms that much of the relevant research has not been published in peer-reviewed journals, has been produced by financially interested scientists connected with truth verification companies, and is significantly less reliable than claimed (Schauer 2009).

Given its clinical and scientific value, there is little call for prohibiting neuroimaging, but some have proposed constraints on nonmedical applications. Tovino (2007), for instance, suggests that we could prohibit the advertising, marketing, or offering of fMRI scanning services for nonclinical or nonresearch uses or restrict the use of fMRI for certain purposes, such as lie detection, or by certain persons or organizations. One possible downside of this option is that it might drive commercial fMRI services underground, perhaps increasing the probability that fraudulent individuals will provide such services illegally, thus lowering the standard of care in the provision of these services. Other suggested options have been to tax or license commercial ventures or to apply consumer protection laws to these practices. Tovino (2007), for instance, recommends that federal and state regulatory agencies be made aware of commercial and other uses of fMRI and enforce truth-in-advertising rules. The Federal Trade Commission Act, state deceptive trade practices acts, and state consumer laws already oblige some advertisers to be truthful and nondeceptive and require them to have scientific evidence to back their claims. At a minimum, the truth-in-advertising principles that underlie these laws could be applied or expanded to apply to fMRI (Tovino 2007).

Another concern over commercial lie detection services is that employers, insurance companies, and other parties might compel or get access to the findings of these products. Legally, except for employees of governmental agencies and companies that provide security, the use of polygraphs on employees is prohibited in the United States under the Federal Employee Polygraph Protection Act of 1988 (FEPPA). Furthermore, to ensure the safety of the clients using fMRI lie detection services, it has been suggested that fMRIs used for lie detection should undergo a complete governmental approval process, perhaps administered by the FDA (Greely 2005). While a lengthy approval process may be extreme, commercial fMRI lie detection companies should be required to make reasonable assurances that clients will be safe. Good faith efforts to protect clients are common practice for commercial enterprises and not a unique ethical issue for fMRI lie detection. To ensure informed consent, then, the equipment used by fMRI lie detection services should be subject to inspection and clients should be made aware of the potential harm involved (White 2010).

Lie detection is the most salient commercial application of neuroscience products but by no means the only one. Another growth area is direct-to-consumer services that offer diagnosis of neurological and psychiatric disorders to detect conditions such as autism (Farah et al. 2009). Although this practice is generally condemned by clinicians, press coverage can promote unrealistic public expectations and demands for services, such as the claim that "for the first time, a quick brain scan that takes just 15 minutes can identify adults with autism with over 90% accuracy" (Medical Research Council 2010). Similarly, the direct marketing of products and health care services to consumers via the Internet and other media is a large growth area, as illustrated by the multi-billion-dollar nutritional supplement market for memory enhancement discussed below.

Ubiquitous advertisements also encourage consumers to request the latest psychotropic drugs from their physicians. For instance, while Ritalin was the market leader in treatments for attention deficit/hyperactivity disorder (ADHD) for many years, recently aggressive marketing campaigns by Eli Lilly and others have increased the use of drugs such as Concerta, Adderall, and Strattera. Not surprisingly, the use of drugs for ADHD in the United States is vastly out of proportion with that of other countries and accounts for about 85 percent of the world's use (Singh 2005). Moreover, new psychopharmaceuticals and cognitive fitness tools, like those being developed by Posit Science, Luminosity, and other firms, will rapidly expand the neurocentric health market. These will be followed by progressively sophisticated neuromodulation, cognitive prosthetics, and neurofeedback technologies that will move from the lab into mainstream use over the next few decades (Dunagan 2010). A comparative analysis of Internet-based advertising for "neuroproducts" shows that existing mechanisms for monitoring the promotion of prescription drugs are being strained and, as technology develops, the information that regulatory bodies will need to be able to oversee will further test their already tenuous oversight mechanisms (Illes and Bird 2006).

Imaging genetics, too, has considerable potential to become an attractive product in emerging high-technology medical markets. Guidance for considering the power of combined technologies may well be gained from past experiences with commercialization. These include considerations of the sale of technology for unintended uses and for which data do not exist, and assessing the impact of aggressive marketing strategies that pose risks, particularly to some of the most vulnerable members of society who suffer from neurological or mental health disorders and cognitive deficits (Illes et al. 2004). Neuroscience databases, including those archiving neurological and genetic data, are being created throughout the world. The resulting challenges for the commercial sector, and for researchers and policy makers, are enormous. A range of legal authorities and potential litigants, such as insurers and employers, may at some point seek this information. Likewise, individuals who are data sources may also try to access information about themselves. As the data can be banked indefinitely, these issues have an indeterminate lifetime for the individual, as well as for family members.

The Case of Neuroenhancement

One area of neuroscience that has created a heated debate about its personal, commercial, and societal ramifications and its treatment by the media is the enhancement of cognitive performance through medical interventions such as drugs and prostheses. Clearly, new developments in neuroscience and psychopharmacology have expanded the possibilities for augmenting mental functioning, such as by improving memory, mood, or even intelligence (de Jongh et al. 2008; Rose 2002). Although the line between enhancement and therapy is often blurry, many applications clearly embody attempts to enhance human traits or performance rather than treat disease or promote health (Miller and Brody 2005; Talbot 2009). According to Repantis and colleagues (2010), the term neuroenhancement refers to improvement in the cognitive, emotional, and motivational functions of healthy individuals through drugs or other means. Singh (2005) adds that enhancement technologies are those treatments that improve human performance, appearance, or behavior where such improvement is not medically warranted. De Jongh and colleagues (2008) distinguish among (1) cognition-enhancing drugs used to improve short- and long-term memory or executive functioning that manages other cognitive processes and is involved in planning, cognitive flexibility, abstract thinking, and inhibiting inappropriate actions; (2) drugs that enhance mood and pro-social behavior; and (3) drugs that prevent the consolidation or reconsolidation of unwanted (traumatic) memories. To date, most attention has focused on the first use (Banjo, Nadler, and Reiner 2010). Neuroenhancement has also been termed cosmetic neurology (Chatterjee 2007, 129).

Despite considerable variation in chemical composition and in the mechanisms by which nootropics (or "smart drugs") act, a common property is their effect on higher integrative brain functions. They are thought to work by altering the availability of the brain's supply of neurotransmitters, enzymes, and hormones; by improving the brain's oxygen supply; or by stimulating nerve growth. The efficacy of most nootropic substances has not been documented, and efforts to do so are complicated by the difficulty of defining and quantifying cognition and intelligence. Although the original rationale for research on these substances was to treat patients with premature dementias such as Alzheimer's disease, increasingly they are being touted as a means of promoting mental agility in healthy persons who want to boost their cognitive abilities. According to Chorover (2005), "basic neuroscientific research has morphed into a branch of bioengineering (neurotechnology) that promises to deliver a remarkable array of practical innovations" (2081). Despite the absence of scientific evidence that these substances actually work in normal persons, an enhancement industry is flourishing (Chatterjee 2007), again manifesting the widespread attractiveness of technological shortcuts to learning (Flaskerud 2010; de Jongh et al. 2008).

In addition to cognitive enhancement, other drugs introduced to treat diseases have the capability to enhance other nervous system functions.

One interesting application involves the use of beta-blocking drugs such as propranolol to reduce stress-induced anxiety (Chatterjee 2007). These drugs compete with adrenaline-like chemicals produced by the sympathetic nervous system that attach to beta-adrenergic receptor sites when the body is under stress, causing rapid heartbeat, muscle tremor, dry mouth, perspiring, and nausea. By occupying the receptor sites, propranolol blocks these physiological responses, thus reducing the symptoms of anxiety. Beta-blockers also appear to alleviate posttraumatic symptoms by cropping disturbing memories of individuals who have been involved, for example, in a car accident (Launis 2010). Musicians and competition shooters use beta-blockers to dampen physiological tremors in order to improve or (in the case of performance anxiety) enable their performance. One study found that 27 percent of members of the International Society of Symphony and Opera Musicians reported using beta-blocking drugs, and of these, 70 percent said they used them without a prescription. Other conceivable users could include surgeons in the operating theater, students before the big exam, and soldiers before battle.

While most attention has focused on drugs, other interventions such as DBS have been discussed to enhance cognitive abilities (Hamani et al. 2008). This use of DBS was endorsed by Abbott (2008) in a highly optimistic article titled "Brain Electrodes Can Improve Learning." Gilbert and Ovadio (2011, 2) note that this "enthusiastic media shock wave was instantly replicated on an international scale" without addressing ethical issues, thereby promoting public acceptance of DBS for enhancement. Although Synofzik and Schlaepfer (2008, 2011) contend that the widespread use of DBS for enhancement purposes is presently questionable, they foresee its potential use to that end after considerably more investigation of its efficacy and ethics. "Given that the ethics of the neuroenhancing application of DBS is less straightforward than it might have appeared at first sight, this responsibility includes, at the very least, a careful ethical assessment of requests for DBS for non-clinical use" (Pacholczyk 2011, 2).

Similarly, there is evidence that transcranial direct current stimulation (tDCS) can improve the ability to learn a simple coordination exercise and that the improvement is still apparent three months later (Reis et al. 2009; Fox 2011). A related study found that several sessions of tDCS applied to the dorsolateral prefrontal cortex improved mood for several weeks (Fregni et al. 2008). Additionally, these authors reported elsewhere that applying tDCS to the same region can make people less likely to take risks (Boggio et al. 2008). With this evidence, one entrepreneur hopes to develop and market "a thinking cap," a tDCS device that corporate executives or advertising copywriters could use to improve their creativity (see Chi and Snyder 2011). According to Fox (2011), the adoption of tCDS for enhancement raises ethical concerns similar to those that surround mind-enhancing drugs such as Adderall and Modafinil, which some students take as study aids. Students could secretly "electrodope" with tDCS before a university entrance exam to inflate their scores. Although none of the studies published so far has shown a type of mind sharpening that would help on such exams, it might simply be a matter of targeting the right brain areas. For Lipsman, Zaner, and Bernstein (2009), neurosurgeons soon will likely be "the purveyors of future enhancement implantable technology" (375).

In their study of the attitudes of neurosurgical staff toward the uses of psychosurgery for enhancement, Mendelsohn, Lipsman, and Bernstein (2010) found little agreement for the future use of DBS and other means of neuromodulation for cognitive enhancement and personality alteration, although most participants felt that physical alteration of nonpathological traits is objectionable. Hildt (2010) and others point out that these uses of DBS, brain implants, or brain-computer interfaces challenge our notions of human nature and the idea of man and how far human functions can be substituted or even enhanced by technical devices (Bell, Mathieu, and Racine 2009; Giordano and Gordijn 2010). Robert (2005) argues that while self-improvement is a noble aim, there is a "dramatic and morally important difference between self-improvement through drugs and neural implants and other forms of enhancement" (29).

Proponents of Enhancement

A 2007 editorial in *Nature* titled "Enhancing, Not Cheating" sparked a debate about the use of cognitive-augmenting drugs by healthy people. The editorial maintained that the use of these drugs was enhancing, not cheating as claimed by the opponents. The editorial argued that one argument in favor of its use was the "pursuit of personal liberty" (320). Such drugs allow people to reach their full potential and should be available. Following up on this theme, a group of scientists and ethicists stated that, in theory, healthy people should have the right to use nootropics and that we should welcome new methods of improving brain function (Greely et al. 2008). Although they suggested a number of cautions and called for more research on the risks, they asserted that enhancing with pills is no more objectionable than eating right or getting a good night's sleep. The authors argued that as more effective brain-boosting pills are developed, demand for them will grow. This demand is bound to be accelerated by the popular media, which have given increased attention to the use of cognitive-enhancing drugs (Flaskerud 2010). The Internet provides thousands of sites that promise significant benefits and immediate shipment, often without prescription. Stimulants like Ritalin and Adderall are being used by people to help them focus their attention, while the narcolepsy drug Provigil is used by people trying to keep their brains alert and awake.

Some libertarians argue that cognition-enhancing drugs should be widely available to anyone who wants them, thus requiring a radical revision of current drug policies that prohibit off-label use beyond their prescription-only status (Capps 2011). Among the most extreme proponents of neuroenhancement are transhumanists. Transhumanism is the loosely defined intellectual and cultural movement that affirms the possibility and desirability of fundamentally improving the human condition through applied reason, especially by developing and making widely available technologies to eliminate aging and to advance human intellectual, physical, and psychological capacities (Bailey 2005; Young 2006). According to Bostrom (2003), the transhumanist vision is to "create the opportunity to live much longer and healthier lives, to enhance our memory and other intellectual faculties, to refine our emotional experiences and increase our subjective sense of well-being, and generally to achieve a greater degree of control over our own lives" (493). Ashcroft and Gui (2005) suggest that this conception of enhancement technologies is all about the extent to which we can correctly consider the human body and individuals' personalities themselves as machines that can be broken down or repaired, and can be considered as designed and improved through redesign.

Transhumanists look forward to posthuman descendants, beings whose basic capacities so radically exceed those of present humans as to be no longer unambiguously human by current standards, who may be resistant to disease, have unlimited youth and vigor, reach intellectual heights far above any current human genius, have increased capacity for pleasure, love, and artistic appreciation, and experience novel states of consciousness that current human brains cannot achieve (Agar 2007, 12). To this end, they favor modifications that produce longer active life spans, better memory, greater intellectual capacities, and the increased ability to make subsequent choices wisely. They predict that parents free to enhance their children's intellects, physical constitutions, and life expectancies will choose to do so (Agar 2007). Furthermore, if there are basic enhancements that would be beneficial for a child but that some parents cannot afford, then society should subsidize them, just as it does with basic education (Bostrom 2003). Ironically, by their unabashed extreme acceptance of enhancement, transhumanists have given fodder to the opponents of enhancement.

Critics of Enhancement

Critics of neuroenhancement focus on the potential safety problems of the long-term use of drugs in healthy individuals, the possibility of direct or indirect coercion to use enhancement drugs, the social justice argument that access to enhancement technologies is likely to be expensive and not available equally, and, that at its core, the use of enhancement poses a threat to social values by undermining the worth and dignity of hard work (Ashcroft and Gui 2005). Often, this opposition is framed in terms of a slippery slope argument (see Launis 2010). In general, then, the concerns of the opponents can be classified into two broad categories, concerns about the harms that may be experienced by those who use the enhancement technologies and concerns about the adverse social impacts of the widespread use and societal embrace of them (Hall, Carter, and Morley 2004; de Jongh et al. 2008; Illes and Bird 2006).

Although there are potential adverse reactions to many therapeutic drugs, these harms are usually outweighed by the relief afforded from the symptoms of the disease. However, when given to disease-free individuals, the trade-off of the adverse effects with the less certain benefits of enhancement are blurred. "Our brain is of such complexity and its neurotransmitter systems are so strongly interlaced that turning a small screw in one system generates unpredictable effects in all other systems with corresponding consequences for behavior" (Quednow 2010, 154). Thus, while access to cognitive enhancers might be desirable in theory, they could have adverse medical consequences (Flaskerud 2010). Prozac, for example, is supposed to make miserable lives tolerable and tolerable lives wonderful, but reality is not that simple. The actual impacts of the drug are less predictable and could lead to personality changes instead of the intended mood improvements (Häyry 2010). Critics also warn that many of these drugs have not been tested for off-label uses and that some, such as stimulants, may be addictive or harmful (Volkow and Swanson 2008; Quednow 2010).

In addition, two concerns have been raised about the societal implications of the widespread use of enhancement technologies. The first is that marked social inequities in access to enhancement technologies will amplify existing social inequities. According to Chatterjee (2010), in modern competitive societies, the social and cultural pressures to secure all the latest enhancements for one's children (so as to give them the best possible start in life) and oneself (so as to become a successful and responsible citizen) will benefit the already best-off. A second and related concern about the social impacts of enhancement technologies is that their widespread use will raise the standards for what counts as normalcy and coerce everyone into using them as a way of "keeping up with the Joneses." Furthermore, it will increase discrimination against disabled persons and people with conditions who decline to be enhanced (Parens 2002). It is also possible that children might be pressured into taking the drugs either directly by their parents who want them to excel or indirectly through peer pressure (Flaskerud 2010).

Carrying this concern further, Martin and Peerzada (2005) argue that we should move with caution in enhancement because it could imply that some people have less intrinsic human worth than others. Although eliminating certain characteristics or increasing certain capacities might express nothing more than a personal preference, it could send the message that some people (the smarter ones, the stronger ones, the more competitive ones) are of greater intrinsic worth than others. Another criticism of neuroenhancement is that it is analogous to doping among athletes—in other words, it is a form of cheating against others who do not use it—or that it is cheating against oneself because it does not represent true achievement (Flaskerud 2010). There are additional questions about whether drugs that enhance concentration might diminish creativity, according to Farah and colleagues (2009).

Kamm (2005), too, has major problems with enhancement, questioning whether we could ever really safely alter people without making disastrous mistakes. He also asks why enhancement should be a top priority in light of scarce medical and societal resources. A deeper issue, he argues, is our lack of imagination as designers. Unfortunately, most people's conception of the varieties of goods is very limited, and if they redesigned people their improvements would likely conform to limited, predictable types (13). However, in his analysis of Kamm's position, Schwartz (2005) asks why interventions that fix biological dysfunctions should have a superior moral status to interventions that modify normal functioning since both aim to fulfill human desires and ease suffering (18). Similarly, the "exclusivity of choice, and an uncritical deployment of enhancement as an unequivocal good, underplay the role of a social and political community, and leave one unable to discriminate between, and solve, conflicting ideas of 'good'" (Capps 2011, 127).

Finally, there is the criticism aimed at the pharmaceutical industry that creates a market for drugs and procedures by convincing people that ordinary states are syndromes that require drug treatment or by selling medications to people who are not ill. Neuroenhancement represents a potentially huge market, not only for pharmaceutical manufacturers but also for physicians who might enter the potentially lucrative specialty of cosmetic neurology (Larriviere and Williams 2010). Many middle-aged people who want youthful memory powers and multitasking workers who need to keep track of numerous demands will want to use these drugs. Of course, drug companies will "gladly have a world where everyone needs to buy their products in order to compete in school and the workforce" (Flaskerud 2010, 63). Similarly, Coors and Hunter (2005) contend that the desire for enhancements by a public ill-equipped to understand the detail of any proposed intervention, coupled with financial incentives on the part of promoters of the technology, has the potential to lead to grievous and potentially irremediable harms to many individuals.

Does Neuroenhancement Work?

Based on a systematic review and meta-analysis, Repantis and colleagues (2010) showed that expectations regarding the effectiveness of these drugs exceed their actual effects. For methylphenidate, an improvement in memory was found, but no consistent evidence for other enhancing effects was uncovered. Modafinil (Provigil), on the other hand, was found to improve attention for well-rested individuals while maintaining wakefulness, memory, and executive functions to a significantly higher degree in sleep-deprived individuals than a placebo did. However, repeated doses of modafinil were unable to prevent deterioration of cognitive performance over a longer period of sleep deprivation, while maintaining wakefulness and possibly even inducing overconfidence in a person's own cognitive performance.

Moreover, as de Jongh and colleagues (2008) note, there are a number of caveats in the development and use of neuroenhancers. First, according to the inverse U-function principle, enhancement is possible only as long as we do not already have an optimal level of arousal, vigilance, or neurotransmitter concentration. Thus, an already optimally tuned brain can hardly be enhanced and, given that usually our brains already perform to the best of their ability, enhancement for most people seems limited (Quednow 2010), To date, cognitive effects in well-rested healthy subjects have been small and hard to detect (Kumar 2008). Second, doses most effective in facilitating one behavior could produce null or even detrimental effects in other cognitive domains. For example, we could enhance our working memory but concurrently decrease our long-term memory, or vice versa, but we will never be able to enhance both simultaneously. Or increases in cognitive stability might come at the cost of a decreased capacity to flexibly alter behavior (de Jongh et al. 2008, 760). Third, individuals with a "low memory span" might benefit from cognition-enhancing drugs, whereas "high span subjects" might become "overdosed."

Even more emphatically, Quednow (2010) contests the pharmacological premises of the enhancement debate as a whole. He contends that the kind of substances presumed by many arguments to make us significantly smarter without serious adverse effects do not exist, and will not exist in the foreseeable future. None of the drugs tested so far has shown replicable and significant effects in healthy human volunteers (de Jongh et al. 2008). Quednow (2010) contends that many ethicists have been led astray by the exaggerated promises of neuroscientists who are either collaborators of the pharmaceutical industry or forced to overstate their own results to get increasingly competitive research funding. "As a matter of fact, we are currently even unable to fully restore disturbed intellectual functioning in psychiatric or neurological diseases and we still do not know how to achieve this goal in the future" (Quednow 2010, 153). Furthermore, he reiterates that because of the complexity of the brain, it is unlikely that we will be able to overcome trade-offs between simultaneous enhancement and impairment by drugs. In addition to the collateral adverse effects on cognitive functions of these drugs, the available substances have many psychiatric and somatic side effects that make them not well suited for use in healthy humans only for the purpose of enhancement (Quednow 2010).

Finally, Trachtman (2005) finds it interesting that those who criticize physicians who dispense treatments that enhance patients' abilities rather than treat a disease assume that there will be a groundswell of people requesting the therapy. Although each advance reported in the press might be greeted by the public with great fanfare, in reality, many treatments are rejected by large segments of the population. "There will always be people in search for the quick fix to treat obesity, prevent dementia, or win an Olympic medal but it is contrary to experience to think that everyone will line up for each new enhancement opportunity" (Trachtman 2005, 32).

National Security and Military Uses of Neurotechnology

The intersection of the cognitive sciences (of which neuroscience may be seen as a subset) and national security raises many ethical, policy, and practical concerns. The Defense Department has invested heavily in the scientific research community, including basic research support to academia. For example, the Defense Advanced Research Projects Agency (DARPA) invested at least $372 million in the cognitive sciences in 2007. Other basic research investments in cognitive sciences across Defense included $13.3 million for the Army, $13.9 million for the Air Force, and $10.4 million for the Navy (Kosal 2008). Although this pales in comparison to the support for biotechnology and even nanotechnology, in light of increased research into human brain function and cognition, some important questions have arisen as to the implications of security-related applications of this research and what impact this research and its applied technologies will have militarily.

Not surprisingly, governments have turned to neuroscience to improve national security and the combat effectiveness of soldiers (Moreno 2004). DARPA launched a $24 million program to develop neuroimaging technology for military applications. A main product of this research has been the Cognitive Technology Threat Warning System, which monitors brain activity for the unconscious detection of threat (O'Connell et al. 2011). Moreover, efforts to create innovative forms of lie detection with fMRI and EEG have received significant funding from DARPA for possible military and security applications (Ashcroft and Gui 2005). DARPA has also been a primary funder of research on tDCS with the hope that it could be used to sharpen soldiers' minds on the battlefield (Fox 2011).

Neuropharmacology
The field of neuropharmacology encompasses research into drugs that influence the chemical balance of the nervous system and the brain. One principal area of research is the development of human enhancement drugs (Huang and Kosal 2008). Ever since militaries began rationing rum to sailors and tobacco to soldiers, they have considered and employed substances that affect troops' brain chemistries. According to Moreno (2008), governments historically have introduced performance-

enhancing drugs to soldiers based on little evidence of their efficacy as compared to the trade-offs, such as artificially extended wakefulness versus impaired judgment and reflexes.

DARPA's program on preventing sleep deprivation, for example, has invested millions of dollars in developing drugs that aim to prevent the harmful effects of sleep deprivation and increase soldiers' ability to function more safely and effectively after the prolonged wakefulness inherent in military operations. Interest in this program kindled further research in performance-enhancing drugs that counter sleep deprivation, such as Ampakine CX717, which enhances attention span and alertness by binding to AMPA-type glutamate receptors in the brain, boosting the activity of the neurotransmitter glutamate, and other improvements from modafinil, a non-amphetamine-based stimulant or "wakefulness-promoting agent" (Huang and Kosal 2008). Other potential neuropharmaceutical applications include improving memory retention and treating posttraumatic stress disorder. In addition to new drugs or new uses for existing drugs, emerging technologies may create new pathways for drug delivery. For instance, nanotechnologies might allow the delivery of drugs across the blood-brain barrier in ways not now possible (Lupia 2011). A larger political and social question is how much enhancement future warfighters can legitimately be expected to accept as part of their preparation for service, especially since the long-term effects of these drugs might not be known until well into the future (Moreno 2008).

A second area of military interest in neuropharmacology concerns the refinement of calmatives that might be used to diffuse hostage situations or affect enemy performance (see box 7.1). While Moreno (2008) agrees that calmative agents in theory represent an attractive option for national security purposes, the time lag between the release of a substance and its effect in targeted individuals makes currently available opiates poor candidates for such uses. For instance, it is not clear why the group holding the Moscow theater hostages did not react when it became apparent that something was going on in the building, but whatever the reason, there is no assurance that future hostage-takers bent on suicide could be controlled in this way. Therefore, governments will need to develop a faster-acting agent before using calmatives becomes an attractive option. Furthermore, the open-air release of an agent is unlikely to be effective because of the dispersal of the aerosol, which severely limits the types of situations in which a calmative could be an effective nonlethal weapon. Even then, one cannot assume that highly effective calmatives could stem violence rather than serving instead to stimulate

Box 7.1
Calmatives for Moscow

A test case for the use of calmatives or other chemicals as a less-than-lethal means in military operations was the 2002 Moscow theater incident, in which the Russian military employed a fentanyl derivative to kill Chechen terrorists who had taken several hundred civilians hostage. Overdoses of the calmative also caused many civilian fatalities. Critics questioned whether the use of fentanyl against terrorists was ethical or violated the Chemical Weapons Convention (CWC). The use of calmative agents in warfare would challenge the CWC, and because they manipulate human consciousness, calmatives could also pose threats to fundamental human rights, including freedom of thought. The questions raised by this incident, however, have not halted research into calmatives. In fact, researchers from University of Göttingen in Germany have found that it is possible to prevent breathing depression from drugs such as fentanyl, opening up further concern about the development of derivative drugs (Huang and Kosal 2008).

aggressive reprisals by adversaries. Therefore, while Moreno (2008) has no hesitancy in supporting the DARPA research, he contends that we should resist the illusion that there can be any neurotechnological fix to the hostilities inherent in combat.

Neuroimaging Applications
The continued development and refinement of neuroimaging technologies is likely to lead to applications that go well beyond those envisioned by current cognitive neuroscience research and clinical medicine. Advanced real-time, continuous readouts of neuroimaging results will become increasingly important for the intelligence community and the Department of Defense to indicate psychological or behavioral states. While predictions about future applications of technology are always speculative, emerging neurotechnologies may help provide insight into intelligence from captured military combatants, enhance training techniques, increase the cognitive abilities and memory of intelligence operatives, screen terrorism suspects at ports of entry, and improve the effectiveness of human-machine interfaces in such applications as remotely piloted vehicles and prosthetics (Lupia 2011). As with drugs, DARPA has assessed the potential of detecting, transmitting, and reconstructing the neural activities of soldiers' brains by including monitoring equipment in a small device, such as headgear. These technologies have

particular defensive applications as well, such as in medical screening and life preservation; in the more effective management of mental stress and workload; and in assessing the physiological status of soldiers. The potential uses of neuroimaging for lie detection are of special interest to Homeland Security and intelligence officials (Wild 2005).

Neuroimaging is also being explored for use in airport security. However, research findings so far suggest that the capabilities of the technology claimed in the media are overstated. DARPA, for instance, highlighted the limits of remote imaging methods such as near-infrared spectroscopy (NIRS) (DARPA 2003). Nevertheless, some in the intelligence and counterintelligence community promote the use of these techniques (see Shamoo 2010), and two Israeli private security firms are offering NIRS airport security services. Although one study reported a capability of EEG methods to identify terrorist intentions, it was criticized for lacking real-life validity since it examined the brain responses of college students, not terrorists (Meixner and Rosenfeld 2011). Similarly, while the impact of using fMRI neuroscience to interrogate terrorist suspects or extract information from actual violators of the law is debatable, some investigators suggest that the future of neuroimaging portends worthwhile applications in counterterrorism or counterinsurgency surveillance strategies and the interrogation of detainees held in custody, either domestically or internationally (Arrigo 2007).

Intelligence

The intelligence community faces the challenging task of analyzing extremely large amounts of information on cognitive neuroscience and neurotechnology, deciding what information has national security implications, and then assigning priorities for decision makers (Lupia 2011). It has been highly challenged to keep pace with rapid scientific advances that can only be understood through close and continuing collaboration with experts from the scientific community, the corporate world, and academia. Not only is the pace of progress swift and interest in research high around the world, but the advances are spreading to new areas of research, including computational biology and distributed human-machine systems with the potential for military and intelligence applications. The intelligence community also confronts massive amounts of pseudoscientific information and journalistic oversimplification related to cognitive neuroscience that it must sort through (Lupia 2011).

As a result of this information overflow, there is considerable interest by intelligence analysts in the use of advanced neuroscience techniques to revolutionize how they handle intelligence imagery, increase the output of imagery, and maximize the accuracy of assessments. The ability to process mass amounts of information in a timely and effective manner can for the military and national security be a matter of life and death (Dunagan 2010). Therefore, the Department of Defense has been working on a series of technologies in augmented cognition that use tools of neurofeedback to curate the information ecology of the user and reduce content when a user's brain is too overwhelmed or overly taxed to process information. Current computer-based target detection capabilities cannot process large volumes of imagery with the speed, flexibility, and precision of the human visual system. Research on visual neuroscience mechanisms indicates that human brains are capable of responding much faster visually than physically. For instance, the goal of DARPA's NIA program is to identify robust brain signals that can be recorded in an operational environment and process these in real time to select images that merit further review, and then to apply these triage methods to static, broad-area and video imagery. The successful development of a neuro-biologically based image triage system should increase the speed and accuracy of image analysis and enable image analysts to process imagery more efficiently (Dunagan 2010).

Research into artificial intelligence spurred early interest in brain-machine interactions. Today, cutting-edge brain-machine applications primarily involve using the brain to control external devices, such as prosthetics and implements that can restore feeling, movement, or sensation to individuals with spinal cord or other paralytic injuries. The most common method of establishing a brain-computer interface, through the external recording of EEG signals, provides limited performance, but more invasive measures, such as implanting microelectrodes into the brain, eventually might yield better results (Scott 2006). This research is of interest to defense and security programs for several reasons. The ability to control a machine directly with a human mind could enable the remote operation of a robot or unmanned vehicle in hostile environments, thereby providing a substantial offensive advantage (Huang 2003). Brain-machine technology also manifests itself in the field of human augmentation or by enhancing human perceptive ability by manipulating the inner workings of the brain. Such a feat could be accomplished through the use of neural prostheses (Huang and Kosal 2008).

Chapter 4 reviewed some of the emerging research on cultural neuro-science that might enlighten our understanding of how culture affects human cognition and even brain functioning and clarify the linkage between culture and brain development. The U.S. military has begun placing greater emphasis on cultural awareness training and education as a critical element in its strategy for engaging in current and future conflicts. Future conflicts will likely involve prolonged interaction with civilian populations in which cultural awareness could be a major factor in outcomes. According to Lupia (2011), research into various aspects of culture can also help the intelligence community understand the current status of cognitive neuroscience research and anticipate the directions it might take in the future. Culturally accurate intelligence and strategic analysis has long been of interest to the intelligence community and national security analysts, but conventional social science models based primarily on Western ideas may be compromised by imperceptible biases. The need is growing to understand hearts and minds at a strategic level because of their potential to exacerbate insurgencies and other problems. Deficiencies in cultural knowledge at the operational level can also adversely affect public opinion. Likewise, ignorance of a culture at the tactical level could endanger both civilians and troops. Advances in inferring cross-cultural intention and meaning are possible with a comparative cultural research agenda. Cross-cultural comparative research can also assess whether the brain function and human behavior are universal, as assumed by Western psychological models (Lupia 2011).

The Defense Department's Minerva Project emphasizes the strategic importance of cross-cultural studies for security. "Where former Defense Secretary Donald Rumsfeld believed that the United States would vanquish its enemies through technological superiority, his replacement Robert Gates has said that cultural expertise in counterinsurgency operations will be crucial in the future wars he anticipates" (Gusterson 2008). Vigorous models of the impact of cognitive science on international security require analysts to consider cutting-edge scientific and technical concepts (Kosal 2008). Directly relevant to the potential strategic role of the cognitive sciences in international security is the question of how one delineates offensive research from defensive research and what metric is used to make this determination. For the neurosciences, experimental psychology, and human performance technology, it is unclear what the parameters will be for differentiating benign research and development from malevolent applications. Biotechnology's dual-use conundrum may hint at the difficulty of "binning" advanced cognitive science research and development into offensive or defensive categories and may challenge traditional international security models (Kosal 2008).

Interrogation

Another conceivable application of neuroscience is to enhance the accuracy and efficiency of interrogation procedures. Recent public outrage over the use of overtly violent techniques such as waterboarding to extract information from persons detained as suspected terrorists has sparked interest in chemical approaches to interrogation. As discussed earlier, the neuropeptide oxytocin has been shown to increase trust in laboratory studies of human and other animal behavior (Kosfeld et al. 2005). Given that this substance is already available in injectable and nasal spray forms, it is possible to envision the potential application of oxytocin as a trust modulator in the military as well as legal spheres (Penney and McGee 2005). It is likely that there will be attempts to develop such substances, similar to the 1950s notion that LSD or other hallucinogens could be "truth serums." Also, some have suggested that brain stimulation might be used to improve the relationship between interrogators and suspected terrorists (Luber et al. 2009, 204).

A committee of the National Research Council predicts that brain scans could eventually aid the acquisition of intelligence from captured unlawful combatants and the screening of terrorism suspects at checkpoints, a development some enthusiasts say would render "enhanced" interrogation techniques unnecessary (see Dossey 2010). Thus, while some contend that the use of neuroimaging might render the "dark art" of interrogation redundant, Marks (2007, 486) argues that, rather than replacing aggressive interrogation, neuroimaging might become a means of selecting detainees for such treatment and feed into narrative constructs that strengthen interrogators' emotional responses to suspected terrorists in a manner likely to result in even harsher mistreatment. Kosal (2008) adds that while less overtly violent than techniques like water boarding, directed chemical or electrical changes in human consciousness under duress are subtle but powerful invasions of one's personality.

There is nothing novel about intelligence operatives seeking the aid of medical science. In the current war on terror, military psychologists and psychiatrists have been designated "behavioral science consultants" and tasked with helping interrogators develop interrogation strategies tailored for individual detainees (Marks 2007, 484). Recently, the intelligence community has recognized that neuroimaging might play a role in assessing the veracity of interrogation responses. However, recent imaging studies have revealed changes in brain function in victims of torture or abuse who suffer from dissociative experiences. Therefore, even if fMRI really was able to detect lies in controlled environments, the application to interrogation subjects who have been captured in stressful circumstances and who may have already been mistreated or abused is, at best, uncertain (Marks 2007, 491). Moreover, there is a risk that intelligence personnel will be seduced by the flamboyant images of fMRI and will overlook the limitations of the technology, the subjectivity of interpretation, and the complexity of brain function outside the realm of playing cards and controlled studies. Therefore, according to Marks (2007, 499), we should be aware of the possibility that the innocent may be tortured or abused by their captors because they have been screened in error with the aid of fMRI.

As noted earlier, considerable funding from DARPA has gone to fMRI research (Thompson 2005). Because the creation of the strong magnetic field for fMRI requires equipment that is cumbersome and nonportable, however, DARPA is now funding work on remote neural monitoring and imaging using wireless near-infrared technology(see box 7.2), techniques that according to Moreno (2006) would provide the added benefit of surreptitious use. Despite technical problems, polygraphy has been used in Iraq and possibly elsewhere in the war on terror. Public documents also make clear the Defense Department's strong interest in these technologies for counterterrorism purposes. Its Polygraph Institute was recently renamed the Defense Academy for Credibility Assessment

Box 7.2
The Preliminary Credibility Assessment Screening System

Research on a portable, handheld lie detection device called the Preliminary Credibility Assessment Screening System, or PCASS, was initiated under the supervision of the DACA in 2008. When attached to the wrist, palm, and a finger, it measures changes in the electrical conductivity of the skin and the interval between heartbeats during an interview designed to detect deception. These signals are immediately analyzed and interpreted by a computer program. If the person is telling the truth, the word "Green" is displayed; if lying, "Red"; or if the result is uncertain, "Yellow." Promoters say PCASS can be used in various scenarios, such as screening groups of local residents who want to work as security guards or interrogators or in service jobs on American bases. Since tests of the device on basic trainees and civilians showed accuracy in only 62 to 79 percent of cases, the Pentagon says it will use it only for triage, to pare down large groups for further scrutiny, and will prohibit its use on Americans. However, some of the criticism of this policy has been "brutal" (Dossey 2010, 277–278).

(DACA), and the program has been extended to encompass "credibility assessment" using other "technical devices," and the director of the Counterintelligence Field Agency is to manage the program and "coordinate, integrate and synchronize [it] with other counterintelligence missions and functions" (Marks 2007, 490).

Moreno (2006) suggests that the likelihood of more potent neurotechnologies, such as brain stimulation–based deception detection, raises a new set of ethical concerns over the role of medical professionals. In practice, physicians must evaluate and monitor subjects receiving brain stimulation. Recently the role of medical professionals in Guantánamo Bay interrogations has attracted considerable attention both within the profession and in the popular press (Bloche and Marks 2005). The ensuing debate has clarified the role of doctors in interrogation as defined by the Geneva Conventions and other instruments of international law, and there is a consensus that physicians should not use drugs or other biological means to subdue enemy combatants or extract information from detainees, or be party to interrogation practices contrary to human rights law or the laws of war (Luber et al. 2009). Whether or not these guidelines speak against the involvement of physicians in the use of brain stimulation technology is open to question and likely depends on context. While the U.S. military has long insisted that the medical profession has a social role that encompasses the needs of the criminal and civil justice system, critics counter by noting that police, intelligence, and military interrogators routinely use deceptive techniques and other coercive measures to obtain information, even though such tactics are contrary to a physician's ethical and professional duties (Rubenstein et al. 2005). Thus, if these technologies are developed for use in the military against enemy combatants or other non-U.S. citizens, the role of physicians would likely come under close scrutiny.

Shamoo (2010) argues that since 9/11, the tension between public good for the United States and its citizens versus the individual rights of suspected foreign personnel and, in some cases, U.S. citizens has increased pressure to resort to innovative methods to obtain information from suspected individuals determined to harm the national security. In addition to the potential harms caused those devices that send waves to the brain, particularly when there are multiple exposures, there are issues over associated psychological stress during testing, as well as issues connected to protecting privacy and the confidentiality of the information obtained from subjects during the conduct of research (Fenton, Meynell, and Baylis 2009). According to Shamoo (2010), the use of innovative methods "raises the specter of how we are perceived by the local culture when conducting such experiments. Was the test conducted sensitive to local cultural needs and sensitivities? Did investigators consult the local community? Answers to these questions are neither trivial to resolve nor without impact on our function abroad" (22).

Conclusions

Sensational and dramatic headlines sell stories and blog hits, and anything about intervention in the brain provides a fertile ground for such stories. As a result, however, reports of neuroscience findings in the media and Internet are frequently oversimplified and misleading. This disconnect between what the research shows and what the media and the blogs report will continue to be a problem. Likewise, there are high economic stakes in brain interventions and strong pressures to move innovations such as neuroenhancement techniques to the marketplace as quickly as possible. This, too, raises important policy issues in the realm of regulation of commercial activity, from consumer protection and truth in advertising to claims of safety and efficacy. Therefore, as the fruits of neuroscience unfold, political debate will intensify around the issues raised in this chapter, particularly with respect to military and national security applications that pit various notions of the collective good against an individual's right to privacy. After discussing the impact of neuroscience on our conceptions of death in chapter 8, we will return to the politics of brain intervention in the concluding chapter.

8

Neuroscience and the Definition of Death

Many death-related policy issues, from treatment abatement to physician-assisted suicide, continue to elicit considerable public and professional debate. This is not surprising because they are among the most intensely emotional and ethically fraught issues. "The seemingly simple concept of death is subject to law and clouded by cultural, religious, and scientific beliefs" (Dinsmore and Garner 2009, 216). One of these social policy issues that will not subside is how we define the death of a human. There are two critical dimensions to this question. The first is the conceptual interpretation of what death means in the context of medical technology, since the traditional understanding of death as the irreversible cessation of cardiopulmonary functions is obscured by having the technological means of prolonging those functions indefinitely (Antommaria 2010). The second dimension concerns the appropriate clinical tests to be used to determine that a patient is in fact dead, especially when the patient's life has been prolonged by technological means. It is argued here that because the technologies in both of these areas are constantly advancing (in the first instance life-sustaining technologies and in the second diagnostic technologies that measure the presence or absence of specific types of brain activity in specific regions), the definition of death will continue to be a contentious issue and one that cannot be resolved on technical grounds alone (Verheijde, Redy, and McGregor 2009; Nair-Collins 2010; Chiong 2005).

All life must end, and all brains die, but the issues surrounding death continue to stir strong emotions and viewpoints. This chapter analyzes the policy issues attending the definition of death within the milieu of our expanding knowledge of the brain and our increased capacity to measure brain activity and function. Ever more precise brain imaging techniques that can measure activity in specific regions of the brain, combined with demographic trends and heightened health budget pressures, are bound to accentuate calls for redefining death in terms of partial- or higher-brain criteria. In addition to questioning our notions of consciousness and of what human life entails, a shift toward higher-brain definitions of death have critical public policy implications. Adding patients in a persistent vegetative state or with end-stage Alzheimer's disease to the ranks of the dead raises many difficult questions, for instance, must the Medicare or private insurers fund continuing care for a legally dead but still breathing patient whose family cannot let go? The implications for the dispositionl of breathing patients who lack brain functions deemed essential to life are extensive. The public is likely to find it difficult to accept burial or cremation of breathing human forms or the use of lethal injections to ready the "dead person" for burial.

Brain Death

In one sense the brain begins to die early in life as large numbers of redundant neurons are eliminated. Over a lifetime, the brain is at risk for losing some of its neurons, although widespread neuron loss is not necessarily a normal process of aging (Society for Neuroscience 2006, 31). However, the normal aging process does include a gradual loss of sensory capacities. For instance, visual acuity declines linearly between ages twenty and fifty and exponentially after age sixty. Depth perception declines at an accelerating rate after age forty-five. By age eighty, speech comprehension may be reduced by more than 25 percent owing to neuronal loss in the superior temporal gyrus of the auditory cortex (Ivy 1996). The aging process, then, is naturally one of decline in the brain, with evidence suggesting a substantial decrease in neuron density as well as in absolute neuron numbers in many of the brain regions, particularly the hippocampus, the subcortical brain regions, and the cerebellum (Selkoe 1992). As a result, "Total brain mass shrinks by approximately 5 to 10% per decade in the normal aged individual, leading to losses of 5% by age 70, 10% by age 80, and 20% by age 90. … The atrophy is most marked over the frontal lobes, although parietal and temporal lobes suffer considerable losses as well" (Ivy 1996, 43).

In addition to the gradual process of brain cell death that accompanies normal aging, an array of neurodegenerative diseases and neurotoxin exposure can cause further demise among certain cell types and regions of the brain, as can injury, cancer, stroke, and other trauma. High blood pressure, for example, has been found to shrink the size of the brain in elderly persons. Also, there is evidence that at least some of the changes associated with aging are related to a decline in hormonal activity (Lamberts, Van den Beld, and Van der Lay 1997), with special emphasis recently placed on estrogen loss. Also, as noted in chapter 4, alcohol and drug abuse can intensify brain cell death at all ages. The concept of brain death discussed here, of course, does not refer to this gradual process of cell death but rather to the cessation of brain activity and function as measured by specific tests, including EEG diagnostics.

The first major step toward defining brain death occurred in 1968, when mounting concern on the part of medical practitioners over how to treat respirator-supported patients led to creation of the "Harvard criteria" for brain death (Ad Hoc Committee 1968). The stated purpose of the Committee was "to define irreversible coma as a new criterion for death" (337). To this end, criteria developed by a Harvard Medical School committee focused on (1) unreceptivity and unresponsiveness, (2) lack of spontaneous movements or breathing, and (3) lack of reflexes. Moreover, a flat EEG showing no discernible electrical activity in the cerebral cortex was recommended as a confirmatory test, when available. Before life-support systems could be terminated, all tests were to be repeated at least twenty-four hours later without demonstrable change. The publication of these criteria led to the mobilization of considerable support to legislate policy standards of brain death and by this means presumably to eliminate the uncertainties faced by hospitals and physicians. According to Truog (1997), however, the Committee gave no explanation as to why brain death should be considered a sufficient criterion for a death determination except that using brain death resolved two practical matters: the care of patients with a diagnosis of brain death and no hope of recovery was seen to be a misuse of valuable ICU resources, and these patients could be a valuable source of organs while their hearts were still beating.

This latter point suggests that another major force for action during this period was the emergence of organ transplantation techniques and the growing need for usable organs. For a major organ transplant procedure to be successful, a viable, intact organ is needed. The suitability of organs, especially the heart, lungs, and liver, for transplantation diminishes rapidly once the donor's respiration and circulation stop. Therefore, the most desirable donors are otherwise healthy persons who have died following traumatic head injuries and whose breathing and blood flow are artificially maintained until organ removal can be completed. Although advocates of the brain death criterion downplay the extent to which the demand for organs influenced this movement, transplant surgery gave the effort to define brain death a new urgency (see Singer 1995, 22ff., for the importance of this factor for the Committee). In agreement with Truog above, Shewmon (2009) contends that the only rationale given by the Committee as to why the irreversible cessation of all brain functions should be equated with death was legal utility: it would free up beds in intensive care units and facilitate organ transplantation. Not surprisingly, in the consideration of alternatives to the use of the brain death diagnosis, facilitating organ procurement and transplantation is at the core of most arguments because it depends heavily on brain death (Zielinski 2011). Moreover, both federal and state laws require physicians to contact an organ procurement organization following a determination of brain death (Wijdicks et al. 2010).

This reluctance to link the brain death criterion directly to transplantation is understandable, however, because the connection could be interpreted as implying that the revised definition of death was implemented solely to facilitate use of an individual's organs. Certainly, organ transplant facilities sought the recognition of brain death and have benefited from the legal clarity provided by statutes that define it. Until then, the lack of consensus regarding when a patient could be declared dead limited organ procurement and "led to a public controversy regarding the ethics of organ procurement," according to Siminoff, Arnold, and Sear (1996, 52). However, if the only rationale for this new definition of death was to facilitate successful organ transplants, support for brain death initially would not have been as persuasive. Thus, "death with dignity" and "patient empowerment" were often cited as reasons behind the change in policy.

In response to these concerns and pressures from the medical community on the inadequacy of existing legal definitions of death, in the 1970s many states passed laws that attempted to incorporate the neurological criteria proposed by the Harvard Ad Hoc Committee. The lack of consistent policy, however, produced confusion and the potential for abuse, especially when patients were transported across state lines for treatment. In 1970 the Kansas state legislature became the first to recognize brain-based criteria for the determination of death. Within several years, four states had passed laws patterned on the Kansas model. The Capron-Kass proposal of 1972 offered the states a more succinct substitute that eliminated some of the Kansas model's ambiguity (Capron and Kass 1972). At least seven states adopted the Capron-Kass model with minor modifications, while three others did so with more substantial changes. Two other model statutes (American Bar Association 1978;

National Conference of Commissioners on Uniform State Laws 1980) were enacted by five and two states, respectively. The American Medical Association's 1979 proposal, which included extensive provisions to limit liability for persons taking actions under the proposal, was never adopted in any state. About ten states enacted nonstandard statutes that included parts of one or more of these models, while about twenty states did not implement clear statutory determinations of death.

This proliferation of similar yet variant models and statutes led the President's Commission for the Study of Ethical Problems in Medicine and Biomedical and Behavorial Research in 1981 to conclude that "in light of ever increasing powers of biomedical science and practice, a statute is needed to provide a clear and socially-accepted basis for making definitions of death" (1981, i) and propose a Uniform Definition of Death Act (UDDA), which still serves as the accepted standard definition of brain death. The act provides for:

(Determination of Death.) An individual who has sustained either 1) irreversible cessation of circulatory and respiratory functions, or 2) irreversible cessation of all functions of the entire brain, including the brain stem, is dead. A determination of death must be made with accepted medical standards.

Before its presentation in the final report, this uniform law was approved by the ABA, the AMA, and the Uniform Law Commissioners as a substitute for their original proposals.

The President's Commission recommended that uniform state statutes address general physiological standards rather than specific medical criteria or tests, since the latter continue to change with advances in biomedical knowledge and refined techniques. In retrospect, this focus on physiological standards might be used as justification for subsequent moves to cortical brain death criteria. It concluded that "death is a unitary phenomenon which can be accurately demonstrated either on traditional grounds of irreversible cessation of heart and lung functions or on the basis of irreversible loss of all functions of the entire brain" (President's Commission 1981, 1). Since that time, most U.S. state legislatures have adopted the UDDA, although several states have added amendments regarding physician qualifications, confirmation by a second physician, or religious exemptions.

Since the UDDA did not define "accepted medical standards," in 1995 the American Academy of Neurology (AAN) published a practice parameter to delineate the medical standards for the determination of brain death (updated with a checklist in Wijdicks et al. 2010). The parameter emphasized three clinical findings necessary to confirm irreversible cessation of all functions of the entire brain, including the brain stem: coma (with a known cause), absence of brain stem reflexes, and apnea. Despite publication of the practice parameter, considerable variation in practice remains, even among leading U.S. hospitals (Greer et al. 2008). Nevertheless, although some patients diagnosed as brain dead may experience complex, non-brain-mediated spontaneous movements that can falsely suggest retained brain function, and although ventilator autocycling may falsely suggest patient-initiated breathing, there are no published reports of recovery of neurological function after a diagnosis of brain death has been made using the criteria reviewed in the AAN practice parameter (Wijdicks et al. 2010).

The clinical criteria of brain death generally include no response to any stimulus; no movement, withdrawal, grimace, or blinking; no breathing efforts when the patient is taken off the ventilator (the apnea test); pupils dilated and not responsive to light; and no gag reflex, corneal reflex (blinking when the surface of the eye is touched), or other specific reflexes. Death is now linked to direct cessation of brain function rather than to indirect cessation after shutdown of the heart and lungs. Because the brain cannot regenerate neural cells, once the entire brain including the brain stem has been seriously damaged, spontaneous respiration can never return, even if breathing is sustained by respirators or ventilators. As discussed in more detail later, the key point is that while machines can maintain certain organic processes in the body, they cannot restore consciousness or other higher brain functioning.

Many technologies are involved in the diagnosis of brain death. CT of the brain is used to ascertain any abnormalities such as hemorrhage, massive stroke, brain injury, or severe brain swelling, while EEGs record electrical brain activity. When brain death is present, the EEG will show no activity. Finally, when a person is brain dead, cerebral radionuclide injection will show no uptake of the radioactive material in the brain. These tests may be performed to confirm brain death and may be used along with the examination-based clinical criteria to show irreversible loss of brain and brain stem function, although not all tests need to be used to declare brain death under prevailing protocols.

In summary, until recent decades, death occurred at the moment of permanent cessation of respiration and circulation. Once the heart and lungs ceased functioning they could not be restored. And once cardiorespiratory function ceased, brain function also ended. Advances in life-support techniques through which the heart can be kept beating with medication and respiration (breathing) can be artificially performed with a ventilator, however, made the conventional notion of death incongruous. In response to these advanced medical techniques and the growing need for viable organs to transplant, the concept of brain death developed. Of the three concepts of brain death—the whole-brain formulation, the brain stem formulation, and the higher-brain formulation—the first is accepted and practiced most widely (Bernat 2005). Whole-brain death occurs when there is no function of the entire brain, including the brain stem. When the brain, including the brain stem, has ceased to function, the individual is dead by medical and legal standards. If a brain-dead individual is not an organ donor, ventilatory and drug support is discontinued and cardiac death is allowed to occur.

The Brain Death Controversy

What does it mean to be human? The present situation, in which all bodily functions need not cease when the heart stops pumping spontaneously, has led to a distinction between human life as a strictly biological existence and human life as an integrated set of social, intellectual, and communicative dimensions. In recent decades, a reasonably strong consensus has developed that recognizes the possibility of social or cognitive death even though the human organism is kept alive biologically by artificial means. However, some observers remain opposed to the recognition of brain death itself, while others object specifically to whole-brain death. Still others who approve of the brain death definition are uncomfortable with the dilemmas it has created. Moreover, any brain-based determination of death in the context of the physical image of a normally functioning body is bound to create emotional and cognitive conflicts for many health care professionals and family members (Long, Sque, and Addington-Hall 2008).

According to Shewmon, however, the much-touted international "consensus" on a neurological standard of death is only skin deep. He argues that the widespread superficial agreement that brain death is death conceals a widespread disagreement over the reason why and even much tacit belief to the contrary. "Brain death as death began as a utilitarian legislative decree and has remained a conclusion in search of a justification ever since: a conclusion clung to at all costs for the sake of the transplantation enterprise that quickly came to depend on it" (Shewmon 2009, 19). Miller and Truog (2008) also contend that the proposition that brain death constitutes the death of the human being is based on dubious and incoherent claims. Lederer (2008) suggests that the deleterious effects on professional integrity from having to maintain the fiction of brain death in order to get patients and families to donate organs are important and argues that "physicians would benefit from a more coherent and defensible ethical account of vital organ donation" (3).

Despite all the committee and commission reports over the decades, there continues to be considerable variability in the determination of brain death and in the requirements for the declaration of brain death throughout the United States and internationally (Busl and Greer 2009). Shewmon (2009) adds that the rationale for brain death also made little headway in the public at large over this time. Reporters still refer to brain-dead patients as being "kept alive" by machines or as "dying" when the ventilator is turned off, and much of the public and a surprising proportion of the medical profession still consider "brain-dead" patients "as good as dead" or "better off dead," but not yet really dead. "Moreover, many who do regard them as dead do so on the grounds of loss of personhood (by virtue of permanent unconsciousness) from a biologically still living human organism, grounds that would also categorize patients in a permanent vegetative state as dead" (Shewmon 2009, 19).

Veatch (2009) finds it interesting that forty years after the launching of the Hastings Center's project on death and dying, "what we thought would be a short-term controversy in which the brain people would win out over the old-fashioned, romantic, heart-based people has turned into a prolonged debate" (16). He argues that we are further from agreement than when we started, and that the debate is no longer between just two or three options. Even within the heart, whole-brain, and higher-brain camps, there are countless nuanced positions. One of the reasons for the sustained debate over brain death is that clinical data made it increasingly clear that patients with total brain failure were not physiologically identical to non-heart-beating corpses and did not necessarily disintegrate if lacking all technological support. Moreover, the rare longer-surviving ones exhibit holistic properties such as homeostasis, teleological repair, and a general ability to survive outside a hospital setting with relatively little support. According to Shewmon (20090, such properties are difficult to reconcile with the assumption that these bodies are "nothing but bags of partially interacting subsystems" (19). Chiong (2005) agrees that "the empirical claim that the loss of whole-brain function is necessary and sufficient for the cessation of integrated functioning has since been cast into doubt, prompting a revival of interest in the cardiopulmonary criterion" (21).

According to Zielinski (2011), there remain two main positions on the question of the reality of the brain death diagnosis. The first position is that the brain is the master supreme regulating organ of the body without which the body is dead. Thus, the loss of integration between the brain and the body constitutes clinical death (McMahan 2006). Therefore, even if all other organs in the body remain intact and functioning, the person is considered to be dead because the mediation of the brain no longer integrates the system as a whole. This way of thinking gives primacy to physiological functioning and essentially reflects the inseparability of the mind, housed in or dependent on the brain, from the body. The second position is held mainly by critics of the brain death diagnosis who claim the body does not lose its integrative abilities when the brain is no longer functional and contend that the brain is indeed separable from the body (Shewmon 2009; Miller and Truog 2008; Lederer 2008).

There are also contrasting views as to whether the mind and body are inseparable and whether brain death equates with clinical death. Some argue a person's consciousness equals life, and clearly, a person cannot maintain consciousness without at least part of a functioning brain (Eberl 2005). On the other side are those who claim that individuals who are brain dead are not clinically dead. Supporters of this perspective may view the mind-body relationship as a separable connection, where the body can still maintain "life" without the mind (Zielinski 2011). For example, Shewmon (2009) argues that the body does not lose its integrative abilities when the brain is no longer functioning, citing research in which, without the function of the brain, children and adults have been able to maintain homeostasis of physiological events, including the absorption of nutrients, elimination of wastes, maintenance of body temperature, wound healing, and even the growth of a fetus. If life is understood from a strictly biological perspective, McMahan (2006) suggests that Shewmon's arguments and evidence force the advocates of brain death to admit defeat.

Redefining Brain Death

Therefore, while the whole-brain definition of death (or brain stem death in Britain, see Dinsmore and Garner 2009) has become the accepted standard of practice in most Western nations, it continues to be enshrouded in controversy. Veatch (1993), for instance, argues that the whole-brain definition of death has become so qualified it can hardly be referring to the death of the whole brain. For example, isolated brain cells continue to live and emit small electrical potentials measurable by EEG even when supercellular brain function is irreversibly destroyed.

Whole-brain death also ignores spinal cord reflexes but requires cessation of lower brain stem activity, thus contradicting the definition that "all functions of the entire brain" be terminated.

Other observers are even more critical of the current reliance on whole-brain death. Truog (2005) contends that the concept is fundamentally flawed, plagued by internal inconsistencies, and confused in theory and in practice: "Brain death and the dead donor rule are widely viewed as the central pillars of our approach to the ethics of organ donation. They are not the views of the public, however" (3). Similarly, Joffe (2010) argues that a "clear and defensible conceptual rationale" to support brain death being death has not been articulated. "This is important because thousands of patients are pronounced with BD every year, and this has been the main way to obtain organs for transplantation without violating the so-called dead donor rule" (47). Nair-Collins (2010) adds that any policy that relies on the whole-brain concept of death "suffers from serious moral failings and so ought to be abandoned" (667).

Other critics contend the specific tests for the determination of brain death are inaccurate and do not always meet the required criteria. Baron and colleagues (2006), for instance, suggest that many controversies arise, not over the existence of brain death per se but rather because of inconsistent application or implementation of the diagnostic criteria. They identify issues to be addressed, including the lack of physician expertise in making the diagnosis, inconsistent clinical criteria, conflicting use of a second brain death examination and a set time interval between the two tests, responses to confounding factors, and the lack of age-specific pediatric guidelines. Furthermore, whole-brain death assumes that the brain is the integrating organ of the body whose functions cannot be replaced, even though ICUs increasingly have become surrogate brain stems replacing respiratory, hormonal, and other regulatory functions. For Truog (1997), "This gradual development of technical expertise has unwittingly undermined one of the central ethical justifications for the whole-brain criterion of death" (31).

The major objection to brain death as now defined, however, comes from those who feel it is not inclusive enough. Whether one believes in life after death or no life after death, there is agreement that we die when we cease to be here, even though our dead bodies remain at least for a while. Furthermore, there is general acceptance that it is necessary and sufficient for one to cease to be here when all possibility of consciousness and mental activity is gone. Until recently, this was congruent with heart and lung stoppage, but, as noted earlier, medical science has separated the two events by providing surrogate blood pumping mechanisms for the brain.

There is growing support now for the view that death comprises the irreversible cessation of the integrated functioning of the various subsystems of the organism as a whole *and*, in humans, the irreversible loss of the capacity for consciousness and mental activity (McMahan 1998). This is clear because the brain functions both to regulate and integrate the systemic functioning of the organism and to generate consciousness. When brain activity ends, both capacities cease without possibility of restoration. Although respiration and heartbeat can be sustained artificially for some time beyond conventional brain death, the organs routinely begin to deteriorate after brain death. According to Singer (1995), brain death was accepted not because of the brain's integrative function, which can be replaced, but because of its association with consciousness or personality. That is why, when higher brain activity ceases without any possibility of restoring consciousness, even though one continues to survive as an organism on machine support, most persons would normally see their life as over.

One significant problem with a whole-brain death definition is that the two essential capacities of the brain are largely localized in different regions of the brain. Generally, consciousness and mental activity are associated with the cerebral cortex while the integration and coordination of somatic functions are associated with the brain stem. As a result, the brain stem can survive in a functioning state after all neocortical capacity ends, therefore continuing to integrate somatic functions, including breathing and heartbeat, that allow the organism to live without life-support systems. According to the conventional whole-brain death definition, neither of the two capacities alone is sufficient for end-of-life determination—only the loss of both. If, however, we find it compelling that the capacity for consciousness is necessary for our continued existence, and its loss congruent with ceasing to exist, then the whole-brain death definition must be amended. If death of the neocortex is both necessary and sufficient for the irreversible loss of consciousness, then, according to McMahan (1998), cortical death should be our criterion for death. "A human organism can remain alive even in the absence of life-support following cortical death. It is, rather, we who go out of existence" (257). What ceases to exist in cerebral death is the mind, which essentially is what in sum and part "we" as an entity are.

One problem in moving toward a cerebral death definition is the current ambivalence concerning brain death itself. As noted earlier, many persons, particularly family members, still find it difficult to equate brain death with death. Adopting a higher-brain definition would complicate this by declaring dead persons with intact brain stems who could breathe on their own. Moreover, even if there is agreement that irreversible loss of consciousness represents death, how can we be certain that the capacity for consciousness is lost? And, if so, how can we be certain its loss can never be reversed at some point in the future? It is on this point that new technologies and knowledge of the brain become important factors in the debate over defining death.

Brain Imaging Techniques Moves to Cerebral Definitions of Death

One of the major forces dictating a reevaluation of the concept of brain death is the rapid advance in the ability to measure activity in very specific regions of the brain. New techniques such as those described in chapter 2 provide vivid images of brains and promise to greatly enhance our understanding of the relationship between the anatomy of the brain and psychological functioning, including consciousness. For Singer (1995), this more precise medical knowledge has "pushed aside" a powerful reason for continuing to treat patients in a persistent vegetative state (PVS) as alive, because if we can demonstrate that there is no blood flow to the cortex, we know that the capacity for consciousness is irreversibly lost. "If blood is not flowing to the cortex, then—even though the brain stem might still be functioning and so the patient would not be brain dead—the patient would be 'cortically dead' and would never regain consciousness" (Singer 1995, 43). This is reiterated by Cranford, who concludes that "permanently unconscious patients have characteristics of both the living and the dead. It would be tempting to call them dead and then retrospectively apply the principles of death, as society has done with brain death" (cited in Shewmon 2009, 19).

The capacity of imaging techniques and ever more sophisticated EEGs to measure levels of activity, or lack thereof, in specific brain regions, in the whole brain or in the normal asymmetry of activity between the two sides of the brain, therefore, gives us for the first time the technical means to measure the end of life in precise brain regions. Further advances in software are likely to match hardware improvements and provide even more detailed imaging of the brain and, thus, more precise measures of the termination of activity in particular regions (Frampas et al. 2009).

As noted earlier, personality, consciousness, memory and reasoning require a functioning cortex. In cases where there is no evidence of blood flow or electrical or biochemical activity in the cortex, some observers feel that a definition of death centered on the loss of these characteristics is warranted (see McMahan 1998). Cerebral death assumes that without consciousness, human life no longer exists. Higher-brain death extends the definition to include patients in a PVS and presumably anencephalic infants. Moreover, some observers would also include patients in end stages of Alzheimer's disease in the category of cortical brain death, using newer imaging techniques such as fMRI and MEG to make the determination. "As with irreversible coma, the parallel with brain death is plain. And a parallel policy of allowing such patients to perish would seem to be in order" (Churchland 1995, 307). If it is these higher brain functions that define us as humans, then higher-brain death, however narrowly defined, could be a more appropriate standard than whole-brain death. Savulescu (2009) warns, however, that since fMRI plays an "increasingly important role in the assessment of patients in a vegetative state, caution is needed in the interpretation of neuroimaging findings" (508).

An extension of brain death to PVS patients would be very problematic. For instance, there continues to be widespread technical and conceptual confusion about PVS. The variety of terms used, such as "comatose," "irreversibly comatose," "patients with locked-in syndrome," "permanent coma," and so forth, complicates the issue. Patients with locked-in syndrome, for instance, have such severe paralysis caused by brain stem damage that they appear unconscious even though they have a fairly normal level and content of consciousness. Patients with permanent coma have eyes-closed unarousability, which might also be caused by extensive damage to the brain stem. However, as noted by Joffe (2009), a person in PVS is not considered dead in any society, and burial, cremation, or organ donation without an anesthetic is unthinkable, thus implying that "loss of cerebral function and consciousness is *not what we mean* by the word death" (53).

Patients with PVS differ from these other patients in that they have a relatively intact brain stem but massive neurological destruction of the cerebral hemispheres, with the complete loss of cerebral cortical functions having occurred because of lack of blood flow (ischemia) or oxygen (hypoxia) to the brain over a period of four to six minutes. As a result, the patient is left amented, in a transient coma for a few days or weeks, and requiring short-term mechanical ventilation. The ensuing prolonged eyes-open unconsciousness can last for years. They experience no pain or suffering and are completely unaware of themselves or their environment. Patients with PVS, however, do have a normal gag reflex and cough reaction and periods of wakefulness and sleep because those are brain stem functions. These latter characteristics, as well as the eyes-open state, would make it very difficult to institute a policy declaring PVS patients brain dead. Although some persons object on grounds of potential reversal of loss of mental activity, McMahan (1998) concludes that "our long experience with this condition, together with advances in techniques for monitoring blood flow to different parts of the brain, make it possible in most cases to determine with virtual certainty when recovery is impossible" (258). Savulescu (2009) cites evidence from fMRI studies that raise the possibility that PVS patients retain some degree of consciousness, but argues that such a finding in very severely brain-damaged patients may provide even more reason to let them die.

Similarly, extending brain death to include patients with end-stage Alzheimer's disease and other dementia would be extremely controversial even if it were a logical move based on an irreversible loss of consciousness and mental capacity. People with dementia have a progressive loss of cortical functions extending over a period of years. In the latter stages of dementia, there is no consciousness of oneself or the environment, though brain stem activity might be unaffected. Redefining death to include these patients, however, would not resolve the problem of doctor-assisted euthanasia. Although Pabst-Battin (1994, 160) views the permitting of active euthanasia of advanced Alzheimer's patients in "conjunction with an antecedently executed living will or personal directive requesting it" as the best possible compromise, she sees problems arising between public perception and philosophical reflection. Cerebral brain death, of course, would undermine the voluntariness that Pabst-Battin and others who support euthanasia demand unless it allowed for antecedent personal choice as to one's acceptance of the official definition.

Although there might be strong economic pressures to terminate treatment for patients with advanced-stage Alzheimer's disease, public objections to any efforts to alter our definition of death to include such persons would be intense. This opposition will come in spite of McMahan's (1998) admonition that even if there is a possibility that some "dim, flickering rudimentary mode of consciousness" might survive cerebral death, this is not a viable objection, because it would be hard to show that whatever "shadowy, semi-conscious mental activity" we imagine might remain would contribute to the good life of that person (259).

Moreover, reliance on cortical death has been criticized because of the difficulty of measuring with precision the loss of higher brain functions and because it focuses on the end of personhood rather than on the death of the organism (Truog 2005). It has been argued that despite advances, diagnostic tests for PVS are unreliable, as evidenced by anecdotal cases of the recovery of some patients who were wrongly diagnosed. Some critics have also charged that if we accept the notion of cerebral brain death, we are vulnerable to sliding down the slippery slope into finding as dead an ever-widening range of marginally functional humans, such as persons with less advanced dementia or those with very low IQs. As a result, Truog (1997) suggests that higher-brain death is bound to remain the domain of philosophers rather than policy makers because of the implications of treating breathing patients as if they were dead.

This objection is congruent with the President's Commission's rejection of any form of partial-brain death on the grounds that to declare dead a person who is spontaneously breathing yet has no higher brain functions would too radically change our definition of death. The public is unlikely to accept the burial or cremation of still breathing humans or the use of lethal injections to terminate cardiorespiratory functions of the brain-dead person prior to burial. Despite these concerns, many observers feel we are now at the stage technologically to move from whole-brain death to cerebral brain death. Veatch (1993) recommends changes in the wording to replace "all functions of the entire brain" with references either to higher brain functions, cerebral functions, or his preferred "irreversible cessation of the capacity for consciousness" (23). He would, however, incorporate a conscience clause that would allow a person to choose through an advance directive his or her preferred option based on personal, religious, or philosophical beliefs. Likewise, Emanuel (1995) proposes a "bounded-zone" definition of death according to which individuals would be allowed to choose higher levels, including PVS, but traditional cardiorespiratory death would be the lower bound for all persons.

Therefore, while cortical brain death has some supporters, Truog rejects any movement in that direction and argues instead for a return to the cardiorespiratory standard. "Instead of searching for better ways of defending brain death, why not consider more seriously whether we need it at all?" (Truog 2005, 3). He contends that while whole-brain death served a useful transition function, brain death is no longer needed to justify withdrawal of life support, has little significance regarding resource allocation, and is not as crucial for organ transplantation as it was in the 1970s. Advance directives and sympathetic court decisions now allow withdrawal without reliance on brain death, although there remain inconsistencies in court decisions in this area.

According to Truog (1997), organ transplantation could be decoupled from brain death by shifting attention from brain death to the principles of consent (or prior consent) of donor and of doing no harm to the PVS patients, although he agrees that it is unlikely that the courts or public would accept this approach because the "process of organ procurement would have to be legitimized as a form of justified killing, rather than just as a dissection of a corpse" (34). McMahan (2006) contends that organ procurement should not rely on the patient being declared brain dead but rather on a patient's prior consent to remove his or her organs at a particular stage. Essentially, the patient would give prior approval for the body to be "killed" to donate organs because "the treatment of a living but unoccupied human organism is governed morally by principles similar to those that govern the treatment of a corpse" (McMahan 2006, 48). For McMahan, it is permissible to kill an "unoccupied organism" if the person who once occupied that organism had given prior consent to the removal of his or her organs, although organ procurement should be permitted only when doing so would not harm the donor: "Individuals who could not be harmed by the procedure would include those who are permanently and irreversibly unconscious (patients in a persistent vegetative state or newborns with anencephaly), and those who are imminently and irreversibly dying" (34).

To some extent, a shift toward this position began in the early 1990s after the supply of organs from relatively healthy but brain-dead persons with suitable organs began to decline, in part because of improved preventive measures such as the legal mandate for motorcycle helmets and improved medical care and rehabilitation of the injured brain (Verheijde 2009, 410). One strategy for making up the difference has been the introduction of organ donation after cardiac death (DCD) protocols, which provide for vital organ donation from dying, hospitalized patients after withdrawal of life support who are declared dead on the basis of cardiac death (Marquis 2010). Typically, under DCD protocols the prospective donor, although not brain dead, has suffered extensive neurological damage and is on life support. Following permission of the person's family, life support is withdrawn and cardiac arrest results. If the heart does not resume beating on its own within two to five minutes, death is declared. With consent for organ donation from the donor or the donor's family in hand, immediately after death is declared the donor's organs are removed for transplantation.

Miller and Truog (2008) contend that because the cessation of circulatory function is not irreversible and the heart could almost certainly be restarted by medical intervention, DCD protocols do not satisfy the standard cardiopulmonary criteria for death. While Bernat (2010) agrees that these patients are not unequivocally dead until cessation of circulation is irreversible, he suggests that from a medical practice standpoint, physicians routinely and rightly declare patients dead at the moment cessation of circulation is permanent. Because this practice is encompassed in a UDDA provision requiring death determination to be "in accordance with accepted medical standards," declaring death in DCD donors does not violate the dead donor rule (Bernat 2010, 3). Miller and Truog (2008, 38) argue that it is time to seriously consider abandoning the dead donor rule and return to the traditional cardiopulmonary standard of death, which would eliminate the need to stop treatment, declare death after a short interval, and then subject the patient (currently considered dead) to interventions designed to preserve vital organs prior to extraction.

Nair-Collins (2010) suggests a different approach and contends we should admit that it is a biological fact that brain-dead individuals are biologically alive, and that we need a forthright public debate as to when it is morally acceptable to remove vital organs and when it is acceptable to remove brain-dead individuals from the ventilator, thus allowing them to die. Although he agrees that brain-dead organ removal (with consent) is morally acceptable, to legally allow this, homicide laws need to be revised to allow exceptions for the case of transplant surgeons, since brain-dead organ donors are in fact killed by the process of organ removal (Nair-Collins 2010, 680). He warns, however, that given the political climate of many countries, it is likely that many people will be dismayed to find out that brain-dead organ donors are actually being killed for the purpose of organ removal, and that such revisions to homicide laws to make brain-dead organ removal legal would be rejected.

Cultural and Religious Issues in the Redefinition of Death

It must be emphasized that however we define death, whether whole brain, cerebral, or cardiorespiratory, there is no benefit for the dead person but rather for the family, society, and potential recipients of organs. Moreover, if we accept the assumption that consciousness is necessary for the human experience, then we cannot harm the person who lacks it by defining that person as dead. Both harm and benefits accrue to the yet living, to the values of society, and to culture. As such, whatever definition is selected, it must be a matter of public policy, not a medical or technological one, even though the changes in medical science introduced here might influence how the public and the experts view death. As noted above, some observers question the necessity or desirability of having a single standard of death and argue that there should be a legal right of patients to choose the criterion of death based on the religious, cultural, or philosophical beliefs of the individual or family (see Sreedhar and Sandjeev 2009). Would it create medical and legal chaos to honor this diversity through inclusion of conscience clauses in death statutes, or is it better for society to devise a single mandatory standard?

For Western observers, brain death as a concept is assumed to be the universal definition of death. In his survey of eighty countries, Wijdicks (2002) found major differences in the procedures for diagnosing brain death and argued for consideration of standardization. But should it be standardized, given the controversy surrounding it? Not according to Veatch (2009), who has long defended the proposition that the "choice of a definition of death, like many other philosophical or theological controversies, will not lend itself to a single, uniform policy and that people should be permitted to choose their own positions within reason" (17). For Sreedhar and Sandjeev (2009), the great variation in the laws and guidelines pertaining to death from country to country and among states in some countries reflects the importance of family and cultural diversity.

In a review of end-of-life decision making in twelve diverse countries, Blank and Merrick (2005) found that the debate over brain death and its connection with and implications for organ transplantation have been at best muted for most of the world's population, particularly in Asia and Africa. Moreover, of the countries where it has reached the policy agenda, only in the UK, the United States, and Brazil is brain death widely accepted and practiced as the basis for organ transplantation. In Kenya, for instance, there is no recognition of brain death and organs cannot be transplanted. Although brain death is defined by law in India, most families refuse to accept it and choose to wait for cardiovascular death to occur. In Japan, there has been an intense debate over brain death and organ transplantation, and although the laws are changing, there remains little public acceptance of brain death, and very few transplants are performed. Brazil, on the other hand, has experienced increased acceptance of brain death and has a transplant rate second only to that of the United States, although a law to require presumed consent designed to increase organ donation rates was repealed after strong public opposition.

But one need not go outside the United States to see the impact of culture and religion on the definition of death. For instance, in their study of Buddhism, Hinduism, Judaism, Catholicism, and Islam, Bresnahan and Mahler (2010) concluded that the claim of organ procurement agencies that all major world religions approve of organ donation does not tell the whole story of the "extensive, heated ethical debate on religious websites about organ donation in the context of brain death" (60). Applbaum and colleagues (2008) discuss a case of Chinese father who asked to maintain his brain-dead daughter on a ventilator to administer traditional medicine and to wait until the mother and other relatives arrived from overseas. This case exemplifies the challenges in handling cross-cultural ethical conflicts in modern medicine. During a time when much attention is being directed toward improving clinicians' cultural competence, this case illustrates the importance of explicitly discussing how health care professionals can work with patients whose health beliefs, practices, and values may be unfamiliar, while at the same time recognizing that accommodation has limits when it comes with real costs to others and society.

Brain death, like other death-related areas that some observers assume are global, such as hospice or palliative care, withholding or withdrawal of care, and advance directives, shows considerable variation across countries. Death-related issues are too passionate and culturally sensitive to expect a convergence to the Western, largely U.S.-driven, norms despite the difficult resource allocation questions surrounding the end of life and aging populations that cross national boundaries. Even more so than in other areas of medicine, end-of-life issues, including the definition of death, elucidate the importance of religion and culture, as well as the role of the family and social structure. In most countries, particularly non-Western ones, there remains fervent opposition among large segments of the population and among health care professions even to the notion of brain death. In many countries there remains the feeling that death is a highly personal matter, not something to be publicly discussed, much less made a matter of public policy. Thus, it is unlikely that moves to advance the criterion of cortical death will be forthcoming despite more precise brain imaging and diagnostic applications.

The Continuing Policy Problem of Brain Death

Whatever standards for determining death are used, they will remain troublesome as a matter of public policy. Verheijde, Rady, and McGregor (2009), for instance, contend that neurological or circulatory parameters currently utilized in heart-beating or non-heart-beating procurement of transplantable organs conceal a medical practice of physician-assisted death. Despite a widespread policy of brain death across jurisdictions, thousands of cortically brain-dead persons are kept alive by artificial means, usually at the request or demand of the family. The difficulty of letting go, the hope for a miracle, and the confusion of values resulting from the new technologies cause many persons to refuse to authorize unplugging the life-support machines even when there is no hope. There is a current debate in the UK as to whether brain-dead transplant donors should be anaesthetized because of the social acceptability of a still death, as well as for the benefit of operating room staff, who find reflex movements disturbing. For Giles Morgan, president of the Intensive Care Society, "In simple terms, if you are dead, you are dead and so dead people don't require anaesthesia. . . . If you aren't dead, you shouldn't be having your organs taken away" (cited in Boseley 2000). Many proponents of organ transplantation, of course, disagree, however, and the issue is expected to intensify as the demand for organs increases.

Any definition of brain death also raises other thorny practical questions. Can third-party payers, including public funding sources, refuse to pay for care or maintenance of a person who is legally dead? On what grounds can private insurers justify such coverage to shareholders? Veatch (1993) suggests incorporation of a clause that standard health insurance not be required to cover medical costs to maintain any person who is "alive with a dead brain" (22). Moreover, can wills be probated in such a case? How can patients be protected from premature termination of helpful treatment under the guise of declaring death? What mechanisms are needed to maintain proper respect for the dignity of a brain-dead person in light of pressures from transplant teams who the need organs to save other patients who are brain-alive?

Although uncommon, brain death determinations have been challenged in courts. According to Burkle, Schipper, and Wijdicks (2011), however, all court rulings to date have upheld the medical practice of death determination using neurological criteria according to state law, irrespective of other elements of the rulings. Nothing in case law suggests a legal justification to alter the current medical standard of brain death determination. U.S. jurisprudence emphasizes that the timing and accurate diagnosis of brain death have important weight in the resolution of conflict among practitioners, hospitals, and family members.

Although a few short years have powerfully transformed the meaning of death, many questions remain. Our very conception of what it means to be human is challenged by rapid advances in medical technology. In light of the expanded knowledge of the brain and its functions, it is likely that some variation of higher-brain death will become the center of debate in the near future. Despite strong opposition, the escalating financial and psychological burden on the living and the intensification of neurodegenerative diseases among an aging population will press in that direction. Although it is tempting to resolve these problems with a simple solution, as noted by Singer (1995), "Solving problems by redefinition rarely works, and this case [brain death] was no exception" (51). The solutions to difficult problems of today, from abortion and euthanasia, cannot be discovered by medical science, or by simply defining them away. They require thoughtful, lengthy dialogue, debate, and utilization of the unique and remarkable human characteristics of consciousness and reasoning.

9

Politics and the Brain

This chapter directs attention to the political implications of neuroscience as manifested by groundbreaking work on decision making, political ideology, and voting behavior. It is argued that the issues raised in this book will not only be prominent in public policy discussions, they will also shape the way in which the dialogue itself is conducted. As noted by Friend and Thayer (2011), since the triumph of behavioralism in the 1950s political science has focused on locating causation anywhere but with the individual. Organizational approaches, the study of bureaucracy, and systemic analyses have almost forced studies of individuals and psychological approaches out of the discipline. However, the advances in neuroscience signal a renaissance of individual approaches and a revival of the individual level of analysis for causation of political behavior. "Rational choice will wane, political psychology will wax as the center of the discipline. The glorification of economics and economic approaches will fade too. It is ironic that psychology was once dismissed as the 'softest' of the social sciences, while economics was triumphed as the most scientific. Yet the advance of the life sciences, evolutionary psychology, cognitive psychology and neuroscience altered this ordering" (Friend and Thayer 2011, 250). To this end, neuroscience will contribute to a transformed view of the human condition that will shift attention back to the individual, where it belongs.

The Brain and Decision Making

At the core of human behavior is the ability to make and execute decisions. The study of decision making spans such varied fields as neuroscience, psychology, economics, political science, and computer science, but despite this diversity of fields and applications, most decisions share the common elements of deliberation, motivation, and commitment

(Gold and Shadlen 2007). Not surprisingly, an incipient area of research is decision neuroscience, the goal of which is to integrate research in neuroscience and behavioral decision making (Shiv et al. 2005). Neuroscience has already made an important contribution by demonstrating that decision making involves the orchestration of multiple neural structures and cognitive systems. Research has shown that areas such as the ventromedial prefrontal cortex, amygdala, insula, somatosensory cortex, dorsolateral prefrontal cortex, and hippocampus are all involved in various aspects of decision making. Gupta, Koscik, and colleagues (2011) note that the amygdala is an important part of an impulsive system involved in decision making that triggers autonomic responses to emotional stimuli, including monetary reward and punishment. Persons with amygdala damage lack these autonomic responses to reward and punishment and are unable to utilize somatic marker–type cues to guide future decision making.

Damage to the orbitofrontal cortex (OFC) has been found to produce a unique deficit that impairs everyday decision making while leaving other cognitive capabilities intact. OFC dysfunction is associated with disorders involving behavior such as obsessive-compulsive disorder, substance abuse, eating disorders, obesity, and pathological gambling, in which subjects report feeling unable to control their behavior, have an immense desire to engage in the behavior, and experience a feeling of release once they do. Wallis (2007) stresses the central role of the OFC for planning and obtaining distant rewards and goals: it integrates multiple sources of information regarding the reward outcome to derive a value signal, thereby calculating how rewarding a reward will be. In turn, this value signal can be held in working memory, where it can be used by the lateral prefrontal cortex to plan and organize behavior toward obtaining the outcome and by the medial prefrontal cortex to evaluate the overall action in terms of its success and the effort that was required. Thus, acting together, these prefrontal areas ensure that our behavior is most efficiently directed toward satisfying our needs (Wallis 2007, 31).

New research has revealed a brain circuit that seems to underlie the ability of humans to resist instant gratification and delay reward for months or even years in order to earn a better payoff. This capacity for "mental time travel," also known as "episodic future thought," enables humans to make choices with high long-term benefits. Thus, while humans normally prefer larger over smaller rewards and discount the value of rewards over time, they exhibit a significant ability to delay gratification. Furthermore, Peters and Büchel (2010) used neuroimaging data to reveal that signals in the anterior cingulate cortex, a part of the brain implicated in reward-based decision making, coupled with the hippocampus, which has been linked with imagining the future, can predict the degree to which forward thinking modulates individual preference functions. Similarly, many situations in life call for us to respond with restraint. Jahfari and colleagues (2010, 1479) suggest that the response delay effect is at least partly explained by active braking, pointing to proactive recruitment of a neurocognitive mechanism usually associated with outright stopping.

According to Pessoa and Engelmann (2010), executive processes such as attention, working memory, and inhibition constitute a set of processes that are particularly important for behavioral planning and production. Because the brain, unlike a computer, has limited processing capacity, to guide behavior it needs to be able to segregate those stimuli that deserve further processing from those that are best ignored, To this end, both bottom-up and top-down processes are involved. By using top-down control, the brain can more economically allocate its resources based on current behavioral goals and prior knowledge. Simultaneously, bottom-up processing resources preferentially shift to salient features of the environment. Based on behavioral evidence, both these processes are intimately linked to reward and motivation. Furthermore, these findings mesh well with previous demonstrations that the motivational dimensions of (top-down) goals rely on the dopamine system and its projection sites, while (bottom-up) stimulus salience is encoded in specific nodes of the reward system, such as the caudate nucleus (e.g., Zink et al. 2006).

Role of Emotions in Decision Making

An emotion is defined as a mental state that arises spontaneously rather than through conscious effort and is often accompanied by physiological changes. We often refer to emotion as a feeling, for instance of joy, sorrow, hate, love, and so forth. The questions as to where such feelings originate and what role they play in human behavior long have been at the core of philosophical debate. Affective neuroscience has begun to identify the neural bases of these mental experiences, and, not surprisingly, one brain structure consistently associated with emotional function is the amygdaloid complex (Gupta, Koscik, et al. 2011). It is recognized that the amygdala plays a key role in the processing of emotional information, regulates emotional responses, and controls fear reactions (Adolphs 2010, 42). Amygdala responses represent a "critical nexus" between the propensity for emotional experience and emotional interactions with perceptual encoding, and its activation is a precondition for negative affective experience (Feldman-Barrett et al. 2007). Similarly, the regulation of negative emotion through reappraisal has been shown to induce increased prefrontal activity and decreased amygdala activity (Modinos, Ormel, and Aleman 2010). This ability to regulate negative emotion is essential for dealing with the distressing experiences we encounter in everyday life (Adolphs 2010).

Similarly, Pessoa (2010b) concludes that the amygdala is critically involved in selective information processing. The amygdala's role in feeling emotions may be tied to enhanced sensitivity toward emotional perception. For instance, the amygdala is part of a vast network of cortical and subcortical structures that allow efficient processing of, and rapid response to, crude visual representations of potentially threatening stimuli (Maratos et al. 2009). Although the amygdala is clearly an important focus for emotional processing research, there is evidence it is not a single static structure; rather, the left and right sides and even separate subregions carry out very different tasks (Patin and Hurlemann 2011, 714). Sergerie, Chochol, and Armony (2008), too, found strong support for a functional dissociation between the left and right amygdala in terms of temporal dynamics. Other studies have found genetic links to amygdala activity related to emotion (see, e.g., Gillihan et al. 2010). Its major involvement in various kinds of integrative sensory and emotional functions makes it a cornerstone for self-relevant appraisals of the environment and also for the processing of autobiographical events (Markowitsch and Staniloiu 2011a). It also serves to allocate processing resources to stimuli, at least in part by modulating the anatomical components that are required to prioritize specific features of information processing in each situation (Pessoa and Adolphs 2010).

The amygdala, therefore, is positioned to make a rapid appraisal of the state of the world and alter the perceptual process by enhancing the encoding of events of potential importance for the individual (Anderson 2007). Neuroimaging studies have shown that emotionally charged scenes vigorously engage early primary and secondary visual cortex. They conclude that viewing unpleasant stimuli might in part reflect the prompting of defensive responses and how they are related to ongoing behaviors (Pereira et al. 2010, 105). Emotions, therefore, skew attention toward events associated with subjective and physiologic arousal, ultimately shaping the contents of awareness (Anderson 2007). The amygdala might also represent embodied attention, the crucial link between central (mental) and peripheral (bodily) resources, and therefore serve as the common primitive foundation of both attention and emotional experience. "If all the world's a stage, then the amygdala may be the emotional spotlight, shedding light on the most dramatic players of our life story" (Anderson 2007, 72).

Another part of the brain that has been found to be critical for emotions is the OFC, a component of the prefrontal cortex that has connections with the amygdala so extensive that it is sometimes considered a limbic structure (Blair 2007b; Bishop 2007). Consistent with its anatomical duality, the OFC has both limbic and prefrontal roles. It is highly involved in emotion and, like other structures of the prefrontal cortex, it mainly serves to regulate emotion, control mood, monitor rewards and punishments, and generally be engaged during planning and decision-making tasks (Kringelbach and Rolls 2004). Together, the OFC and the amygdala promote stimulus-reinforcement learning and decision making in healthy individuals, particularly during associative learning (Schoenbaum, Saddoris, and Stalnaker. 2007). In support, Finger and colleagues (2008) found increased OFC activity during operant extinction learning. According to Adolphs (2010), this prefrontal activity is involved in down-regulating activity in emotion-generating regions by its inverse association with the amygdala's response to negative stimuli.

In their meta-analysis of emotion, Vytal and Hamann (2010) found that each of the basic emotional states (happiness, sadness, anger, fear, and disgust) was associated with consistent and discriminable patterns of neural activation, and that the signature patterns of neural activation that characterized each emotion consistently differentiated that emotion from other emotions:

Happiness consistently activated rostral ACC and right STG, and activity in both regions differentiated happiness from sadness, anger, fear, and disgust (ACC only). Sadness consistently activated MFG and head of the caudate/subgenual ACC, and activity in both regions reliably differentiated sadness from happiness, anger, fear, and disgust. Anger consistently activated IFG and PHG, and both regions differentiated anger from all other emotion states. Fear consistently activated amygdala and insula, and these regions differentiated fear from happiness, sadness, anger (insula only), and disgust (posterior insula). Disgust consistently activated IFG/anterior insula, and these regions reliably differentiated disgust from all other emotion states. (2879)

In sum, these findings demonstrate that basic emotional states are associated with discernible patterns of brain activation that vary substantially among emotions. Moreover, Keightly and colleagues (2011) examined the influence of emotional valence and type of item to be remembered on brain activity during recognition of faces and scenes. They concluded that both emotional valence and type of visual stimulus modulate brain activity at recognition and influence multiple networks mediating visual, memory, and emotion processing. Emotion, like cognition, then, is not a single entity but a multicomponent process.

Emotion and Cognition

Although emotion and motivation have crucial roles in governing human behavior, how they interact with cognitive control functions is less well understood. For Pessoa (2009, 160), the "dual competition" framework suggests that emotion and motivation affect both perceptual and executive competition. In particular, the anterior cingulate cortex is thought to be engaged in attention control mechanisms and to interact with other brain structures in integrating emotionally significant signals with control signals in prefrontal cortex. This implies that emotion and motivation can either enhance or impair behavioral performance, depending on how they interact with executive control functions. Thus, the impact of an emotional stimulus on behavior depends on how it affects the flow of executive functions. Pessoa (2009, 161) hypothesizes that the results will depend on the level of threat, which will determine whether emotional content enhances or impairs behavioral performance.

Historically, emotion and cognition have been viewed as largely separate processes implemented by different regions of the brain. Often emotion has been contrasted to cognition by associating it with "irrational" or "suboptimal" processes that are more basic, that is, more linked to survival, than cognitive ones (Pessoa 2010a). Under this framework, functional interactions between the amygdala and the prefrontal cortex mediate emotional influences on cognitive processes such as decision making, as well as the cognitive regulation of emotion. More recent data suggest, however, that such a view is misleading and that, to understand how complex behaviors are carried out in the brain, it is necessary to understand the interaction between emotion and cognition (Pessoa 2010a). For instance, it has become evident that large portions of both cortex and subcortex are engaged during emotional information analyses. Neurons in these structures often have entangled representations whereby single neurons encode multiple cognitive and emotional variables.

In their meta-analysis of research on the representation and utilization of cognitive and emotional parameters, Salzman and Fusi (2010) contend that these mental state parameters are "inextricably linked" and represented in dynamic neural networks composed of interconnected prefrontal and limbic brain structures. Thus, cognition and emotion are much more effectively integrated in the brain than previously assumed. Emotional processes can influence cognitive processes and cognitive processes can regulate or modify our emotions. Both of these actions can be implemented by changing mental state variables (either emotional or cognitive ones), and emotions and thoughts will shift together, corresponding to the new mental state. Although the amygdala is most clearly tied to fear-related functions and is an effective alarm system, important roles for the amygdala in cognitive operations, such as attention and decision making, indicate that it is counterproductive to separate emotion and cognition (Salzman and Fusi 2010).

Summary: Emotions, Cognition, and Decision Making
It is clear even from the relatively embryonic state of neuroscience research summarized above that the neural functions involved in making a decision are complicated and include both emotional and cognitive dimensions. In their study, Camerer, Loewenstein, and Prelec (2005) provide a valuable contribution by conceptualizing two dimensions that produce four possible combinations, I-IV, as illustrated in table 9.1. The first dimension is divided into two types of processes. *Controlled* processes are serial, using step-by-step logic or computations; tend to be invoked deliberately by the agent; and are often associated with a

Table 9.1
Two Dimensions of Neural Functioning

	Cognitive	Affective
Controlled processes Serial Effortful Evoked deliberately Good introspective access	I. Controlled Cognitive	II. Controlled Affective
Automatic processes Parallel Effortless Reflexive No introspective access	III. Automatic Cognitive	IV. Automatic Affective

Source: Camerer, Loewenstein, and Prelec (2005, 16).

subjective feeling of effort. A person can typically provide a good intro-spective account of controlled processes. In contrast, *automatic* processes operate in parallel, are not accessible to consciousness, and are relatively effortless.

The second dimension includes affective and cognitive processes, often seen as passion and reason. Although we often associate affect with feeling states, and indeed, most affect states do produce feeling states when they reach a threshold level of intensity, Camerer and colleagues (2005) contend that most affect probably operates below the threshold of conscious awareness. The central feature of affect, therefore, is not the feeling states associated with it but its role in human motivation. As discussed above, all emotions have either a positive or negative valence and many also carry action tendencies, such as anger and aggression or fear and escape. This means that cognition by itself cannot produce action; to influence behavior, the cognitive system must operate via the affective system. While it is often presumed that cognition is controlled and affect is automatic, in reality, much cognitive processing is automatic as well. Moreover, research on "emotional regulation" demonstrates many ways that cognition can influence emotion, implying a capacity for controlling emotion.

To conclude, emotion is a disposition to action and is part biological and universal and part sociologically and culturally specific. Emotion and cognition evolved together and function together, and neither can be understood without understanding the other. Emotion galvanizes, informs, and motivates cognition, and cognition, in turn, refines, strat-egizes, and sometimes overrides affective input. As noted above, research suggests that emotion, in the right measure, facilitates, and indeed might be necessary for, rational decision making, assigning priorities to sensory data and sustaining attention, identifying problems and preferences, as well as motivating, directing, and accelerating strategic reasoning and helping to store and retrieve memories. In ways thus far only partially understood, emotion animates and helps coordinate problem-solving techniques and their application (Long and Brecke 2003, 27). Therefore, continuing to ignore or marginalize the role of emotions in decision making does not serve us well in our attempts to understand social action and policy making.

What Does This All Mean for Policy Making?

Why should this research on the neurological roots of emotions and decision making have any relevance for the study of politics and public policy? Although there many reasons why these findings regarding brain function are crucial in discovering how the brain processes political situations, two stand out. The first is that it demonstrates that many individual decisions are made automatically, without conscious reflection or calculation. At the least, this knowledge calls into question the rational decision maker postulated in rational choice models. Moreover, collectively it challenges traditional policy-making institutions by raising questions about group behavior and dynamics. A second and related reason why this work is important is that it consistently demonstrates that the affective system is intricately involved in decision making. Contrary to the old view that "passion is the enemy of reason," emotions not only have an influence on all decisions, they also have corresponding neurological dynamics that interact with cognitive processes in the brain (Marcus, Neuman, and McKuen 2000; Marcus 2000, 2002). Unlike economic self-interested behavior, political self-interested behavior is not necessarily defined by utility functions, and concern about others is not explicitly related to considerations of the welfare of others (Kato et al. 2009).

The common theme that emerges from this research is that many decisions occur at an automatic level until the brain is faced with an unexpected or novel stimulus. When this occurs, the emotional system assists in arousing higher-level conscious systems, which are used to analyze and deal with the novelty at hand. In other words, making decisions requires concerted activity across those regions of the brain associated with basic impulses, as well as those associated with cognition and reason, although not in any constant proportion. Thus, many political decisions, especially those that might be perceived as routine, are not necessarily rational, at least not in the sense that prevailing policy-making models assume. For a more detailed discussion of these issues and their implications for democratic theory, see Tingley (2006), who states that the "quest to be able to generate a more complete understanding of political behavior by bridging 'multiple levels of analysis' and connecting political science to neuroscience is exceedingly complicated for a variety of reasons" (6), but necessary if we are to understand political behavior.

The biological paradigm introduced in chapter 1, backed by considerable research findings from neuroscience, directly challenges the prevailing models of policy making. Over the last half century, many models have been promulgated, most prominently variations of the *rational* and *incremental* models, where the debate has coalesced over how logical and systematic the policy-making process actually is. Public choice

advocates argue that approximations of rational conduct through calculations of cost and benefit are sufficient to explain the behavior of individuals in organizations, and thus even psychological factors are of little
value (Losco 1994, 49). Although intuitively attractive, the rational actor
model is vulnerable because of conflicting objectives that are found in
any organization. Furthermore, in practice, decisions are often made
without adequate information and comparable benefits.

Alternately, *bureaucratic* and *organizational* models have been proposed that highlight respectively the importance of impact of bargaining
among personnel and agencies as they pursue their own perceived interests and the impact of values and regular patterns of behavior that are
common to any large organization. While still other models have placed
greater emphasis on beliefs and ideology and begin to address the degree
to which decision making is structured by social and political values that
are not rational or impartial, notably absent from all these conventional
policy-making models are biological factors or any reference to the brain.
Even *belief system* models that explicitly challenge rational models by
suggesting that decision makers are not rational, rigorous, and objective
agents and that decisions are shaped by often unconscious perceptions
and concepts generally fail to account for the evolutionary dimensions
of these preconceptions and beliefs. In other words, they too neglect the
biological foundations of decision making and the intricate contributions
of both emotion and cognition, despite the mounting neuroscience evidence reviewed here to the contrary.

Rationality has come to mean the conscious, goal-oriented, reasoned
process by which an individual, expressing and thus also revealing his
or her preferences, chooses a utility-maximizing action from among an
array of alternative actions. Although rational choice theory does not
specify that people must be "conscious" or fully in "control" of their
decision-making process in order to be rational, it is most often applied
in situations where the posited actors are assumed to be conscious and
acting intentionally (Tingley 2006, 27). Also, while a few cognitive theo-
reticians have considered a role for emotion in decision making and some
organization theory and agent-level theories allow for the incorporation
of emotional variables, most either ignore it or see its role as marginal
and view its formal consideration as counterproductive.

Alternatively, *social constructivism* explains phenomena by reference
to social interactions rather than anything innate to human beings and
claims that society and societal phenomena can be understood solely as
the product of social interaction. Thus, emotion and the behavior associ-

ated with it are merely social constructs that are learned and reinforced through social discourse. Long and Brecke (2003), however, argue that emotion is not reducible to a sociocultural construct, as constructivism assumes. Undoubtedly, our emotion is provoked by sociocultural circumstances and in part expressed according to sociocultural conventions, but in between provocation and expression is nothing but brain and body. Although the human mind is capable of recognizing and describing subtle emotional states that reflect one's particular society and culture, basic emotions such as fear, anger, sadness, and joy transcend culture and represent universals.

Political Ideology and Political Attitudes

Although political ideology and attitudes are commonly assumed to have solely environmental causes, recent studies have identified biological influences on an individual's political orientation (Alford, Funk, and Hibbing 2005). Kanai and colleagues (2011) found that high-level concepts of political attitudes are reflected in the structure of focal regions of the brain. Moreover, brain structure can exhibit systematic relationships with an individual's experiences and skills altered through extensive training (Scholz et al. 2009) and is related to different aspects of conscious perception (Fleming et al. 2010; Kanai and Rees 2011). Since comprehending changes in political attitudes is critical to understanding political behavior, stable preferences toward others, such as partisanship and membership in social groups, as well as preferences that are more susceptible to change constitute important subjects for study. Neuropolitical studies have found that activation of emotion-related areas in the brain is connected to resilient political preferences (Kato et al. 2009). Additionally, the study of ideology and attitudes is one in which the interaction of genetic and neural factors is most discernible.

Conservatives have been found to be more organized and persistent in their judgments and approaches to decision making, as indicated by higher average scores on psychological measures of personal need for order, structure, and closure (Amodio et al. 2007). Liberals, by contrast, report higher tolerance for ambiguity and complexity, and display greater openness to new experiences on psychological measures (Amodio et al. 2007). Research has sought to determine whether political liberalism (versus conservatism) is associated with differences in gray matter volume in the anterior cingulate cortex. Apart from the anterior cingulate cortex, other brain structures also confirm patterns of neural activity that reflect political attitudes. One study found that conservatives respond to threatening situations with more aggression than liberals and are more sensitive to threatening facial expressions (Kanai et al. 2011). This heightened sensitivity to emotional faces suggests that individuals with a conservative orientation might exhibit differences in brain structures associated with emotional processing such as the amygdala. In fact, structural MRI data suggest that conservatism is associated with increased gray matter volume in the right amygdala (Kanai et al. 2011, 678).

Moreover, the results of a study by Amodio and colleagues (2007) are consistent with previous findings that political orientation reflects individual differences in the functioning of a general mechanism related to cognitive control and self-regulation (Jost et al. 2003; Alford, Funk, and Hibbing 2005). Amodio and colleagues found that greater liberalism was associated with stronger conflict-related anterior cingulate activity (a part of the middle surface in the front of the brain), suggesting greater neurocognitive sensitivity to cues for altering a habitual response pattern. In contrast, stronger conservatism was associated with less neurocognitive sensitivity to response conflicts. Moreover, the thickness of the gray matter in the anterior cingulate cortex is greater the more people describe themselves as liberal or left wing and thinner the more they describe themselves as conservative or right wing. At the behavioral level, a liberal orientation was associated with better performance on the response-inhibition task, implying that conservatives would perform better on tasks in which a more fixed response style is optimal.

In their controversial study, Kanai and colleagues (2011) found that conservatives tend to have a larger amygdala. This suggests either that there is something about political attitudes that is entrenched in our brain structure through experience or that our brain structure determines or produces our political attitudes. Political orientation is often associated with psychological processes for managing fear and uncertainty. Individuals with a larger amygdala are more sensitive to fear, which may suggest they are more inclined to integrate conservative views into their belief system (Kanai et al. 2011, 678). Similarly, it is striking that conservatives are more sensitive to disgust. On the other hand, an association between anterior cingulate cortex volume and political attitudes may be linked to tolerance of uncertainty. It is possible that individuals with a larger anterior cingulate cortex have a higher capacity to tolerate uncertainty and conflicts, allowing them to accept more liberal views. Although speculative (and very controversial), such propositions provide a basis for theorizing about the psychological constructs underlying political attitudes.

A recent study concludes that political attitudes are influenced much more heavily by genetics than by parental socialization (Kanai et al. 2011). Some believe that for the overall index of political conservatism, genetics accounts for approximately half of the variance in ideology, while shared environment, including parental influence, accounts for only 11 percent. With respect to variance in people's tendencies to possess political opinions at all, regardless of their ideological direction, genetics is claimed to represent one-third of the variance, while shared environment is inconsequential. These findings have implications for political science since acknowledging a role for heritability in politics affects our understanding of political issues, political learning, and political cleavages (Alford, Funk, and Hibbing 2005). There is evidence that inherited attitudes are very different from acquired tastes. They are manifested more quickly, are more resistant to change, and increase the likelihood that people will be attracted to those who share those particular attitudes. To the extent that political ideologies are inherited and not learned, they become more difficult to manipulate (Alford, Funk, and Hibbing 2005).

Most observers believe that genes are unlikely to affect attitudes directly. Rather, they affect neurological systems, which in turn affect cognitive processing tendencies, which in turn affect personality and value traits, which in turn affect political ideology, which in turn affects stances on issues of the day. Rather than connecting political preferences to genetics directly, psychologists have been more likely to connect political beliefs to personality, moral foundations, life choices, tastes, and values. Current research on political ideologies suggests that people hold a political preference if they have a consistent view on issues of the day, if their stance on one issue constrains their stances on other issues, and if they are aware of the meaning of commonly employed ideological labels. To be considered ideological, an individual would need to understand what it means to be liberal or conservative and would need to have a collection of political beliefs that is consistent and can logically be placed under one of those labels (Beckwith and Morris 2008).

In direct contrast, some observers contend that whatever constitutes ideology, its ultimate cause is environmental, and any particular ideology can only make sense within the context of a unique political culture. According to Smith and colleagues (2011), the dominant view of political scientists that ideology is optional (and, in fact, largely absent in most people) and is merely a collection of issue positions, not reliant on deeper, more universal, psychological tendencies, has resulted in a shallow and narrowly political conceptualization of the term and hindered the ability to integrate their findings with those of psychologists working on the same topic. If political ideology is to become useful and meaningful, it is necessary to accept that ideology is not a superficial label or bundle of topical positions but rather a central component of an individual's general life orientations. In the past, political scientists measured ideology in ways that ensured it could only be seen as emanating from particular cultures, and, for Smith and colleagues (2011), this unfortunate vision of ideology must end. Acceptance that there are biological and even genetic precursors to political attitudes and behavior is a tough sell, however, because the traditional view of politics places great weight on smaller issues arising only within a certain cultural context, thereby making it much more difficult to imagine how broad biological forces could be at work (Smith et al. 2011). Moreover, most political scientists know very little about either genetics or neuroscience, resulting in an insensitivity to or ignorance of the biological bases of political behavior.

Recent studies have suggested that the prevailing environmental-based theories of political behavior—rational choice and behaviorism—do not adequately explain the variance in political behavior because neither theory takes into account where preferences come from but instead focuses entirely on individuals' reactions to their environment (Hatemi et al. 2007; Fowler and Schreiber 2008). While the choice of a particular candidate or party does not appear to be heritable, it remains an open question whether or not the act of voting is heritable (Fowler, Baker, and Dawes 2008). Future work should explore the interaction effects of genes and environment on participation. Studying this will help us understand what the causal mechanisms are that link genes which have taken millions of years to evolve from small-scale to large-scale political behavior, an extremely recent phenomenon on the scale of human evolution.

It is extremely unlikely that such efforts will uncover a voting gene, however there is some suggestion that there is some set of genes whose expression – in combination with environmental factors – regulates political participation. The task of finding out which genes they are and what physical function they have will improve our understanding of the biological processes that underlie these complex social behaviors and may also shed light on their evolutionary origin. (Fowler, Baker, and Dawes 2008, 244)

The realization that participation is heritable has already helped to generate additional theories of turnout, partisanship, and pro-social behavior. Although it may not surprise geneticists that participation is heritable, it is premature to argue that heritability studies will not bear fruit in political science. These studies will provide the first step needed to excite the imagination in a discipline not used to thinking about the role of biology in human behavior. The study of brains and genes potentially promises better understanding of the constraints imposed on political psychology. Genes are the institutions of the human body, and we ignore them at our peril if we want to develop a full understanding of human psychology and political behavior (Settle, Dawes, and Fowler 2009; Fowler and Christakis 2008).

Traditional voting choice theories make important assumptions regarding voting behavior that differ from typical assumptions made by behavioral geneticists or neuroscience because they focus principally on external and environmental influences on vote choice, providing no explicit role for biological factors (Hatemi et al. 2007). Previous studies examining political traits have been limited to attitudes, and the act of voting is only part of the complex interdependent and context-dependent social attitude factors that are both genetically and environmentally influenced. Over the past few years the claims that establish a link between genes and political behavior have been reinforced by twin research. This has led to the emergence of the nascent field of "genopolitics" (Joseph 2010). Recent twin studies have concluded that genes are a very important part of what it means to be a human. The twin method compares the trait resemblance of reared-together MZ (monozygotic, identical) twin pairs, who share a 100 percent genetic similarity, with the trait resemblance of reared-together same sex DZ (dizygotic, fraternal) twin pairs, who average a 50 percent genetic similarity. Twin researchers and most of their critics agree that MZ twins resemble each other more than same-sex DZ pairs for most behavioral and psychological traits (Joseph 2010).

Those political scientists influenced by behavioral genetics argue that this finding provides concrete evidence that genetic factors play an important role in politics. Thus, political behavior joins a long list of other intuitively nongenetic behaviors now claimed for genetics on the basis of twin research, including anorexia, loneliness, perfectionism, and religiousness. However, twin studies have many outspoken critics. Some scholars have argued that heritability claims based on twin studies generally cannot be trusted because of the confounding influence of the greater environmental similarity of MZ twins. According to Joseph (2010), "Because the twin method is no more able than a family study to disentangle potential genetic and environmental influences, twin studies of political attitudes and behavior provide no scientifically acceptable evidence in support of genetic factors" (213). However, Alford, Funk, and Hibbing (2005) argue that these concerns with twin studies have been raised and soundly rebutted.

Voting Behavior

Voting is one of the most important decisions that individuals can make in a democratic society; therefore it is no surprise that the factors influencing voting behavior and turnout have long been of interest to political scientists and have spawned many models. Traditional approaches to the processes in which voters identify with political candidates tend to ignore the effect of emotions and instead conclude that voters are more responsive to the candidates' position on the issues, their party affiliation, and physical traits such as race, gender, and age. Although these factors undeniably influence to varying degrees voter choice, neglecting the influence of emotions from the study of voter preference ignores a wealth of evidence of the fundamental role they play in mediating attitude change (see box 9.1). Recently, brain imaging studies investigating the neural

Box 9.1
One Finds What One Looks For

Much of modern political science assumes that voters consciously select candidates based on their estimates of how the candidate will satisfy their values and material needs. This is a controlled and cognitive process, and more likely to activate an executive region of the brain, such as the lateral prefrontal cortex, that implements the strategic pursuit of goals. Quite plausibly, however, voters might also prefer candidates in part because they are physically attractive, which would correspond to automatic or affective processing because it elicits a noncontrolled response that is highly affective and likely processed in part by elements of the limbic portion of the brain. Explanations of voter behavior thus might have two different theories and two different predictions of neurological activity (Todorov et al. 2005). "Of course, elements of both processes occur. The point is twofold: first, the type of political explanation being offered affects the type of neurological evidence that would be suggested; second, if both processes occur, fMRI could be used to study their relationship" (Tingley 2006, 11–12).

bases of partisan political judgment, candidate preference, and voting have provided some of the most innovative and controversial research on political behavior. Much of this research strongly suggests that neural substrates and biological mechanisms shape our political orientations, preferences, and decisions regarding candidates (Friend and Thayer 2011).

While conventional behavioral research on voting behavior presumes that a candidate's appearance can affect voter opinion, recent cognitive neuroimaging studies provide us with a better understanding of how emotion and motivated reasoning influence political judgment, partisan preference, and voting behavior. Behavioral studies have shown that judgments about candidates' physical appearance correlate with election outcomes (Todorov et al. 2005; Ballew and Todorov 2007), suggesting that information derived from visual appearance alone (so-called "thin-slice" information) affects voting behavior. Much of the research on how appearance affects voting behavior focuses on the facial expressions of candidates and what regions of the brain respond to the emotional displays. Despite the significant consequences electoral choice, research has shown that electoral outcomes can be predicted on the basis of individuals' snap judgments of the faces of the candidates (Todorov et al. 2005; Chiao, Bowman, and Gill. 2008). One study found that judgments of competence based on seeing American political candidates' faces for only 100 milliseconds accurately predicted 59.6 percent of the actual electoral outcomes (Ballew and Todorov 2007). The finding that people automatically make valence/trustworthiness judgments from facial appearance was replicated by Todorov, Pakrashi, and Oosterhof (2009), who note, "Trustworthiness judgments are made within a single glance of a face" (830).

Moreover, brain imaging studies, in combination with behavioral studies on the psychological mechanisms involved in appearance-based judgments, have begun to offer valuable insights into the underlying neural processes of voter preference. A fundamental question in politics is the extent to which voters' decisions are driven by positive motives, which induce them to vote for candidates that they like, or by negative ones, which induce them to vote for the candidate that they do not dislike (i.e., negative voting). To determine whether the effect of appearance on voting is more strongly influenced by positive or negative emotions, Spezio and colleagues (2008) conducted neuroimaging studies in which two separate groups of participants viewed pairs of unfamiliar real politicians (both Republican and Democrat) from previous elections. In the first study, images of losing candidates elicited greater activation in the insula and ventral anterior cingulate cortex than images of winning candidates, suggesting that negative attributions from appearance exert greater influence on voting than do positive ones. In a second study, when negative attribution processing was enhanced, images of losing candidates again elicited greater activation in the same brain regions. The authors concluded that these findings support the view that negative emotions play a critical role in mediating the effects of appearance on voter decisions, an effect that may be of special importance when other information is absent (Spezio et al. 2008). In other words, these thin-slice attributions play a larger role in influencing voters who have little knowledge of the candidates, thereby reinforcing already strong evidence that, when voters know little about a candidate, perceived negative aspects of a candidate exert a stronger influence on voter turnout than positive aspects (Stevens et al. 2008).

Similarly, Knutson et al. (2006) assessed political attitudes using the Implicit Association Test (IAT) in which participants under fMRI were presented faces and names of well-known Democrat and Republican politicians along with positive and negative words. They found a significant behavioral IAT effect for the face, but not the name, condition. Amygdala and fusiform gyrus were activated during perceptual processing of familiar faces with amygdala activation also associated with measures of strength of emotion. Their findings of ventromedial PFC activation suggest that stereotypic knowledge is activated while processing associative knowledge concerned with politicians, but additionally, the anterior prefrontal activations imply that more elaborative, reflective knowledge about the politician is also triggered.

In another fMRI study, Kato and colleagues (2009) imaged subjects after viewing negative campaign videos and found that those with stronger activation in the dorsolateral prefrontal cortex lowered their ratings for the candidate they originally supported more than did those with weaker fMRI activation in the same cortical area. Moreover, subjects with stronger activation in the medial prefrontal cortex tended to increase their ratings for the candidates attacked in the negative campaign videos more than those with weaker activation. Interestingly, the same regions were not activated while subjects viewed negative advertisements for cola, which were used for purposes of comparison. These results imply that neural activity after exposure to negative information about previously supported political candidates was linked to cognitive control of socially relevant stimuli. The activation of distinct prefrontal areas indicates that different kinds of cognitive controls were associated with opposite responses to negative information about the previously supported candidates, that is, they were associated with increasing and decreasing political support (Kato et al. 2009).

Although these studies have confirmed that a candidate's appearance can induce both positive and negative effects, it remains less certain what attributes of a person's appearance signal negative traits that could threaten voters and result in election loss. A cross-cultural fMRI study of the influence of the amygdala on political judgments and subsequent voting decisions may offer some insight into the involvement of these "political intangibles" (Rule et al. 2010). American and Japanese university students viewed and made judgments on facial images of candidates from the 2004 and 2006 U.S. Senate elections and from the 2000 election of the Japanese Diet, respectively. The study found that among both American and Japanese participants, "the bilateral amygdala was significantly more responsive to the faces of politicians for whom participants chose to vote" and "participants showed a greater overall amygdala response to the faces of out-group candidates than they did the faces of in-group candidates" (Rule et al. 2010, 353–354). These findings suggest that the amygdala may be important in evaluating candidates for political office and that the neural basis for electoral decisions extends across cultures. One explanation for the stronger amygdala response to the faces of cultural out-group candidates could be that the amygdala is highly responsive to novel stimuli, such as the faces of out-group members (Van Bavel, Packer and Cunningham 2008). Furthermore, because these findings are consistent with studies of the neural underpinnings of prejudice and bias involved in evaluating out-group members, it appears that voter judgment and decision making are not only influenced by rational cost-benefit evaluations of a candidate's issues and policies but are also, and possibly more so, the product of emotional reactions to adaptive biological mechanisms associated with maintaining in-group cohesion and survival (Friend and Thayer 2011).

Along similar lines, Westen (2007) argues that a neuronal approach to emotion-biased motivated reasoning provides a more accurate understanding of the formation of partisan political judgment and voter choice than conventional theories. His book, *The Political Brain,* in which he argues that the prevailing theory that the voter's mind is a cool calculator that makes decisions by weighing the evidence bears no relation to how the brain actually works, caused considerable debate. Westen suggests that when political candidates assume that voters dispassionately make decisions based on the issues, they lose. In politics, when reason and emotion collide, emotion invariably wins. Elections are decided in the marketplace of emotions, a marketplace filled with values, images, analogies, moral sentiments, and moving oratory, in which logic plays only a supporting role. He shows, through the evolution of the passionate brain and a summary of fifty years of American presidential and national elections, why campaigns succeed and fail, and argues that the evidence is overwhelming that three things determine how people vote. In order, they are: their feelings toward the parties and their principles, their feelings toward the candidates, and, if they haven't decided by then, their feelings toward the candidates' policy positions. Although one cannot change the structure of the brain, one can change the way you appeal to it.

In an earlier article, Westen and colleagues (2006) reported the results of their fMRI study, which showed that self-described Democrats and Republicans responded to negative remarks about their political candidate of choice in systematically biased ways. Specifically, when Republican test subjects were shown self-contradictory quotes by George W. Bush and when Democratic test subjects were shown self-contradictory quotes by John Kerry, both groups tended to explain away the apparent contradictions in a manner biased to favor their candidate of choice. Similarly, none of the circuits involved in conscious reasoning was markedly engaged, while areas of the brain controlling emotions, particularly the amygdala and or cingulate gyrus, showed increased activity as compared to the subject's responses to politically neutral statements associated with politically neutral people. In addition, the dorsolateral prefrontal cortex and anterior cingulate cortex are activated when subjects view images of a candidate of a political party different from their own (Kaplan, Freedman, and Iacoboni 2007). Furthermore, when subjects are presented with information that exonerates their candidate of choice, areas of the brain involved in reward processing (the OFC or the striatum/nucleus accumbens, or both) display increased activity. Accordingly, "Essentially, it appears as if partisans twirl the cognitive kaleidoscope until they get the conclusions they want ... emotionally biased judgments when they have a vested interest in how to interpret 'the facts'" (Westen et al. 2006).

In a related area of investigation about elections, some researchers have examined the hormonal effect of winning and losing for the voters. For instance, in their study of the 2008 presidential election, Stanton and colleagues (2009) found that vicarious victory and defeat via democratic elections have the same effect on men's testosterone levels as does winning or losing an interpersonal dominance contest. They found that men who voted for the winner, Obama, had stable post-outcome levels of testosterone, and men who voted for losers McCain or Barr had decrements in their testosterone levels. In other words, male voters experienced the outcome of winning or losing vicariously through their candidate even though they did not personally win or lose the election.

Persuasion

A key element of election campaigning is persuasion through which candidates and parties attempt to convince potential voters to turn out and vote for them. Reasoning, emotion, and the characteristics of the message source have been central factors in modern models of persuasion and attitude change. Here again, imaging studies are being used to provide insights into this phenomenon. For instance, in their three cross-cultural fMRI studies, Falk and colleagues (2009) found that across linguistically and culturally diverse groups, as well as across different media, neural activations associated with feeling persuaded were almost exclusively associated with a neural network involved in mentalizing and perspective taking. Persuasion was associated with increased activity in the medial temporal lobes and visual cortex in the first two studies, in which participants viewed text-based messages and made ratings following the scanner session, but not in the third study, in which participants viewed video-based messages and made ratings directly following each message, where persuasion was associated with increased activity in the left ventrolateral prefrontal cortex (VLPFC) (Falk et al. 2009, 2457). These results are consistent with prior behavioral research that suggests a close relationship between social cognition and persuasion, although most other studies of persuasion have not focused directly on perspective taking as a mechanism of persuasion (Campbell and Babrow 2004).

The overlap between the brain regions associated with persuasion effects and mentalizing might show that thinking about the beliefs, desires, and intentions of a specific person represents a special case of thinking about them more generally, whether they are tied to a particular individual or presented as part of a more general argument. All persons are surrounded by signs and other artifacts that suggest particular beliefs (e.g., smoking is bad, thin is good) without referring to a particular person who promotes this belief. Although we typically associate perspectives and points of view with individuals, it has long been known that content often has an impact long after its association with the source is forgotten. Since the left VLPFC was found to be the only region that was more active in response to persuasive compared to unpersuasive messages, it is plausible that this region plays a role in selecting among competing beliefs and memory representations regarding the persuasion topic. This subregion of the VLPFC has been regularly observed in studies of memory selection and emotional reappraisal (Badre and Wagner 2007). Since persuasion involves adopting a new interpretation over an existing one, the VLPFC may play a key role in this process. In any case, as with other aspects of politics it is clear from neuroimaging studies that persuasion can be tied to particular brain regions, and that future research will clarify both how we are persuaded and how we persuade others.

The Brain and Society

The brain is a major resource for society. Both individually and collectively, the health of the brain and the maximization of its potential should be a high priority. Moreover, the increased emphasis on technical skills in the global economy means that brain power will heighten in importance. All evidence suggests that the era of the need for a large unskilled labor force is now history and that individuals without technical or professional skills increasingly risk being socially excluded. Furthermore, in the competitive global economy, countries unable or unwilling to cultivate a highly trained workforce will suffer economically, and resources will go disproportionately to those with the necessary abilities to function in a more mental, less physical environment.

Despite the emphasis in this book on technological interventions, in the end there must be a renewed focus on the social environment to maximize brain potential. There is convincing evidence that the development of the brain is constantly influenced by the outside world. The prenatal environment and the family setting, especially in early childhood, are critical for proper maturation and stimulation of the brain. Furthermore, there is considerable corroboration that childhood socioeconomic status (SES) is an important predictor of neurocognitive performance, particularly of language and executive function, and that SES differences are found in neural processing even when performance levels are equal (Hackman and Farah 2009, 71). Moreover, the sensitive chemical balance of the brain makes it susceptible to environmental hazards, including malnutrition, neurotoxins, violence, substance abuse, and accidents. Ironically, one set of environmental insults that we should be attentive to in the future includes the potential misuse or overuse of neurotechnologies themselves. Each time we intervene in the brain, for treatment or enhancement, unanticipated consequences could be triggered that cause harm to this formidable but fragile organ. The complexity of each human brain and its unique wiring means that iatrogenic harm is always a possibility, one that can be mitigated only by making certain that we intervene only with caution and strong justification.

The framework of politics and public policy is constantly changing. As seen here, new developments in neuroscience are a major source of this change, both in terms of the policy issues raised and in terms of the policy process itself. The technologies reviewed here, particularly the more sophisticated neuroimaging and brain-computer innovations, have the potential either to democratize politics or to undermine democracy, depending on one's perspective and how the technology unfolds. Likewise, discoveries about the brain and its relationship to behavior are altering how we view political behavior and, thus, how we actually practice politics and make public policy. To the extent that our new understanding of the brain clarifies the neurological bases of conflict and cooperation and the behavior of leaders and followers, it has important ramifications for the conduct of domestic as well as international policy making. At the individual and collective levels, neuroscience has the potential to undermine empirical theories, including public choice theories and rational voter models. This, in turn, makes suspect any methodologies that exclude biological variables, even down to the actions of neurotransmitters, the importance of which is reflected in the research summarized here.

Conclusions: The Emergence of Neuroscience Policy

The recent attention placed on neuroscience has raised awareness of its importance to all aspects of human life, but it has not yet placed it high on the policy agenda. This, however, will change. The wide array of new intervention capacities and the tremendous costs of CNS-related problems for society, along with the emergence of the view of inseparability of the mental and physical dimensions of health, will demand considerably more attention from policy makers in the coming decades. The societal and personal benefits of more specifically designed psychotropic drugs, neurogenetic treatments for addictions and neurodegenerative diseases, and a heightened understanding of brain function in general are multiplying.

Regrettably, the emergence of these new intervention techniques comes at a time when health care resources are scarce and competition for funding is tight. Many interventions, particularly medications, are likely to be cost-effective, and in fact might lead to cost savings. However, other emerging treatment strategies, such as brain implants and neurogenetic procedures, will be very costly on a per-case basis. Moreover, because total cost involves frequency of use as well as per-case cost, even less expensive procedures such as deep brain stimulation and brain implants can add significantly to overall health care spending when applied to large populations such as Alzheimer's patients. Seldom, to date, have even rough estimates been given for prospective innovative neurotechnolgy procedures. Although this is understandable at this early stage of development, analysis of cumulative costs is crucial before the new procedures come into routine use. While high cost alone is not justification to block or slow the diffusion of various neural intervention techniques, as with other medical innovations it must be taken into consideration.

One research area that has not received the attention it warrants is the impact of environmental influences on the brain. The role of neurotoxins in altering health status remains poorly understood, not because we are unaware of the risks but because we have placed a relatively low priority on prevention in general. Unlike the research areas discussed here, the study of neurotoxins lacks the urgent, dramatic context found in curative interventions. Despite this relative neglect, the biggest benefits for the health of the population are likely to come from efforts to prevent neural diseases. Rather than placing highest priority on developing new and costly therapies, in the long term we are likely to be better served by finding the causes of mental disorders and diseases and developing policies to reduce their incidence. While prevention goes counter to the individual-oriented medical model because it places emphasis on statistical as opposed to identifiable patients, any policy that continues to minimize the proximate causes of neurological disorders will fail to maximize health no matter how many innovative treatments are invented. Moreover, while some observers argue that preventive efforts interfere with individual privacy and require behavioral changes, as pointed out in chapter 3, all brain interventions, preventive or therapy, inherently raise issues of privacy, autonomy, and confidentiality.

In addition to placing a higher priority on searching for environmental contributions to neurological disorders, we must also incorporate our new knowledge of the brain-behavior association into health care in general. As we come to better understand the neurogenetic roots of risk-taking behavior, addiction, and aggression, we should use this information to design appropriate social strategies for reducing the massive social problems these behaviors cause. Likewise, in light of the high incidence of brain injury and trauma, more emphasis on accident prevention is warranted, particularly among young people, for whom the long-term economic and personal costs can be staggering. Likewise, reductions in domestic abuse, handgun injuries, and violence should be integral elements in a broad-based strategy to trim the need for dramatic methods of brain intervention to treat the results of such actions.

A related issue has to do with the impact of these technologies on our perceptions of candidates for their use. As discussed in chapter 3, the stigma attached to victims of mental disorders is manifest in most countries. Whether this stigma is exacerbated or abated by new interventions depends less on the innovations themselves and more on how society perceives them. Knowledge of the genetic bases of mental disorders, for instance, could lead either to increased sympathy for those dealt a poor genetic hand or to rigid screening programs designed to reduce the number of individuals with those traits in the population. Similarly, our views of violent offenders or those who exhibit other antisocial behaviors will be affected by neuroscience research that leads to the discovery of the biological bases of such behavior. On the one hand, this evidence could produce an empathy with affected individuals and intensified efforts to help them cope with what are basically biochemical problems. Eventually, this could alter the way the justice system handles offenders, by shifting emphasis from punishment to prevention and treatment. On the other hand, this knowledge could result in heightened use of involuntary neurotropic interventions for individuals identified at special risk based on genetic testing, imaging genetics, and brain imaging. Again, the response to this knowledge is political and social.

The same situation pertains to neuroscience findings regarding sex differences, sexual orientation, and addictive personalities or behaviors. While technological advances in neuroscience can provide the basis for better understanding of human variation and an acceptance of differences, alternatively, they can work to exaggerate inequalities and lead to repression of the most vulnerable members of society. How they actually affect society depends on the motivations behind their use. While there is little evidence at present that developments in neuroscience are motivated by anything other than beneficence, ultimately the political climate will determine to what ends they are put.

We must guard against a dual system in which interventions are used voluntarily to improve the lives of the already well-off but involuntarily to control the behavior of the least well-off. Another tiered system to be avoided is one that denies needed treatment to the least well-off members of society. This is especially critical for those interventions without which a person is unable to function as a full member or for enhancement interventions. For the former, the person will remain disadvantaged. For the latter, there is a danger that those persons who already enjoy advantages in capabilities will have the means to extend their advantage, thus widening the gap. These issues in distribution and access become more important in a society where mental capacity is of highest importance.

As in other areas of biomedicine, innovative developments in neuroscience will emerge at an accelerating rate in the coming decades. Furthermore, because we are still at an early stage of understanding the nervous system, the speed of discovery in neuroscience is likely to outpace that in other areas of medical science. This book has demonstrated that despite the recency of knowledge in neuroscience, the technological fruits of the Decade of the Brain represent an impressive start to a more complete understanding of brain function and dysfunction, the genetic and neurological basis of human behavior, and the biochemical foundations of mental disorders. Less impressive to date have been our gains in knowledge of the impact of the environment on the brain.

This book also has raised a broad range of policy issues that accompany our unfolding knowledge of the brain and the application of a widening array of techniques for intervention in the brain. In coming years these issues are likely to elicit heightened research attention from social scientists and bioethicists. Although brain policy reflects problems generic to other biomedical areas, because the impact of the brain is pervasive in all areas of human life and death, it raises unique challenges to the study of political behavior and the human condition in general. Neuroscience policy, therefore, represents a new conceptual area of study that warrants urgent attention from policy makers, policy analysts, and informed citizens. Moreover, the linkage between genetics and neuroscience is intractable. As we come to understand the mechanisms and role of neurotransmitters and receptors and their relationship to genes, policy concerns inherent in these seemingly disparate substantive areas will merge. The lines between genetic policy and brain policy drawn to date are most likely a reflection of the relatively primitive state of knowledge about the brain—in other words, a temporary condition—but already much genetic research is directed at ameliorating mental disorders or deficits or at behavioral problems.

Furthermore, as the discussion in chapter 8 illustrated, our conception of the brain is deeply embedded in the full range of death-related issues. Our knowledge about the brain and the ability to measure more precisely brain activity will strongly influence our very definition of death. Similarly, policy on organ transplantation is dependent on the concept of brain death, as are the issues concerning the use of organs from anencephalic infants. Less directly, brain policy is interrelated with policy concerning fetal workplace hazards, fetal research, and the treatment of severely ill newborns with Down syndrome or other brain-centered problems. Given these linkages, brain policy takes on even more significance as an important new area for investigation by those trained in public policy.

References

Abbott, A. 2008. Brain electrodes can improve learning. *Nature.* doi:10.1038/news.2008.538.

Abbott, A. 2009a. Brain imaging skewed: Double dipping of data magnifies errors in functional MRI scans. *Nature* 458:1087–1090.

Abbott, A. 2009b. Brain imaging studies under fire: Social neuroscientists criticized for exaggerating links between brain activity and emotions. *Nature* 457:245–246.

Abbott, A. 2009c. Neuroscience: Opening up brain surgery. *Nature* 461: 866–868.

Abbott, A. 2011. Brain implants have long-lasting effect on depression: Technique can alleviate symptoms for six years. *Nature.* doi:10.1038/news.2011.76.

Ad Hoc Committee of the Harvard Medical School. 1968. A definition of irreversible coma. *Journal of the American Medical Association* 205 (6): 337–340.

Adolphs, R. 2003. Cognitive neuroscience of human social behaviour. *Nature Reviews. Neuroscience* 4:165–177.

Adolphs, R. 2010. What does the amygdale contribute to social cognition? *Annals of the New York Academy of Sciences* 1191:42–61.

Agar, N. 2007. Whereto transhumanism? The literature reaches a critical mass. *Hastings Center Report* 37 (3): 12–17.

Aguilar, J. A., T. S. Vesagas, R. D. Jamora, L. Ledesma, et al. 2011. The promise of deep brain stimulation in X-linked dystonia parkinsonism. *International Journal of Neuroscience* 121 (Suppl. 1): 57–63.

Alexander, G. M., and M. Hines. 2002. Sex differences in response to children's toys in nonhuman primates (Cercopithecus aethiops sabaeus). *Evolution and Human Behavior* 23:467–479.

Alford, J. R., C. Funk, and J. R. Hibbing. 2005. Are political orientations genetically transmitted? *American Political Science Review* 99:153–167.

Alford, J. R., and J. R. Hibbing. 2004. The origin of politics: An evolutionary theory of political behavior. *Perspectives on Politics* 2 (4): 707–723.

Alia-Klein, N., R. Z. Goldstein, A. Kriplani, J. Logan, D. Tomasi, B. Williams et al. 2008. Brain monoamine oxidase A activity predicts trait aggression. *Journal of Neuroscience* 28 (19):5099–5104.

Allen, L. S., and R. A. Gorski. 1991. Sexual dimorphism of the anterior commissure and massa intermedia of the human brain. *Journal of Comparative Neurology* 312:97–104.

American Bar Association. 1978. *ABA Annual Report* 100 (February 1975): 231–323.

American Psychiatric Association (APA). 1990. *The Practice of ECT: Recommendations for Treatment, Training, and Privileging.* Washington, DC: American Psychiatric Press Inc.

Amodio, D. M., J. T. Jost, S. L. Master, and C. M. Yee. 2007. Neurocognitive correlates of liberalism and conservatism. *Nature Neuroscience* 10:1246–1247.

Anderson, A. K. 2007. Feeling emotional: The amygdala links emotional perception and experience. *Social Cognitive and Affective Neuroscience* 2 (2): 71–72.

Anderson, B. M., M. C. Stevens, S. A. Meda, K. Jordan, V. D. Calhoun, and G. Pearlson. 2011. Functional imaging of cognitive control during acute alcohol intoxication. *Alcoholism, Clinical and Experimental Research* 35:156–165.

Anderson, C. A., and B. J. Bushman. 2002. Human aggression [altruism and aggression]. *Annual Review of Psychology* 53: 27–51.

Anderson, J. E. 1990. *Public Policy-Making: An Introduction.* Boston: Houghton Mifflin.

Andrews, M. M., A. M. Shashwath, A. D. Thomas, M. N. Potenza, J. H. Krystal, et al. 2011. Individuals family history positive for alcoholism show functional magnetic resonance imaging differences in reward sensitivity that are related to impulsivity factors. *Biological Psychiatry* 69 (7): 675–683.

Annas, G. J. 2007. Forward: Imaging in a new era of neurimaging, neuroethics and neurolaw. *American Journal of Law & Medicine* 33:163–170.

Anonymous 2009. Brain difference in psychopaths identified. *Telemedicine Law Weekly,* August 22, 168.

Antommaria, A. H. M. 2010. Conceptual and ethical issues in the declaration of death. *Pediatrics* 31:427–430. doi:10.1542/pir.31-10-427

Appelbaum, P. 2005. Psychopharmacology and the power of narrative. *American Journal of Bioethics* 5 (3):48–49.

Applbaum, A. I., J. C. Tilburt, M. T. Collins, and D. Wendler. 2008. A family's request for complementary medicine after patient brain death. *Journal of the American Medical Association* 299 (18): 2188–2193.

Archer, J. 2006. Testosterone and human aggression: An evaluation of the challenge hypothesis. *Neuroscience and Biobehavioral Reviews* 30 (3): 319–345.

Archer, J., N. Graham-Kevan, and M. Davies. 2005. Testosterone and aggression: A reanalysis of Book, Starzyk, and Quinsey's (2001) study. *Aggression and Violent Behavior* 10:241–261.

Archives of General Psychiatry. 1996. Violence, crime, and mental illness: How strong a link? *Archives of General Psychiatry* 53 (6): 481–486.

Ariely, D., and G. S. Berns. 2010. Neuromarketing: The hope and hype of neuroimaging in business. *Nature Reviews. Neuroscience* 11:284–292.

Armony, J. L., and R. J. Dolan. 2002. Modulation of spatial attention by fear-conditioned stimuli: An event-related fMRI study. *Neuropsychologia* 40 (7): 817–826.

Árnason, G. 2010. Neuroimaging, uncertainty, and the problem of dispositions. *Cambridge Quarterly of Healthcare Ethics* 19:188–195.

Arrigo, B. A. 2007. Punishment, freedom, and the culture of control: The case of brain imaging and the law. *American Journal of Law & Medicine* 33 (2–3): 457–482.

Arrow, K. J. 1974. *The Limits of Organization*. New York: W. W. Norton.

Ashcroft, R. E., and K. P. Gui. 2005. Ethics and world pictures in Kamm on enhancement. *American Journal of Bioethics* 5 (3): 21–22.

Badre, D., and A. D. Wagner. 2007. Left ventrolateral prefrontal cortex contributions to the control of memory. *Neuropsychologia* 45 (13): 2883–2901.

Bailey, R. 2005. *Liberation Biology: The Scientific and Moral Case for the Biotech Revolution*. Amherst, MA: Prometheus Books.

Ballew, C. C., and A. Todorov. 2007. Predicting political elections from rapid and unreflective face judgments. *Proceedings of the National Academy of Sciences of the United States of America* 104:17948–17953.

Bandettini, P. A. 2009. What's new in neuroimaging methods? *Annals of the New York Academy of Sciences* 1156:260–293.

Banjo, O. C., R. Nadler, and P. B. Reiner. 2010. Physician attitudes towards pharmacological cognitive enhancement: Safety concerns are paramount. *PLoS ONE* 5 (12): e14322.

Bao, A.-M., and D. F. Swaab. 2011. Sexual differentiation of the human brain: Relation to gender identity, sexual orientation and neuropsychiatric disorders. *Frontiers in Neuroendocrinology*. doi:10.1016/j.yfrne.2011.02.007.

Baron, L., S. D. Shemie, J. Teitelbaum, and C. J. Doig. 2006. Brief review: History, concept and controversies in the neurological determination of death. *Neuroanesthesia and Intensive Care* 53 (6):602–608.

Baron-Cohen, S. 2010. Empathizing, systemizing and the extreme male brain theory of autism. *Progress in Brain Research* 186:167–175.

Baron-Cohen, S. 2011. *Zero Degrees of Empathy: A New Theory of Human Cruelty*. New York: Allen Lane/Basic Books.

Barth, A. S. 2007. A double-edged sword: The role of neuroimaging in federal capital sentencing. *American Journal of Law & Medicine* 33 (2–3): 501–522.

Bartholow, B. D., B. J. Bushman, and M. A. Sestir. 2006. Chronic violent video game exposure and desensitization to violence: Behavioral and event-related brain potential data. *Journal of Experimental Social Psychology* 42 (4): 532–539.

Baskin, J. H., J. G. Edersheim, and B. H. Price. 2007. Is a picture worth a thousand words? Neuroimaging in the courtroom. *American Journal of Law & Medicine* 33 (2–3): 239–269.

Batts, S. 2009. Brain lesions and their implications in criminal responsibility. *Behavioral Sciences & the Law* 27:261–272.

Baumgartner, F. R., and B. D. Jones. 1991. Agenda dynamics and policy subsystems. *Journal of Politics* 53:1047.

Baumgartner, T., M. Heinrichs, A. Vonlanthen, U. Fischbacher, and E. Fehr. 2008. Oxytocin shapes the neural circuitry of trust and trust adaptation in humans. *Neuron* 58:639–650.

Baxter, M. G., and E. A. Murray. 2002. The amygdala and reward. *Nature Neuroscience* 3:563–574.

Beasley, D. 2003. Researchers compile "atlas" of the brain. Reuters News. http://news.yahoo.com/news?tmpl=story2&cid=570&u=%2Fnm%2F20030808%2Fsc_nm%2Fhealth_brain.html.

Becker, J. B., and M. Hu. 2008. Sexual differences in drug abuse. *Frontiers in Neuroendocrinology* 29:36–47.

Beckwith, J., and C. Morris. 2008. Twin studies of political behavior: Untenable assumptions? *Perspectives on Politics* 6:785–791.

Bell, E. C., G. Mathieu, and E. Racine. 2009. Preparing the ethical future of deep brain stimulation. *Surgical Neurology* 72:577–586.

Bell, E. C., M. C. Willson, A. H. Wilman, S. Dave, and P. H. Silverstone. 2006. Males and females differ in brain activation during cognitive tasks. *NeuroImage* 30:529–538.

Benabid, A. L. 2007. What the future holds for deep brain stimulation. *Expert Review of Medical Devices* 4 (6): 895–903.

Bennett, C. M., and M. B. Miller. 2010. How reliable are the results from functional magnetic resonance imaging? *Annals of the New York Academy of Sciences* 1191:133–155.

Bennett, M., and P. Hacker. 2011. Criminal law as it pertains to patients suffering from psychiatric diseases. *Journal of Bioethical Inquiry* 8 (1): 45–58.

Bernat, J. L. 2005. The concept and practice of brain death. *Progress in Brain Research* 150:369–379.

Bernat, J. L. 2010. The Debate over Death Determination in DCD. *Hastings Center Report* 40 (3): 3.

Bickart, K. C., C. I. Wright, R. J. Dautoff, B. C. Dickerson, and L. Feldman Barrett. 2011. Amygdala volume and social network size in humans. *Nature Neuroscience* 14:163–164.

Birbaumer, N., R. Veit, M. Lotze, M. Erb, C. Hermann, W. Grodd, and H. Flor. 2005. Deficient fear conditioning in psychopathy: A functional magnetic resonance imaging study. *Archives of General Psychiatry* 62:799–805.

Bishop, S. J. 2007. Neurocognitive mechanisms of anxiety: An integrative account. *Trends in Cognitive Sciences* 11 (7):307–316.

Biswal, B. B., M. Mennes, X.-N. Zuo, S. Gohel, C. Kelly, et al. 2010. Toward discovery science of human brain function. *Proceedings of the National Academy of Sciences of the United States of America* 107 (10): 4734–4739.

Blair, R. J. R. 2003. Neurobiological basis of psychopathy. *British Journal of Psychiatry* 182:5–19.

Blair, R. J. R. 2005. Applying a cognitive neuroscience perspective to the disorder of psychopathy. *Development and Psychopathology* 17:865–891.

Blair, R. J. R. 2007a. The amygdala and ventromedial prefrontal cortex in morality and psychopathy. *Trends in Cognitive Sciences* 11 (9): 387–392.

Blair, R. J. R. 2007b. Dysfunctions of medial and lateral orbitofrontal cortex in psychopathy. *Annals of the New York Academy of Sciences* 1121:461–479.

Blair, R. J. R., K. S. Peschardt, S. Budhani, D. G. V. Mitchell, and D. S. Pine. 2006. The development of psychopathy. *Journal of Child Psychology and Psychiatry, and Allied Disciplines* 47 (3–4):262–276.

Blank, R. H. 1999. *Brain Policy: How the New Neuroscience Will Change Our Lives and Our Politics*. Washington, DC: Georgetown University Press.

Blank, R. H. 2006. The brain, aggression and public policy. *Politics and the Life Sciences* 24 (1–2): 12–21.

Blank, R. H. 2011. Brain sciences and politics: Some linkages. In *Biology and Politics: The Cutting Edge*, ed. S. Peterson and A. Somit, 205–229. Bingley, UK: Emerald Books.

Blank, R. H., and J. C. Merrick. 2005. *End-of-Life Decision Making: A Cross-National Study*. Cambridge, MA: MIT Press.

Bloch, K. E., and K. C. McMunigal. 2005. *Criminal Law: A Contemporary Approach : Cases, Statutes, and Problems*. New York: Aspen Publishers.

Bloche, M. G., and J. H. Marks. 2005. Doctors and interrogators at Guantanamo Bay. *The New England Journal of Medicine* 353 (1): 6–8.

Boggio, P. S., N. Sultani, S. Fecteau, L. Merabet, T. Mecca, et al. 2008. Prefrontal cortex modulation using transcranial DC stimulation reduces alcohol craving: A double-blind, sham-controlled study. *Drug and Alcohol Dependence* 92: 55–60.

Boire, R. G. 2005. Searching the brain: The Fourth Amendment implications of brain-based deception detection devices. *American Journal of Bioethics* 5:62–63.

Bonnicksen, A. L. 1992. Human embryos and genetic testing: A private policy model. *Politics and the Life Sciences* 11 (1): 53–62.

Bootzin, R. R., J. R. Acocella, and L. B. Alloy. 1993. *Abnormal Psychology: Current Perspectives*. New York: McGraw-Hill.

Boseley, S. 2000. Transplant row over pain rule. *Guardian*, August 19. http://www.guardian.co.uk/uk/2000/aug/19/sarahboseley1.

Bostrom, N. 2003. Human genetic enhancements: A transhumanist perspective. *Journal of Value Inquiry* 37:493–506.

Boutrel, B., N. Cannella, and L. De Lecea. 2010. The role of hypocretin in driving arousal and goal-oriented behaviors. *Brain Research* 1314:103–111.

Brain Fingerprinting Laboratories. 2003. Scientific Procedure, Research, and Applications. http://www.scribd.com/doc/49079040/BRAIN-FINGERPRINTING.

Brakemeier, E.-L., R. Berman, J. Prudic, K. Zwillenberg, and H. A. Sackeim. 2011. Self-evaluation of the cognitive effects of electroconvulsive therapy. *Journal of ECT* 27 (1): 59–66.

Brass, M., and P. Haggard. 2008. The What, When, Whether model of intentional action. *Neurocientist* 14:319–325.

Bresnahan, M. J., and K. Mahler. 2010. Ethical debate over organ donation in the context of brain death. *Bioethics* 24 (2): 54–60.

Brief, E., and J. Illes. 2010. Tangles of neurogenetics, neuroethics, and culture. *Neuron* 68:174–177.

Broome, M. R., L. Bortolotti, and M. Mameli. 2010. Moral responsibility and mental illness: A case study. *Cambridge Quarterly of Healthcare Ethics* 19 (2):179–187.

Brown, R., and H. Hewstone. 2005. An integrative theory of intergroup contact. In *Advances in Experimental Social Psychology*. vol. 37. ed. M. Zanna, 255–343. San Diego, CA: Academic Press.

Bruneau, E. G., and R. Saxe. 2010. Attitudes towards the outgroup are predicted by activity in the precuneus in Arabs and Israelis. *NeuroImage* 52 (4): 1704–1711.

Buchman, D. Z., J. Illes, and P. B. Reiner. 2010. The paradox of addiction neuroscience. *Neuroethics* 22 (June). doi:10.1007/s12152-010-9079-z.

Buchman, D. Z., W. Skinner, and J. Illes. 2010. Negotiating the relationship between addiction, ethics, and brain science. *AJOB Neuroscience* 1 (1): 36–45.

Buckholtz, J. W., and A. Meyer-Lindenberg. 2008. MAOA and the neurogenetic architecture of human aggression. *Trends in Neurosciences* 31:120–129.

Buckholtz, J. W., M. T. Treadway, R. L. Cowan, N. D. Woodward, S. D. Benning, et al. 2010. Mesolimbic dopamine reward system hypersensitivity in individuals with psychopathic traits. *Nature Neuroscience* 13 (4): 419–421.

Bueno de Mesquita, B. 1981. *The War Trap*. New Haven, CT: Yale University Press.

Buller, T. 2005. Can we scan for truth in a society of liars? *American Journal of Bioethics* 5 (2): 58–60.

Buller, T. 2010. Rationality, responsibility and brain function. *Cambridge Quarterly of Healthcare Ethics* 19:196–204.

Burdea, G., and P. Coiffet. 1994. *Virtual Reality Technology*. New York: John Wiley and Sons.

Burkle, C. M., A. M. Schipper, and E. F. M. Wijdicks. 2011. Brain death and the courts. *Neurology* 76 (9): 837–841.

Burns, K., and A. Bechara. 2007. Decision making and free will: A neuroscience perspective. *Behavioral Sciences & the Law* 25:263–280.

Bushman, B. J., and L. R. Huesmann. 2006. Short-term and long-term effects of violent media on aggression in children and adults. *Archives of Pediatrics & Adolescent Medicine* 160 (4): 348–352.

Busl, K. M., and D. M. Greer. 2009. Pitfalls in the diagnosis of brain death. *Neurocritical Care* 11 (2): 276–287.

Cacioppo, J. T. 2006. The neuroscience of social interaction: Decoding, imitating, and influencing the actions of others. *American Journal of Psychiatry* 19:664–670.

Cacioppo, J. T., and J. Decety. 2011. Social neuroscience: Challenges and opportunities in the study of complex behavior. *Annals of the New York Academy of Sciences* 1224:162–173.

Cahill, L. 2006. Why sex matters for neuroscience. *Nature Reviews. Neuroscience.* doi:10.1038/nrn1909.

Cahill, L. 2010. Sex influences on brain and emotional memory: The burden of proof has shifted. *Progress in Brain Research* 186:29–40.

Cairns, E., J. B. Kenworthy, A. Campbell, and M. Hewstone. 2006. The role of in-group identification, religious group membership, and intergroup conflict in moderating in-group and out-group affect. *British Journal of Social Psychology* 45:701–716.

Camerer, C., G. Loewenstein, and D. Prelec. 2005. Neuroeconomics: How neuroscience can inform economics. *Journal of Economic Perspectives* 43: 9–64.

Campbell, N. D. 2010. Toward a critical neuroscience of "addiction." *Biosocieties* 5 (1): 89–104.

Campbell, R., and A. Babrow. 2004. The role of empathy in responses to persuasive risk communication: Overcoming resistance to HIV prevention messages. *Health Communication* 16:159–182.

Capps, B. 2011. Libertarianism, legitimation, and the problems of regulating cognition-enhancing drugs. *Neuroethics* 4:119–128.

Capron, A. M., and L. R. Kass. 1972. A statutory definition of the standards for determining human death: An appraisal and a proposal. *University of Pennsylvania Law Review* 121 (87):101–104.

Carnagey, N., C. Anderson, and B. Bartholow. 2007. Media violence and social neuroscience: New questions and new opportunities. *Current Directions in Psychological Science* 16 (4): 178–182.

Carney, S., and J. Geddes. 2003. Electroconvulsive therapy. *British Medical Journal* 326:43.

Carré, J. M., and C. M. McCormick. 2008. Aggressive behavior and change in salivary testosterone concentrations predict willingness to engage in a competitive task. *Hormones and Behavior* 54 (3): 403–409.

Carter, A., E. Bell, E. Racine, and W. Hall. 2010. Ethical issues raised by proposals to treat addiction using deep brain stimulation. *Neuroethics* 2 (September). doi:10.1007/s12152-010-9091-3.

Carter, A., and W. Hall. 2011. Proposals to trial deep brain stimulation to treat addiction are premature. *Addiction (Abingdon, England)* 106 (2): 235–237.

Cartwright, G. F. 1994. Virtual or real? The mind in cyberspace. *Futurist* 28 (2): 22–26.

Caspi, A., A. R. Hariri, A. Holmes, R. Uher, and T. E. Moffitt. 2010. Genetic sensitivity to the environment: The case of the serotonin transporter gene and its implications for studying complex diseases and traits. *American Journal of Psychiatry* 167 (5): 509–527.

Chandler, J. 2011. Autonomy and the unintended legal consequences of emerging neurotherapies. *Neuroethics*. doi:10–1007/s12152–011–9109–5.

Chance, B., E. Anday, S. Nioka, et al. 1998. A novel method for fast imaging of brain function, non-invasively with light. *Optics Express* 2:411.

Changeux, J.-P. 1997. *Neuronal Man: The Biology of the Mind*. Princeton, NJ: Princeton University Press.

Chatterjee, A. 2007. Cosmetic neurology and cosmetic surgery: Parallels, predictions, and challenges. *Cambridge Quarterly of Healthcare Ethics* 16:129–137.

Chatterjee, A. 2010. Neuroaesthetics: A coming of age story. *Journal of Cognitive Neuroscience* 23 (1): 53–62.

Cheon, B. K., D.-M. Im, T. Harada, J.-S. Kim, V. A. Mathur, J. M. Scimeca, T. B. Parrish, H. W. Park, and J. Y. Chiao. 2011. Cultural influences on neural basis of intergroup empathy. *NeuroImage* 57 (2): 642–650.

Cheshire, W. P. 2007. Can grey voxels resolve neurethical dilemmas? *Ethics & Medicine* 23 (3): 135–140.

Chi, R. P., and A. W. Snyder. 2011. Facilitate insight by non-invasive brain stimulation. *PLoS ONE* 6 (2): e16655. doi:10.1371/journal.pone.0026655.

Chiao, J. Y. 2010. At the frontier of cultural neuroscience: Introduction to the special issue. *Social Cognitive and Affective Neuroscience* 5 (2–3): 109–110.

Chiao, J. Y., and N. Ambady. 2007. Cultural neuroscience: Parsing universality and diversity across levels of analysis. In *Handbook of Cultural Psychology*, ed. S. Kitayama and D. Cohen, 237–254. New York: Guilford Press.

Chiao, J. Y., N. E. Bowman, and H. Gill. 2008. The political gender gap: Gender bias in facial inferences that predict voting behavior. *PLoS ONE* 3:3666.

Chiao, J. Y., T. Iidaka, H. L. Gordon, et al. 2008. Cultural specificity in amygdala response to fear faces. *Journal of Cognitive Neuroscience* 20:2167–2174.

Childress, A. R., P. D. Mozley, W. McElgin, J. Fitzgerald, M. Reivich, and C. P. O'Brien. 1999. Limbic activation during cue-induced cocaine craving. *American Journal of Psychiatry* 156 (1):11–18.

Chiong, W. 2005. Brain death without definitions. *Hastings Center Report* 35 (6): 20–30.

Chorover, S. L. 2005. Who needs neuroethics? *Lancet* 365:2081–2082.

Choudhury, S., S. K. Nagel, and J. Slaby. 2009. Critical neuroscience: Linking neuroscience and society through critical practice. *Biosocieties* 4 (1): 61–77.

Christen, M., and S. Müller. 2011. Single cases promote knowledge transfer in the field of DBS. *Frontiers in Integrative Neuroscience* 5. doi:10.3389/fnint .2011.00013.

Churchland, P. M. 1995. *The Engine of Reason, the Seat of the Soul: A Philosophical Journey into the Brain.* Cambridge, MA: MIT Press.

Churchland, P. S. 2011. *Braintrust: What Neuroscience Tells Us about Morality.* Princeton, NJ: Princeton University Press.

Ciaramelli, E., M. Muccioli, E. Ladavas, and G. di Pellegrino. 2007. Selective deficit in personal moral judgment following damage to ventromedial prefrontal cortex. *Social Cognitive and Affective Neuroscience* 2 (2): 84–92.

Cima, M., F. Tonnaer, and M. D. Hauser. 2010. Psychopaths know right from wrong but don't care. *Social Cognitive and Affective Neuroscience.* doi:10.1093/ scan/nsp051.

Clark, L., S. R. Chamberlain, and B. J. Sahakian. 2009. Neurocognitive mechanisms in depression: Implications for treatment. *Annual Review of Neuroscience* 32:57–74.

Clausen, J. 2011. Conceptual and ethical issues with brain-hardware interfaces. *Current Opinion in Psychiatry* 24 (6): 495–501.

Cobb, R. W., and C. D. Elder. 1983. *Participation in American Politics: Dynamics of Agenda-Building.* 2nd ed. Baltimore, MD: Johns Hopkins University Press.

Coccaro, E. F., M. S. McCloskey, D. A. Fitzgerald, and K. L. Phan. 2007. Amygdala and orbitofrontal reactivity to social threat in individuals with impulsive aggression. *Biological Psychiatry* 62:168–178.

Colletti, V., R. V. Shannon, M. Carner, S. Veronese and L. Colletti. 2009. Progress in restoration of hearing with the auditory brainstem implant. *Progress in Brain Research* 175: 333–345.

Comings, D. E. 1996. Both genes and environment play a role in antisocial behavior. *Politics and the Life Sciences* 15 (1):84–85.

Coors, M. E., and L. Hunter. 2005. Evaluation of genetic enhancement: Will human wisdom properly acknowledge the value of evolution? *American Journal of Bioethics* 5 (3): 21–22.

Coplan, M. J., S. C. Patch, R. D. Masters, and M. S. Bachman. 2007. Confirmation of and explanations for elevated blood lead and other disorders in children exposed to water disinfection and fluoridation chemicals. *Neurotoxicology* 28 (5): 1032–1042.

Coricelli, G., and R. Nagel. 2009. Neural correlates of depth of strategic reasoning in medial prefrontal cortex. *Proceedings of the National Academy of Sciences of the United States of America* 106:9163–9168.

Corwin, R. L., and P. S. Grigson. 2009. Symposium overview--Food addiction: fact or fiction? *Journal of Nutrition* 139 (3):617–619.

Couppis, M. H., C. H. Kennedy, and G. D. Stanwood. 2008. Differences in aggressive behavior and in the mesocorticolimbic DA system between A/J and BALB/cJ mice. *Synapse* 62 (10): 715–724.

Courtwright, D. T. 2010. The NIDA brain disease paradigm: History, resistance and spinoffs. *Biosocieties* 5 (1): 137–147.

Craig, I. W., and K. E. Halton. 2009. Genetics of human aggressive behaviour. *Human Genetics* 126 (1): 101–113.

Crawford, M. B. 2008. The limits of neuro-talk. *New Atlantis* 19:65–78.

Crews, D. 2008. Epigenetics and its implications for behavioral neuroendocrinology. *Frontiers in Neuroendocrinology* 29 (3): 344–357.

Crombag, H. S., C. R. Ferrario, and T. E. Robinson. 2008. The rate of intravenous cocaine or amphetamine delivery does not influence drug-taking and drug-seeking behavior in rats. *Pharmacology, Biochemistry, and Behavior* 90:797–804.

Crowe, S. L., and R. J. R. Blair. 2008. The development of antisocial behaviour: What can we learn from functional neuroimaging studies? *Development and Psychopathology* 20:1145–1159.

Cunningham, W. A., C. L. Raye, and M. K. Johnson. 2004. Implicit and explicit evaluation: fMRI correlates of valence, emotional intensity, and control in the processing of attitudes. *Journal of Cognitive Neuroscience* 16:1717–1729.

Cunningham, W. A., J. J. Van Bavel, and I. R. Johnsen. 2008. Affective flexibility: Evaluative processing goals shape amygdale activity. *Psychological Science* 19:152–160.

Curtis, A. L., T. Bethea, and R. J. Valentino. 2005. Sexually dimorphic responses of the brain norepinephrine system to stress and corticotropin-releasing factor. *Neuropsychopharmacology* 31:544–554.

Custers, R., and H. Aarts. 2010. The unconscious will: How the pursuit of goals operates outside of conscious awareness. *Science* 329 (5987): 47–50.

DARPA. 2003. *Augmented Cognition Technical Integration Experiment (TIE)*. San Diego: SPAWAR System Center.

Davidson, R. J., and S. Begley. 2012. *The Emotional Life of Your Brain: How Its Unique Patterns Affect the Way You Think, Feel, and Live—And How You Can Change Them*. New York: Hudson Street Press.

Davidson, R. J., K. M. Putnam, and C. L. Larson. 2000. Dysfunction in the neural circuitry of emotion regulation: A possible prelude to violence. *Science* 289 (5479): 591–594.

Davis, C., and J. C. Carter. 2009. Compulsive overeating as an addiction disorder: A review of theory and evidence. *Appetite* 53:1–8.

Davis, C., R. D. Levitan, J. Carter, et al. 2008. Personality and eating behaviors: A case-control study of binge eating disorder. *International Journal of Eating Disorders* 41:243–250.

Davis, C., S. Strachan, and M. Berkson. 2004. Sensitivity to reward: Implications for overeating and overweight. *Appetite* 42:131–138.

Dayan, P., and Q. J. M. Huys. 2009. Serotonin in affective control. *Annual Review of Neuroscience* 32:95–126.

Deary, I. J., W. Johnson, and L. M. Houlihan. 2009. Genetic foundations of human intelligence. *Human Genetics* 126 (1): 215–232.

De Biasi, M., and J. A. Dani. 2011. Reward, addiction, withdrawal to nicotine. *Annual Review of Neuroscience* 34:105–130.

Decety, J. 2010. The neurodevelopment of empathy in humans. *Developmental Neuroscience* 32 (4): 257–268.

Decety, J., S. Echols, and J. Correll. 2010. The blame game: The effect of responsibility and social stigma on empathy for pain. *Journal of Cognitive Neuroscience* 22 (5): 985–997.

De Dreu, C. K. W., L. L. Greer, M. J. J. Handgraaf, S. Shalvi, G. A. Van Kleef, et al. 2010. The neuropeptide oxytocin regulates parochial altruism in intergroup conflict among humans. *Science* 328 (5984): 1408–1411.

Dehaene, S. 2005. Evolution of human cortical circuits for reading and arithmetic: the neuronal recycling hypothesis. In *From Monkey Brain to Human Brain*, ed. S. Dehaene, J.-R. Duhamel, M. D. Hauser, and G. Rizzolatti, 137–157. Cambridge, MA: MIT Press.

de Jongh, R., I. Bolt, M. Schermer, and B. Olivier. 2008. Botox for the brain: Enhancement of cognition, mood and pro-social behavior and blunting of unwanted memories. *Neuroscience and Biobehavioral Reviews* 32 (4): 760–776.

De Oliveira-Souza, R., R. D. Hare, I. E. Bramati, G. J. Garrido, F. A. Ignácio, F. Tovar-Moll, and J. Moll. 2008. Psychopathy as a disorder of the moral brain: Fronto-temporo-limic grey matter reductions demonstrated by voxel-based morphometry. *NeuroImage* 40 (3):1202–1213.

Derks, B., M. Inzlicht, and S. Kang. 2008. The neuroscience of stigma and stereotype threat. *Group Processes & Intergroup Relations* 11 (2): 161–181.

Derntl, B., C. Windischerger, S. Robinson, I. Kryspin-Exner, R. C. Gur, E. Moser, and U. Habel. 2009. Amygdala activity to fear and anger in healthy young males is associated with testosterone. *Psychoneuroendocrinology* 34 (5):687–693.

DeWall, C. N., T. Deckman, M. T. Gailliot, and B. J. Bushman. 2011. Sweetened blood cools hot tempers: Physiological self-control and aggression. *Aggressive Behavior* 37:73–80.

Díaz, J. L. 2010. The psychobiology of aggression and violence: Bioethical implications. *International Social Science Journal* 61:233–245.

Di Chiara, G., and V. Bassareo. 2007. Reward system and addiction: What dopamine does and doesn't do. *Current Opinion in Pharmacology* 7 (2): 69–76.

Dickie, E. W., and J. L. Armony. 2008. Amygdala responses to unattended fearful faces: Interaction between sex and trait anxiety. *Psychiatry Research* 162 (1):51–57.

Dillow, C. 2010. New optogenetic neural implants use precise beams of light to manipulate the brain. http://www.popsci.com/science/article/2010-02/new-optical-neural-implants-use-light-manipulate-brain?page=.

Dinsmore, J., and A. Garner. 2009. Brain stem death. *Surgery* 27 (5): 216–220.

Dmitrieva, J., C. Chen, E. Greenberger, O. Ogunseitan, and Y.-C. Ding. 2011. Gender-specific expression of the DRD4 gene on adolescent delinquency, anger and thrill seeking. *Social Cognitive and Affective Neuroscience* 6 (1): 82–89.

Dolan, M., W. J. F. Deakin, N. Roberts, and I. Anderson. 2002. Serotonergic and cognitive impairment in impulsive aggressive personality disordered offenders: Are there implications for treatment? *Psychological Medicine* 32 (1):105–117.

Dolan, S. L., A. Bechara, and P. E. Nathan. 2008. Executive dysfunction as a risk marker for substance abuse: The role of impulsive personality traits. *Behavioral Sciences & the Law* 26:799–822.

Dominguez Duque, J. F., R. Turner, E. D. Lewis, and G. Egan. 2010. Neuroanthropology: A humanistic science for the study of the culture–brain nexus. *Social Cognitive and Affective Neuroscience* 5 (2–3): 138–147.

Dossey, L. 2010. Neurolaw or Frankelaw? The thought police have arrived. *Explore (New York)* 6 (5): 275–286.

Double, R. 1991. *The Non-Reality of Free Will*. New York: Oxford University Press.

Double, R. 1996. *Metaphilosophy and Free Will*. New York: Oxford University Press.

Draca, S. 2010. Gender-specific functional cerebral asymmetries and unilateral cerebral lesion sequelae. *Reviews in the Neurosciences* 21 (6): 421–425.

Drewnowski, A., and N. Damon. 2005. The economics of obesity: Dietary energy density and energy cost. *American Journal of Clinical Nutrition* 82:265S–273S.

Dunagan, J. 2010. Six eurocentric ndustries. The Institute for the Future. http://www.good.is/post/six-neurocentric-industries.

Dunbar, D., H. I. Kushner, and S. Vrecko. 2010. Drugs, addiction and society. *Biosocieties* 5 (1): 2–7.

Dupont, R. L. 1997. *The Selfish Brain: Learning from Addiction*. Washington, DC: American Psychiatric Press.

Eagleman, D. 2008. Neuroscience and the Law. *Houston Lawyer* 45:36–38.

Eaton, M. L., and J. Illes. 2007. Commercializing cognitive neurotechnology: The ethical terrain. *Nature Biotechnology* 25:393–397.

Eberhardt, J. L. 2005. Imaging race. *American Psychologist* 60 (2):181–190.

Eberl, J. T. 2005. A Thomistic understanding of human death. *Bioethics* 19 (1): 29–48.

Ebstein, R. P. 2006. The molecular genetic architecture of human personality: Beyond self-report questionnaires. *Molecular Psychiatry* 11:427–445.

Eisenberger, N. I., B. M. Way, S. E. Taylor, W. T. Welch, and M. D. Lieberman. 2007. Understanding genetic risk for aggression: Clues from the brain's response to social exclusion. *Biological Psychiatry* 61:1100–1108.

Ellis, L. 2011. Evolutionary neuroandrogenic theory and universal gender differences in cognition and behaviour. *Sex Roles* 64:707–722.

El-Zein, R. A., S. Z. Abdel-Rahmanb, M. J. Hayb, M. S. Lopeza, M. L. Bondya, D. L. Morris, and M. S. Legator. 2005. Cytogenetic effects in children treated with methylphenidate. *Cancer Letters* 230 (2): 284–291.

Emanuel, L. L. 1995. Re-examining death: The asymptotic model and a bounded zone definition. *Hastings Center Report* 25 (4): 27–35.

Endicott, P. G., H. A. Kennedy, A. Zangen, and P. B. Fitzgerald. 2011. Deep repetitive transcranial magnetic stimulation associated with improved social functioning in a young woman with an autism spectrum disorder. *Journal of ECT* 27 (1): 41–43.

England, R. 1995. Sensory-motor systems in virtual manipulation. In *Simulated and Virtual Realities: Elements of Perception*, ed. K. Carr and R. England. London: Taylor and Francis.

Enserink, M. 2000. Searching for the mark of Cain. *Science* 289:575–579.

Erickson, S. K. 2010. Blaming the brain. *Minnesota Journal of Law, Science and Technology* 11 (1): 27–77.

Erickson, S. K., and A. R. Felthous. 2009. Introduction to this issue: The neuroscience and psychology of moral decision making and the law. *Behavioral Sciences & the Law* 27:119–121.

Erler, A. 2011. Does memory modification threaten our authenticity? *Neuroethics* 4 (3): 235–249.

Etkin, A., T. Egner, and R. Kalisch. 2010. Emotional processing in anterior cingulate and medial prefrontal cortex. *Trends in Cognitive Sciences* 15 (2): 85–93.

Fakra, E., L. W. Hyde, A. Gorka, P. M. Fisher, K. E. Muñoz, M. Kimak, I. Halder, R. E. Ferrell, S. B. Manuck, and A. R. Hariri. 2009. Effects of HTR1A C(-1019) G on amygdala reactivity and trait anxiety. *Archives of General Psychiatry* 66 (1):33–40.

Falk, E. B., L. Rameson, E. T. Berkman, B. Liao, Y. King, T. Inagaki, and M. D. Lieberman. 2009. The neural correlates of persuasion: A common network across cultures and media. *Journal of Cognitive Science* 22 (11): 2447–2459.

Farah, M. J. 2002. Emerging ethical issues in neuroscience. *Nature Neuroscience* 5 (11): 1123–1129.

Farah, M. J. 2005. Neuroethics: The practical and the philosophical. *Trends in Cognitive Sciences* 9 (1): 34–40.

Farah, M. J., ed. 2010. *Neuroethics: An Introduction with Readings*. Cambridge, MA: MIT Press.

Farah, M. J., J. Illes, R. Cook-Deegan, H. Gardner, E. Kandel, P. King, E. Parens, B. Sahakian, and P. R. Wolpe. 2004. Neurocognitive enhancement: What can we do and what should we do? *Nature Reviews. Neuroscience* 5: 421–425.

Farah, M. J., M. E. Smith, C. Gawuga, D. Lindsell, and D. Foster. 2009. Brain imaging and brain privacy: A realistic concern? *Journal of Cognitive Neuroscience* 21 (1): 119–127.

Farrer, T. J., and D. W. Hedges. 2011. Prevalence of traumatic brain injury in incarcerated groups compared to the general population: A meta-analysis. *Progress in Neuro-Psychopharmacology & Biological Psychiatry* 35:390–394.

Farris, S., and M. Giroux. 2011. Deep brain stimulation: A review of the procedure and the complications. *Official Journal of the American Academy of Physician Assistants* 24 (2): 39–45.

Farrow, T. F. D., and P. W. R. Woodruff. 2005. Neuroimaging of forgivability. In *Handbook of Forgiveness*, ed. E. L. Worthington, Jr., 259–272. New York: Brunner-Routledge.

Fehr, E., and C. Camerer. 2007. Social neuroeconomics: The neural circuitry of social preferences. *Trends in Cognitive Sciences* 11 (10): 419–427.

Feigenson, N. 2006. Brain imaging and courtroom evidence: On the admissibility and persuasiveness of fMRI. *International Journal of Law in Context* 2: 233–248.

Feldman-Barrett, L., E. Bliss-Moreau, S. L. Duncan, S. L. Rauch, and C. I. Wright. 2007. The amygdala and the experience of affect. *Social Cognitive and Affective Neuroscience* 2 (2): 73–83.

Felthous, A. R. 2010. Psychopathic disorders and criminal responsibility in the USA. *European Archives of Psychiatry and Clinical Neuroscience* 260 (Suppl. 2): S137–S141.

Fenker, D. B., D. Heipertz, C. N. Boehler, M. A. Schoenfeld, T. Noesselt, H-J. Heinze, E. Duezel, and J-M. Hopf. 2010. Mandatory processing of irrelevant fearful face features in visual search. *Journal of Cognitive Neuroscience* 22 (12): 2926–2938.

Fenton, A., L. Meynell, and F. Baylis. 2009. Ethical challenges and interpretive difficulties with non-clinical applications of pediatric fMRI. *American Journal of Bioethics* 91:3–13.

Filley, C.M., B.H. Price, V.D. Nell et al. 2001. Toward an understanding of violence: Neurobehavioral aspects of unwarranted physical aggression. *Neuropsychiatry, Neuropsychology, and Behavioral Neurology* 14 (1): 1.

Finger, E. C., A. A. Marsh, D. G. Mitchell, M. E. Reid, C. Sims, S. Budhani, et al. 2008. Abnormal ventromedial prefrontal cortex function in children with psychopathic traits during reversal learning. *Archives of General Psychiatry* 65 (5): 586–594.

Fink, M. 2011. Transcranial magnetic stimulation is not a replacement for electroconvulsive therapy in depressive mood disorders. *Journal of ECT* 27 (1): 3–4.

Fischbach, R. L., and G. D. Fischbach. 2005. The brain doesn't lie. *American Journal of Bioethics* 5 (2): 54–55.

Fischer, H., C. I. Wright, P. J. Whalen, S. C. McInerney, L. M. Shin, and S. L. Rauch. 2003. Brain habituation during repeated exposure to fearful and neutral faces: A functional MRI study. *Brain Research Bulletin* 59 (5): 387–392.

Fisher, C. E., L. Chin, and R. Klitzman. 2010. Defining neuromarketing: Practices and professional challenges. *Harvard Review of Psychiatry* 18 (4): 230–237.

Fiske, S. T. 2011. *Envy Up, Scorn Down: How Status Divides Us*. New York: Russell Sage Foundation.

Fiske, S. T. 2002. What we know about bias and intergroup conflict: The problem of the century. *Current Directions in Psychological Science* 11 (4): 123–128.

Flaskerud, J. H. 2010. American culture and neuro-cognitive enhancing drugs. *Issues in Mental Health Nursing* 31 (1): 62–63.

Fleming, S. M., R. S. Weil, Z. Nagy, R. J. Dolan, and G. Rees. 2010. Relating introspective accuracy to individual differences in brain structure. *Science* 329:1541–1543.

Focquaert, F. 2011. Pediatric deep brain stimulation: A cautionary approach. *Frontiers in Integrative Neuroscience* 5. doi:10.3389/fnint.2011.00009.

Foddy, B. 2011. Addicted to food, hungry for drugs. *Neuroethics* 4:79–89.

Forbes, C. E., and J. Grafman. 2010. The role of the human prefrontal cortex in social cognition and moral judgment. *Annual Review of Neuroscience* 33:299–324.

Ford, P. J. 2009. Vulnerable brains: Research ethics and neurosurgical patients. *Journal of Law, Medicine & Ethics* 37 (1):73–83.

Foster, D., and J. F. Meech. 1995. Social dimensions of virtual reality. In *Simulated and Virtual Realities: Elements of Perception*, ed. K. Carr and R. England. London: Taylor and Francis.

Fountas, K. N., and J. R. Smith. 2007. Historical evolution of stereotactic amygdalotomy for the management of severe aggression. *Journal of Neurosurgery* 106:710–713.

Fowler, J. H., L. A. Baker, and C. T. Dawes. 2008. Genetic variation in political participation. *American Political Science Review* 102:233–248.

Fowler, J. H., and N. A. Christakis. 2008. Dynamic spread of happiness in a large social network: Longitudinal analysis over 20 years in the Framingham Heart Study. *British Medical Journal* 337:a2338.

Fowler, J. H., and D. Schreiber 2008. Biology, politics, and the emerging science of human nature. *Science* 322 (November 7): 912–914.

Fox, D. 2011. Neuroscience: Brain buzz. *Nature* 472:156–159.

Frampas, E., M. Videcoq, E. de Kerviler, F. Ricolfi, et al. 2009. CT angiography for brain death diagnosis. *AJNR. American Journal of Neuroradiology* 30: 1566–1570.

Freedman, D., and D. Hemenway. 2000. Precursors of lethal violence: A death row sample. *Social Science & Medicine* 50:1757–1770.

Fregni, F., F. Orsati, W. Pedrosa, S. Fecteau, F. A. Tome, M. A. Nitsche, et al. 2008. Transcranial direct current stimulation of the prefrontal cortex modulates the desire for specific foods. *Appetite* 51 (1): 34–41.

Frick, P. J., and S. F. White. 2008. The importance of callous-unemotional traits for developmental models of aggressive and antisocial behavior. *Journal of Child Psychology and Psychiatry, and Allied Disciplines* 49 (4):359–375.

Friend, J. M., and B. A. Thayer. 2011. Brain imaging and political behavior: A survey. In *Biology and Politics: The Cutting Edge*, ed. S. Peterson and A. Somit, 231–255. Bingley, UK: Emerald Books.

Fukuyama, F. 2002. *Our Posthuman Future: Consequences of the Biotechnology Revolution*. New York: Farrar, Straus and Giroux.

Galván, A., R. A. Poldrack, C. M. Baker, K. M. McGlennen, and E. D. London. 2011. Neural correlates of response inhibition and cigarette smoking in late adolescence. *Neuropsychopharmacology* 36 (5): 970–978.

Gao, Y., A. Raine, F. Chan, P. H. Venables, and S. A. Mednick. 2010. Early maternal and paternal bonding, childhood physical abuse and adult psychopathic personality. *Psychological Medicine* 40:1007–1016.

Garland, B. 2004. Neuroscience and the law. *Professional Ethics Report* 17 (1): 1–3.

Garnett, A., L. Whiteley, H. Piwowar, E. Rasmussen, and J. Illes. 2011. Neuroethics and fMRI: Mapping a fledgling relationship. *PLoS ONE* 6 (4): e18537.

Garrett, A., and K. Chang. 2008. The role of the amygdale in bipolar disorder development. *Development and Psychopathology* 20:1285–1296.

Gazzaniga, M. S. 2005. *The Ethical Brain*. New York: Dana Press.

Gazzaniga, M. S. 2011. Neuroscience in the courtroom. *Scientific American* 304 (4): 54–59.

George, O., and G. F. Koob. 2010. Individual differences in prefrontal cortex function from drug use to drug dependence. *Neuroscience and Biobehavioral Reviews* 35:232–247.

Gelernter, J., and H. R. Kranzler. 2009. Genetics of alcohol dependence. *Human Genetics* 126 (1): 1–2.

Gilbert, F., and D. Ovadia. 2011. Deep brain stimulation in the media: Over-optimistic portrayals call for a new strategy involving journalists and scientists in ethical debates. *Frontiers in Integrative Neuroscience* 5:16. doi:10.3389/fnint.2011.00016.

Gillett, G. 2011. The gold-plated leucotomy standard and deep brain stimulation. *Journal of Bioethical Inquiry*. doi:10–1007/s11673–010–9281z.

Gillihan, S. I., H. Rao, J. Wang, J. A. Detre, J. Breland, et al. 2010. Serotonin transporter genotype modulates amygdala activity during mood regulation. *Social Cognitive and Affective Neuroscience* 5 (1): 1–10.

Giordano, J. J. 2010. Neuroethical issues in neurogenetic and neuro-transplantation technology: The need for pragmatism and preparedness in practice and policy. *Studies in Ethics, Law, and Technology* 4 (3): Article 4.

Giordano, J. J., and B. Gordijn, eds. 2010. *Scientific and Philosophical Perspectives in Neuroethics*. Cambridge, UK: Cambridge University Press.

Glannon, W. 2009. Stimulating brains, altering minds. *Journal of Medical Ethics* 35 (5): 289–292.

Glannon, W. 2010. Consent to deep brain stimulation for neurological and psychiatric disorders. *Journal of Clinical Ethics* 21 (2): 104–111.

Glenn, A. L., R. Iyer, J. Graham, S. Koleva, and J. Haidt. 2009. Are all types of morality compromised in psychopathy? *Journal of Personality Disorders* 23:384–398.

Glenn, A. L., R. Kurzban, and A. Raine. 2011. Evolutionary theory and psychopathy. *Aggression and Violent Behavior* 16:371–380.

Glenn, A. L., and A. Raine. 2009. Psychopathy and instrumental aggression: Evolutionary, neurobiological, and legal perspectives. *International Journal of Law and Psychiatry* 32:253–258.

Gold, J. I., and M. N. Shadlen. 2007. The neural basis of decision making. *Annual Review of Neuroscience* 30:535–574.

Goldberg, S. 2007. MRIs and the perception of risk. *American Journal of Law & Medicine* 33 (2–3): 229–237.

Goldman-Rakic, P. 1997. Working memory and the mind. *Scientific American* (September):111–117.

Goldstein, R. Z., A. D. Craig, A. Bechara, H. Garavan, A. R. Childress, et al. 2009. The neurocircuitry of impaired insight in drug addiction. *Trends in Cognitive Sciences* 13 (9): 372–380.

Goodman, W. K. 2011. Electroconvulsive therapy in the spotlight. *New England Journal of Medicine* 364 (19): 1785–1787.

Goodwin, J., J. M. Jasper, and F. Polletta, eds. 2001. *Passionate Politics*. Chicago: University of Chicago Press.

Greely, H. T. 2005. Pre-market approval regulation for lie detections: An idea whose time may be coming. *American Journal of Bioethics* 5 (2): 50–52.

Greely, H. T., and J. Illes. 2007. Neuroscience-based lie detection: The urgent need for regulation. *American Journal of Law & Medicine* 33 (2–3): 377–431.

Greely, H. T., B. Sahakian, J. Harris, R. C. Kessler, M. Gazzaniga, P. Campbell, and M. J. Farah. 2008. Towards responsible use of cognitive-enhancing drugs by the healthy. *Nature* 456:702–705.

Green, R. M. 2005. Spy versus spy. *American Journal of Bioethics* 5 (2):53–54.

Greene, J. D. 2007. Why are VMPFC patients more utilitarian?: A dual-process theory of moral judgment explains. *Trends in Cognitive Sciences* 11 (8):322–323.

Greene, J. D., and J. Cohen. 2004. For the law, neuroscience changes nothing and everything. *Philosophical Transactions of the Royal Society of London. Series B, Biological Sciences* 359:1775–1785.

Greene, J. D., S. A. Morelli, K. Lowenberg, L. E. Nystrom, and J. D. Cohen. 2008. Cognitive load selectively interferes with utilitarian moral judgment. *Cognition* 107 (3):1144–1154.

Greene, J. D., L. E. Nystrom, A. D. Engell, J. M. Darley, and J. D. Cohen. 2004. The neural bases of cognitive conflict and control in moral judgment. *Neuron* 44 (2): 389–400.

Greer, D. M., P. N. Varelas, S. Haque, and E. F. M. Wijdicks. 2008. Variability of brain death determination guidelines in leading US neurologic institutions. *Neurology* 70:284–289.

Griggins, C. 2005. Dosing dilemmas: Are you rich and white or poor and black? *American Journal of Bioethics* 5 (3): 55–57.

Grover, S. K., S. Chakrabarti, N. Khehra, and R. Rajagopal. 2011. Does the experience of electroconvulsive therapy improve awareness and perceptions of treatment among relatives of patients? *Journal of ECT* 27 (1): 67–72.

Gupta, R., T. R. Koscik, A. Bechara, and D. Tranel. 2011. The amygdala and decision-making. *Neuropsychologia* 49:760–766.

Gupta, S., C. Hood, and R. Chaplin. 2011. Use of continuation and maintenance electroconvulsive therapy: UK national trends. *Journal of ECT* 27 (1): 77–80.

Gurley, J. R., and D. K. Marcus. 2008. The effects of neuroimaging and brain injury on insanity defenses. *Behavioral Sciences & the Law* 26:85–97.

Güroğlu, B., W. van den Bos, S. A. R. B. Rombouts, and E. A. Crone. 2010. Unfair? It depends: Neural correlates of fairness in social context. *Social Cognitive and Affective Neuroscience* 5 (4): 414–423.

Gusterson, H. 2008. The U.S. military's quest to weaponize culture. *Bulletin of the Atomic Scientists,* June 20. http://thebulletin.org/web-edition/columnists/hugh-gusterson/the-us-militarys-quest-to-weaponize-culture.

Gutchess, A. H., R. C. Welsh, A. Boduroglu, and D. C. Park. 2006. Cultural differences in neural function associated with object processing. *Cognitive, Affective & Behavioral Neuroscience* 6 (2):102–109.

Gutchess, A. H., T. Hedden, S. Ketay, A. Aron, and J. D. E. Gabrieli. 2010. Neural differences in the processing of semantic relationships across cultures. *SCAN* 5:254–263.

Hackman, D. A., and M. J. Farah. 2009. Socioeconomic status and the developing brain. *Future of Children* 15:71–89.

Haier, R. J., R. E. Jung, R. A. Yeo, K. Head, and M. T. Alkire. 2005. The neuroanatomy of general intelligence: Sex matters. *NeuroImage* 25:320–327.

Hall, W., L. Carter, and K. I. Morley. 2004. Neuroscience research on the addictions: A prospectus for future ethical and policy analysis. *Addictive Behaviors* 29:1481–1495.

Hamani, C., M. P. McAndrew, M. Cohn, M. Oh, D. Zumsteg, C. M. Shapiro, R. A. Wennberg, and A. Lozano. 2008. Memory enhancement induced by hypothalamic/fornix deep brain stimulation. *Annals of Neurology* 63 (1): 119–123.

Hamann, S. 2011. Introduction to special issue on the human amygdala and emotional function. *Neuropsychologia* 49:585–588.

Hammock, E. 2011. NSF Workshop on Genes, Cognition, and Social Behavior. Washington, DC: National Science Foundation's Political Science Program (SBE-1037831), 115–124.

Han, S., and G. Northoff. 2008. Culture-sensitive neural substrates of human cognition: A transcultural neuroimaging approach. *Nature Reviews. Neuroscience* 9:646–654.

Hare, R. D., and C. S. Neumann. 2008. Psychopathy as a clinical and empirical construct. *Annual Review of Clinical Psychology* 4:217–246.

Hare, T. A., J. O'Doherty, C. F. Camerer, W. Schultz, and A. Rangel. 2008. Dissociating the role of the orbitofrontal cortex and the striatum in the computation of goal values and prediction errors. *Journal of Neuroscience* 28:5623–5630.

Harenski, C. L., S. H. Kim, and S. Hamann. 2009. Neuroticism and psychopathy predict brain activation during moral and nonmoral emotion regulation. *Cognitive, Affective & Behavioral Neuroscience* 9 (1):1–15.

Hariri, A. R. 2009. The neurobiology of individual differences in complex behavioral traits. *Annual Review of Neuroscience* 32:225–247.

Hariri, A. R., E. M. Drabant, and D. R. Weinberger. 2006. Imaging genetics: Perspectives from studies of genetically driven variation in serotonin function and corticolimbic affective processing. *Biological Psychiatry* 59:888–897.

Hariz, M. I., P. Blomstedt, and L. Zrinzo. 2010. Deep brain stimulation between 1947 and 1987: The untold story. *Neurosurgical Focus* 29 (2): e1.

Harkavy, R. E. 2000. Defeat, national humiliation and the revenge motif in international politics. *International Politics* 37 (3): 345–368.

Harris, L. T., and S. T. Fiske. 2006. Dehumanizing the lowest of the low: Neuroimaging responses to extreme out-groups. *Psychological Science* 17 (10): 847–853.

Harris, L. T., and S. T. Fiske. 2007. Social groups that elicit disgust are differentially processed in mPFC. *Social Cognitive and Affective Neuroscience* 2 (1): 45–51.

Harrison, P. J., and E. M. Tunbridge. 2008. Catechol-O-methyltransferase (COMT): A gene contributing to sex differences in brain function, and to sexual dimorphism in the predisposition to psychiatric disorders. *Neuropsychopharmacology* 33:3037–3045.

Hart, H. L. A. 1968. Legal responsibility and excuses. In *Punishment and Responsibility*. New York: Oxford University Press.

Hart, A. J., P. J. Whalen, L. M. Shin, S. C. McInerney, H. Fischer, and S. L. Rauch. 2000. Differential response in the human amygdala to racial outgroup vs. ingroup face stimuli. *NeuroReport* 11:2351–2355.

Harth, E. 1993. *The Creative Loop: How the Brain Makes a Mind*. Reading, MA: Addison-Wesley.

Hatemi, P. K. 2011. NSF Workshop on Genes, Cognition, and Social Behavior. Washington, DC: National Science Foundation's Political Science Program (SBE-1037831), 125–137.

Hatemi, P. K., S. E. Medland, K. I. Morley, A. C. Heath, and N. G. Martin. 2007. The genetics of voting: An Australian twin study. *Behavior Genetics* 37:435–448.

Hayashi, A., N. Abe, A. Ueno, Y. Shigemune, E. Mori, et al. 2010. Neural correlates of forgiveness for moral transgressions involving deception. *Brain Research* 1332:90–99.

Häyry, M. 2010. Neuroethical theories. *Cambridge Quarterly of Healthcare Ethics* 19:165–178.

Hecht, D. 2011. An inter-hemispheric imbalance in the psychopath's brain. *Personality and Individual Differences* 51 (1): 3–10.

Heckman, K. E., and M. D. Happel. 2006. Mechanical detection of deception: A short review. In *Educing Information—Interrogation: Science and Art. Foundations for the Future. Intelligence Science Board, Phase 1 Report, National Defense Intelligence College (NDIC).* Washington, DC: NDIC Press.

Heilbronn, R., and S. Weger. 2010. Viral vectors for gene transfer: Current status of gene therapeutics. *Handbook of Experimental Pharmacology* (197): 143–170.

Heinrichs, B. 2010. A new challenge for research ethics: Incidental findings in neuroimaging. *Journal of Bioethical Inquiry*. doi:10.1007/s11673-010-9268-9.

Heinrichs, M., T. Baumgartner, C. Kirschbaum, and U. Ehlert. 2003. Social support and oxytocin interact to suppress cortisol and subjective responses to psychosocial stress. *Biological Psychiatry* 54:1389–1398.

Herman-Stahl, M. A., C. P. Krebs, L. A. Kroutil, and D. C. Heller. 2006. Risk and protective factors for methamphetamine use and nonmedical use of prescription stimulants among young adults aged 18 to 25. *Addictive Behaviors* 32 (5): 1003–1015.

Hewstone, M., M. Rubin, and H. Willis. 2002. Intergroup bias. *Annual Review of Psychology* 53:575–604.

Hildt, E. 2010. Brain-computer interaction and medical access to the brain: Individual, social and ethical implications. *Studies in Ethics, Law, and Technology* 4 (3): Article 5.

Hines, M. 2010. Sex-related variation in human behaviour and the brain. *Trends in Cognitive Sciences* 14 (10): 448–456.

Hines, M. 2011. Gender development and the human brain. *Annual Review of Neuroscience* 34. doi:10.1146/annrev-neuro-061010-113654.

Hinton, V. J. 2002. Ethics of neuroimaging in pediatric development. *Brain and Cognition* 50:455–468.

Holtzheimer, P. E. 2011. Deep brain stimulation for psychiatric disorders. *Annual Review of Neuroscience* 34. doi:10.1146/annurev-neuro-061010-113638.

Hölzel, B. K., J. Carmody, K. C. Evans, E. A. Hoge, J. A. Dusek, L. Morgan, R. K. Pitman, and S. W. Lazar. 2010. Stress reduction correlates with structural changes in the amygdala. *Social Cognitive and Affective Neuroscience* 5 (1): 11–17.

Hommer, D. W., J. M. Bjork, and J. M. Gilman. 2011. Imaging brain response to reward in addictive disorders. *Annals of the New York Academy of Sciences* 1216:50–61.

Hoptman, M. J., and D. Antonius. 2011. Neuroimaging correlates of aggression in schizophrenia: An update. *Current Opinion in Psychiatry* 24 (2): 100.

Hotz, R. L. 1997. Estrogen may play key role in brain. *Los Angeles Times*. January 26.

Howard, M. 2001. The causes of war. In *Turbulent Peace*, ed. C. A. Crocker, F. O. Hampson, and P. Aall, 29–38. Washington, DC: United States Institute of Peace Press.

Hsu, M., M. Bhatt, R. Adolphs, D. Tranel, and C. F. Camerer. 2005. Neural system responding to degrees of uncertainty in human decision-making. *Science* 310:1681–1683.

Huang, G. T. 2003. Mind-machine merger. *Technology Review* 106:39–45.

Huang, J. Y., and M. E. Kosal. 2008. The security impact of the neurosciences. *Bulletin of the Atomic Scientists*, June 8.

Hughes, J. 2005. Beyond "real boys" and back to parental obligations. *American Journal of Bioethics* 5 (3): 61–62.

Hughes, V. 2010. Science in court: Head case. *Nature* 464:340–342. doi: 10.1038/464340a.

Hurley, R. A., and K. H. Taber. 2008. *Windows to the Brain: Insights from Neuroimaging*. New York: American Psychiatric Publishing.

Husain, M. M., and S. H. Lisanby. 2011. Repetitive transcranial magnetic stimulation (rTMS): A noninvasive neuromodulation probe and intervention. *Journal of ECT* 27 (1): 2.

Huster, R. J., R. Westerhausen, and C. S. Herrmann. 2011. Sex differences in cognitive control are associated with midcingulate and callosal morphology. *Brain Structure & Function* 215 (3–4): 225–235.

Hyde, L. W., A. Gorka, S. B. Manuck, and A. R. Hariri. 2011. Perceived social support moderates the link between threat-related amygdala reactivity and trait anxiety. *Neuropsychologia* 49:651–656.

Hyde, J. S., and M. C. Linn. 1998. Gender differences in verbal ability: A meta-analysis. *Psychological Bulletin* 104 (1):53–69.

Hyman, S. E. 2007. Neuroethics: At age 5, the field continues to evolve. The Dana Foundation. http://www.dana.org.

Iacoboni, M. 2009. Imitation, empathy, and mirror neurons. *Annual Review of Psychology* 60:653–670.

Iadaka, T., D. N. Saito, H. Komeda, Y. Mano, N. Kanayama, T. Osumi, N. Ozaki, and N. Sadato. 2010. Transient neural activation in human amygdala involved in aversive conditioning of face and voice. *Journal of Cognitive Neuroscience* 22 (9): 2074–2085.

Illes, J., and S. J. Bird. 2006. Neuroethics: A modern context for ethics in neuroscience. *Trends in Neurosciences* 30 (10): 1–7.

Illes, J., and V. N. Chin. 2008. Bridging philosophical and practical implications of incidental findings in brain research. *Journal of Law, Medicine & Ethics* 36 (2): 298–301.

Illes, J., and E. Racine. 2005. Imaging or imagining? A neuroethics challenge informed by genetics. *American Journal of Bioethics* 5 (2): 5–18.

Illes, J., A. C. Rosen, L. Huang, et al. 2004. Ethical consideration of incidental findings on adult brain MRI in research. *Neurology* 62:888–890.

Isaias, I. U., R. L. Alterman, and M. Tagliati. 2008. Outcome predictors of pallidal stimulation in patients with primary dystonia: The role of disease duration. *Brain* 131:1895–1902.

Ito, T. A., and B. Bartholow. 2008. The neural correlates of race. *Trends in Cognitive Sciences* 13 (12): 524–531.

Ivy, G. O. 1996. The aging nervous system. In *Brain Mechanisms and Psychotropic Drugs*, ed. A. Baskys and G. Remington. Boca Raton, FL: CRC Press.

Izuma, K., D. N. Saito, and N. Sadato. 2010. Processing of the incentive for social approval in the ventral striatum during charitable donation. *Journal of Cognitive Neuroscience* 22 (4): 621–631.

Jahfari, S., C. M. Stinear, M. Claffey, F. Verbruggen, and A. R. Aron. 2010. Responding with restraint: What are the neurocognitive mechanisms? *Journal of Cognitive Neuroscience* 22 (7): 1479–1492.

Joffe, A. 2010. Are recent defenses of the brain death concept adequate? *Bioethics* 24 (2): 47–53.

Johansen-Berg, H. 2009. Imaging the relationship between structure, function and behaviour in the human brain. *Brain Structure & Function* 213:499–500.

Johansson, V., M. Garwicz, M. Kanje, J. Schouenborg, et al. 2011. Authenticity, depression and deep brain stimulation. *Frontiers in Integrative Neuroscience* 5:21. doi:10.3389/fnint.2011.00021.

Joseph, J. 2010. The genetics of political attitudes and behavior: Claims and refutations. *Ethical Human Psychology and Psychiatry* 12 (3): 200–217.

Jost, J. T., J. Glaser, A. W. Kruglanski, and F. J. Sulloway. 2003. Political conservatism as motivated social cognition. *Psychological Bulletin* 129:339–375.

Juni, P., F. Holenstein, J. Sterne, C. Bartlett, and M. Egger. 2002. Direction and impact of language bias in meta-analyses of controlled trials: Empirical study. *International Journal of Epidemiology* 31 (1):115–123.

Kaisera, A., S. Hallerb, S. Schmitzd, and C. Nitscha. 2009. On sex/gender related similarities and differences in fMRI language research. *Brain Research. Brain Research Reviews* 61 (2): 49–59.

Kahane, G., K. Wiech, N. Shakel, M. Farias, J. Savulescu, and I. Tracey. 2011. The neural basis of intuitive and counterintuitive moral judgment. *Social Cognitive and Affective Neuroscience* 6 (1). doi:10.1093/scan/nsr005.

Kalnin, A. J., C. R. Edwards, Y. Wang, W. G. Kronenberger, T. A. Hummer, K. M. Mosier, D. W. Dunn, and V. P. Mathews. 2011. The interacting role of media violence exposure and aggressive-disruptive behavior in adolescent brain activation during an emotional Stroop task. *Psychiatry Research: Neuroimaging* 192:12–17.

Kamm, F. M. 2005. Is there a problem with enhancement? *American Journal of Bioethics* 5 (3): 5–14.

Kamm, F. M. 2009. Neuroscience and moral reasoning: A note on recent research. *Philosophy & Public Affairs* 37 (2): 330–345.

Kanai, R., T. Feilden, C. Firth, and G. Rees. 2011. Political orientations are correlated with brain structure in young adults. *Current Biology* 21:677–680.

Kanai, R., and G. Rees. 2011. The structural basis of inter-individual differences in human behaviour and cognition. *Nature Reviews. Neuroscience* 12:231–242. doi:10.1038/nrn3000.

Kandel, E. R., and R. D. Hawkins. 1992. The biological basis of learning and individuality. *Scientific American* 267 (3): 79–86.

Kane, R. 1996. *The Significance of Free Will*. New York: Oxford University Press.

Kanwisher, N., and G. Yovel. 2006. The fusiform face area: A cortical region specialized for the perception of faces. *Philosophical Transactions of the Royal Society B, Biological Science* 361:2109–2128.

Kaplan, J. T., J. Freedman, and M. Iacoboni. 2007. Us versus them: Political attitudes and party affiliation influence neural response to faces of presidential candidates. *Neuropsychologia* 45:55–64.

Kato, J., H. Ide, I. Kabashima, H. Kadota, K. Takano, and K. Kansaku. 2009. Neural correlates of attitude change following positive and negative advertisements. *Frontiers of Behavioral Neuroscience* 3:6.

Katz, L. Y., A. L. Kozyrskyj, H. J. Prior, M. W. Enns, B. J. Cox, and J. Sareen. 2008. Effect of regulatory warnings on antidepressant prescription rates, use of health services and outcomes among children, adolescents and young adults. *Canadian Medical Association Journal* 178 (8): 1005–1011.

Keane, H., and K. Hamill. 2010. Variations in addiction: The molecular and the molar in neuroscience and pain medicine. *Biosocieties* 5 (1): 52–69.

Keckler, C. N. W. 2006. Cross-examining the brain: A legal analysis of neural imaging for credibility impeachment. *Hastings Law Journal* 57:509–556.

Keightley, M. L., K. S. Chiew, J. A. E. Anderson, and C. L. Grady. 2011. Neural correlates of recognition memory for emotional faces and scenes. *Social Cognitive and Affective Neuroscience* 6 (1): 24–37.

Kellner, C. H. 2011. The FDA Advisory Panel on the Reclassification of ECT Devices. *Psychiatric Times* 28 (7): 1.

Kempf, L., and A. Meyer-Lindenberg. 2006. Imaging genetics and psychiatry. *Focus (San Francisco, Calif.)* 4 (3): 327–338.

Kessler, D. 2009. *The End of Overeating: Taking Control of the Insatiable North American Appetite*. Toronto: McClelland and Stewart.

Khoshbin, L. S., and S. Khoshbin. 2007. Imaging the mind, minding the image: An historical introduction to brain imaging and the law. *American Journal of Law & Medicine* 33 (2–3): 171–192.

Kiehl, K. A., A. M. Smith, A. Mendrek, B. B. Forster, R. D. Hare, and P. F. Liddle. 2004. Temporal lobe abnormalities in semantic processing by criminal psychopaths as revealed by functional magnetic resonance imaging. *Psychiatry Research: Neuroimaging* 130:297–312.

Kilpatrick, L. A., D. H. Zald, J. V. Pardo, and L. F. Cahill. 2006. Sex-related differences in amygdala functional connectivity during resting conditions. *NeuroImage* 30:452–461.

Kim, H. J., and S. A. Thayer. 2009. Lithium increases synapse formation between hippocampal neurons by depleting phosphoinositides. *Molecular Pharmacology* 75 (5): 1021–1030.

Kim, M. J., R. A. Loucks, M. Neta, F. C. Davis, J. A. Oler, E. C. Mazzulla, and P. J. Whelen. 2010. Behind the mask: The influence of mask-type on amygdala response to fearful faces. *Social Cognitive and Affective Neuroscience* 5 (4): 363–368.

Kimonis, E. R., P. J. Frick, J. L. Skeem, M. A. Marsee, K. Cruise, L. C. Munoz, K. J. Aucoin, and A. S. Morris. 2008. Assessing callous–unemotional traits in adolescent offenders: Validation of the Inventory of Callous-Unemotional Traits. *International Journal of Law and Psychiatry* 31:241–252.

Kimura, D. 1992. Sex differences in the brain. *Scientific American* 267 (3):118–125.

Kimura, D. 2007. Sex differences in the brain. *Scientific American,* http://www.scientificamerican.com/article.cfm?articleID=00018E9D-879D-1D06 -8E49809EC588EEDF.

Kingdon, J. W. 1995. *Agendas, Alternatives, and Public Policies.* 2nd ed. New York: Longman.

Kinnunen, L. H., H. Moltz, J. Metz, and M. Cooper. 2004. Differential brain activation in exclusively homosexual and heterosexual men produced by the selective serotonin reuptake inhibitor, fluoxetine. *Brain Research* 1024:251–254.

Kitayama, S., and J. Park. 2010. Cultural neuroscience of the self: Understanding the social grounding of the brain. *Social Cognitive and Affective Neuroscience* 5 (2–3): 111–129.

Klein, C. 2011. The dual track theory of moral decision-making: A critique of the neuroimaging evidence. *Neuroethics* 4 (2): 143–162.

Knabb, J. J., R. K. Walsh, J. G. Ziebell, and K. S. Reimer. 2009. Neuroscience, moral reasoning and the law. *Behavioral Sciences & the Law* 27:219–236.

Knutson, K. M., J. N. Wood, M. V. Spampinato, and J. Grafman. 2006. Politics on the brain: An fMRI investigation. *Social Neuroscience* 1:25–40.

Koenigs, M., L. Young, R. Adolphs, D. Tranel, F. Cushman, M. Hauser, and A. Damasio. 2007. Damage to the prefrontal cortex increases utilitarian moral judgements. *Nature* 446:908–911.

Konrad, P., and T. Shanks. 2010. Implantable brain computer interface: Challenges to neurotechnology translation. *Neurobiology of Disease* 38 (3): 369–375.

Koob, G. F., and M. Le Moal. 2006. *Neurobiology of Addiction.* London: Academic Press.

Kosal, M. E. 2008. Basic research is needed to better understand the effects of drugs that affect the brain. *Bulletin of the Atomic Scientists.* http://thebulletin .org/web-edition/roundtables/the-military-application-of-neuroscience-research.

Koscik, T. R., and D. Tranel. 2011. The human amygdala is necessary for developing and expressing normal interpersonal trust. *Neuropsychologia* 49 (4): 602–611.

Kosfeld, M. 2007. Trust in the brain: Neurobiological determinants of human social behaviour. *EMBO Reports* 8:s44–s47.

Kosfeld, M., M. Heinrichs, P. J. Zak, U. Fischbacher, and E. Fehr. 2005. Oxytocin increases trust in humans. *Nature* 435 (7042): 673–676.

Kosslyn, S. M., and O. Koenig. 1992. *Wet Mind: The New Cognitive Science.* New York: Free Press.

Kotulak, R. 1996. *Inside the Brain: Revolutionary Discoveries of How the Brain Works.* Kansas City: Andrews and McMeel.

Kozel, F.A., L.J. Revell, J.P. Lorberbaum, A. Shastri, J.D. Elhai, M.D. Horner, A. Smith, Z. Nahas et al. 2004. A pilot study of functional magnetic resonance imaging brain correlates of deception in healthy young men. *Journal of Neuropsychiatry and Clinical Neurosciences* 163:295–305.

Kozel, F. A., K. A. Johnson, E. L. Grenesko, S. J. Laken, S. Kose, X. Lu, D. Pollina, A. Ryan, and M. S. George. 2008. Functional MRI detection of deception after committing a mock sabotage crime. *Journal of Forensic Sciences* 54 (1): 220–231.

Kraemer, F. 2011. Me, myself and my brain implant: Deep brain stimulation raises questions of personal authenticity and alienation. *Neuroethics* (12 May), 1–15. doi:10.1007/s12152-011-9115-7.

Kragh, J. V. 2010. Shock therapy in Danish psychiatry. *Medical History* 54:341–364.

Kringelbach, M. L., A. L. Green, and T. Z. Aziz. 2011. Balancing the brain: Resting state networks and deep brain stimulation. *Frontiers in Integrative Neuroscience* 5. doi:10.3389/fnint.2011.00008.

Kringelbach, M. L., and E. T. Rolls. 2004. The functional neuroanatomy of the human orbitofrontal cortex: Evidence from neuroimaging and neuropsychology. *Progress in Neurobiology* 72 (5): 341–372.

Kronenberger, W., V. Mathews, D. Dunn, Y. Wang, E. Wood et al. 2005. Media violence exposure and executive functioning in aggressive and control adolescents. *Journal of Clinical Psychology* 61 (6): 725–737.

Kuhar, M. 2010. Contributions of basic science to understanding addiction. *Biosocieties* 5 (1): 25–35.

Kuhn, J., W. Gaebel, J. Klosterkoetter, and C. Woopen. 2009. Deep brain stimulation as a new therapeutic approach in therapy-resistant mental disorders: Ethical aspects of investigational treatment. *European Archives of Psychiatry and Clinical Neuroscience* 259 (Suppl. 2): S135–S141.

Kulynych, J. J. 2007. The regulation of MR neuroimaging research: Disentangling the Gordian knot. *American Journal of Law & Medicine* 33 (2–3): 295–317.

Kumar, R. 2008. Approved and investigational uses of modafinil: An evidence-based review. *Drugs* 68 (13):1803–1839.

Kurzban, R., J. Tooby, and L. Cosmides. 2001. Can race be erased? Coalitional computation and social categorization. *Proceedings of the National Academy of Sciences of the United States of America* 98 (26): 15387–15392.

Kushner, H. I. 2010. Toward a cultural biology of addiction. *Biosocieties* 5:8–24.

Lamberts, S. W. J., A. W. Van den Beld, and A.-J. Van der Lay. 1997. The endocrinology of aging. *Science* 278:419–424.

Lamm, C., A. N. Meltzoff, and J. Decety. 2010. How do we empathize with someone who is not like us? A functional magnetic resonance imaging study. *Journal of Cognitive Neuroscience* 22 (2): 362–376.

Lang, P. J., and M. Davis. 2006. Emotion, motivation, and the brain: Reflex foundations in animal and human research. *Progress in Brain Research* 156:3–29.

Langleben, D. D., L. Schroeder, J. A. Maldjian, R. C. Gur, S. McDonald, J. D. Ragland, C. P. O'Brien, and A. R. Childress. 2002. Brain activity during simulated deception: An event-related functional magnetic resonance study. *NeuroImage* 15:727–732.

Larriviere, D., and M. A. Williams. 2010. Neuroenhancement: Wisdom of the masses or "false phronesis"? *Clinical Pharmacology and Therapeutics* 88 (4): 459–461.

Launis, V. 2010. Cosmetic neurology: Sliding down the slippery slope? *Cambridge Quarterly of Healthcare Ethics* 19:218–229.

Leben, D. 2010. Cognitive neuroscience and moral decision-making: Guide or set aside? *Neuroethics*. doi:10.1007/s12152-010-9087-z.

Lederer, S. E. 2008. Putting death in context. *Hastings Center Report* 38 (6): 3.

LeDoux, J. 2002. *Synaptic Self: How Our Brains Become Who We Are.* New York: Viking Press.

LeDoux, J. E. 2007. The amygdala. *Current Biology* 17:R868–R874.

Lee, N., A. J. Broderick, and L. Chamberlain. 2007. What is neuromarketing? A discussion and agenda for future research. *International Journal of Psychophysiology* 63:199–204.

Lee, T. M. C., S. C. Chan, and A. Raine. 2008. Strong limbic and weak frontal activation to aggressive stimuli in spouse abusers. *Molecular Psychiatry* 13:655–660.

Lenroot, R. K., and J. N. Giedd. 2010. Sex differences in the adolescent brain. *Brain and Cognition* 72:46–55.

Lenz, K. M., and M. M. McCarthy. 2010. Organized for sex: Steroid hormones and the developing hypothalamus. *European Journal of Neuroscience* 32 (12): 2096–2104.

Leonard, C. M., S. Towler, S. Welcome, L. K. Halderman, R. Otto, et al. 2008. Size matters: Cerebral volume influences sex differences in neuroanatomy. *Cerebral Cortex* 18:2920–2931.

Lesch, K.-P. 2007. Linking emotion to the social brain: The role of the serotonin transporter in human social behavior. *EMBO Reports* 8:s24–s29.

LeVay, S. 1991. A difference in hypothalamic structure between heterosexual and homosexual men. *Science* 253:1034–1070.

LeVay, S., and D. H. Hamer. 1994. Evidence for a biological influence in male homosexuality. *Scientific American* 270:44–49.

Leyens, J., B. Cortes, S. Demoulin, J. Dovidio, S. Fiske, R. Gaunt, et al. 2003. Emotional prejudice, essentialism, and nationalism: The 2002 Tajfel lecture. *European Journal of Social Psychology* 33:703–717.

Lieberman, M. D., E. T. Berkman, and T. D. Wager. 2009. Correlations in social neuroscience aren't voodoo: Commentary on Vul et al. *Perspectives on Psychological Science* 4 (3): 299–307.

Lindquist, M. A., and A. Gelman. 2009. Correlations and multiple comparisons in functional imaging: A statistical perspective. *Perspectives on Psychological Science* 4 (3): 310–313.

Lipsman, N., M. Ellis, and A. M. Lozano. 2010. Current and future indications for deep brain stimulation in pediatric populations. *Neurosurgical Focus* 29 (2): E2.

Lipsman, N., R. Zaner, and M. Bernstein. 2009. Personal identity, enhancement and neurosurgery: A qualitative study in applied ethics. *Bioethics* 23 (6): 375–383.

Long, T., M. Sque, and J. Addington-Hall. 2008. Conflict rationalisation: How family members cope with a diagnosis of brain stem death. *Social Science & Medicine* 67 (2): 253–261.

Long, W. J., and P. Brecke. 2003. The emotive causes of recurrent international conflicts. *Politics and the Life Sciences* 22 (1): 24–35.

Looney, J. W. 2010. Neuroscience's new techniques for evaluating future dangerousness: Are we returning to Lombroso's biological criminality? *University of Arkansas Little Rock Law Review* 32:301.

Lopez-Larson, M. P., J. S. Anderson, M. A. Ferguson, and D. Yurgelun-Todd. 2011. Local brain connectivity and associations with gender and age. *Developmental Cognitive Neuroscience* 1 (2): 187–197.

Losco, J. 1994. Biology and public administration. In *Research in Biopolitics*. vol. 2. ed. A. Somit and S. Peterson, 47–67. London: JAI Press.

Losin, E. A. R., M. Dapretto, and M. Iacoboni. 2010. Culture and neuroscience: Additive or synergistic? *Social Cognitive and Affective Neuroscience* 5 (2–3): 148–158.

Lovett, R. A. 2010. Reproducibility of brainscan studies questioned: Some magnetic resonance imaging studies could be less reliable than has been presumed. *Nature.* doi:10.1038/news.2010.129.

Luber, B., C. Fisher, P. S. Appelbaum, M. Ploesser, and S. H. Lisanby. 2009. Noninvasive brain stimulation in the detection of deception: Scientific challenges and ethical consequences. *Behavioral Sciences & the Law* 27 (2): 191–208.

Luders, E., C. Gaser, K. L. Narr, and A. W. Toga. 2009. Why sex matters: Brain size independent differences in gray matter distributions between men and women. *Journal of Neuroscience* 29 (45): 14265–14270.

Luders, E., and A. W. Toga. 2010. Sex differences in brain anatomy. *Progress in Brain Research* 186:2–12.

Lupia, A. 2011. NSF Workshop on Genes, Cognition, and Social Behavior. Washington, DC: National Science Foundation's Political Science Program (SBE-1037831).

Lv, B., J. Li, H. He, M. Li, M. Zhao, et al. 2010. Gender consistency and difference in healthy adults revealed by cortical thickness. *NeuroImage* 53 (2): 373–382.

Ma, Q., Q. Shen, Q. Xu, D. Li, L. Shu, and B. Weber. 2011. Empathetic responses to others' gains and losses: An electrophysiological investigation. *NeuroImage* 54 (3):2472–2480.

Mackenzie, R. 2011. The neuroethics of pleasure and addiction in public health strategies moving beyond harm reduction: Funding the creation of non-addictive drugs and taxonomies of pleasure. *Neuroethics* 4 (2): 103–117.

Maratos, F. A., K. Mogg, B. P. Bradley, G. Rippon, and C. Senior. 2009. Coarse threat images reveal theta oscillations in the amygdala: A magnetoencephalography study. *Cognitive, Affective & Behavioral Neuroscience* 9 (2): 133–143.

Marcus, G. E. 2000. Emotions in politics. *Annual Review of Political Science* 3:221–250.

Marcus, G. E. 2002. *The Sentimental Citizen: Emotion in Democratic Politics.* University Park: Pennsylvania State University Press.

Marcus, G. E., W. R. Neuman, and M. McKuen, eds. 2000. *Affective Intelligence and Political Judgment.* Chicago: University of Chicago Press.

Markowitsch, H. J., and A. Staniloiu. 2011a. Amygdala in action: Relaying biological and social significance to autobiographical memory. *Neuropsychologia* 49:718–733.

Markowitsch, H. J., and A. Staniloiu. 2011b. Neuroscience, neuroimaging and the law. *Cortex* 47 (10): 1248–1251.

Marks, J. H. 2007. Interrogation neuroimaging in counterterrorism: A "nobrainer" or a human rights hazard? *American Journal of Law & Medicine* 33 (2–3): 483–500.

Marks, W. A., J. Honeycutt, F. Acosta, and M. Reed. 2009. Deep brain stimulation for pediatric movement disorders. *Seminars in Pediatric Neurology* 16 (2): 90–98.

Marquis, D. 2010. Are DCD donors dead? *Hastings Center Report* 40 (3): 24–31.

Marsh, A. A., E. C. Finger, D. G. Mitchell, M. E. Reid, C. Sims, D. S. Kosson, et al. 2008. Reduced amygdala response to fearful expressions in children and adolescents with callous-unemotional traits and disruptive behavior disorders. *American Journal of Psychiatry* 165 (6): 712–720.

Martell, D. A. 2009. Neuroscience and the law: Philosophical differences and practical constraints. *Behavioral Sciences & the Law* 27:123–136.

Martin, A. M., and J. Peerzada. 2005. The expressive meaning of enhancement. *American Journal of Bioethics* 5 (3): 25–27.

Martin, P., and R. Ashcroft. 2005. Background paper prepared for the 2005 Wellcome Trust Summer School on Neuroethics.

Masten, C. L., E. H. Telzer, and N. I. Eisenberger. 2011. An fMRI investigation of attributing negative social treatment to racial discrimination. *Journal of Cognitive Neuroscience* 23 (5): 1042–1051.

Masters, R. D., and M. Coplan. 2001. Toxins, brain chemistry, and behavior. *Optimal Center Newsletter* 262: 6–9.

Mattai, A. K., J. L. Hill, K. Rhoshel, and M. D. Lenroot. 2010. Treatment of early-onset schizophrenia. *Current Opinion in Psychiatry* 23:304–310.

Mathew, S. J., J. M. Amiel, and H. A. Sackeim. 2005. Electroconvulsive therapy in treatment-resistant depression. *Primary Psychiatry* 12 (2): 52–56.

Mathias, C. W., D. M. Marsh-Richard, and D. M. Dougherty. 2008. Behavioral measures of impulsivity and the law. *Behavioral Sciences & the Law* 26: 691–707.

McCarthy, M. M., G. J. de Vries, and N. G. Forger. 2009. Sexual differentiation of the brain: Mode, mechanisms and meaning. In *Hormones, Brain and Behavior*. 2nd ed., ed. A. P. Arnold, A. M. Etgen, S. E. Fahrbach, R. T. Rubin, and D. W. Pfaff. New York: Academic Press.

McDermott, R. 2009. Mutual interests: The case of increasing dialogue between political science and neuroscience. *Political Research Quarterly* 62 (3): 571–583.

McDermott, R., D. Johnson, J. Cowden, and S. Rosen. 2007. Testosterone and aggression in a simulated crisis game. *Annals of the American Academy of Political and Social Science* 614:15–33.

McDermott, R., D. Tingley, J. Cowden, G. Frazzetto, and D. Johnson. 2009. Monoamine oxidase A gene (MAOA) predicts behavioral aggression following provocation. *Proceedings of the National Academy of Sciences of the United States of America* 106 (7): 2118–2123.

McMahan, J. 1998. Brain death, cortical death and persistent vegetative state. In *A Companion to Bioethics*, ed. Helga Kuhse and Peter Singer. Oxford: Blackwell.

McMahan, J. 2006. An alternative to brain death. *Journal of Law, Medicine & Ethics* 34 (1): 44–48.

Medical Research Council. 2010. *Autism in Adults Diagnosed by Quick, New Brain Scan.* London: Medical Research Council.

Meegan, D. V. 2008. Neuroimaging techniques for memory detection: Scientific, ethical, and legal issues. *American Journal of Bioethics* 8 (1): 9–20.

Mehrkens, J. H., K. Boetzel, U. Steude, K. Zeitler, A. Schnitzler, V. Sturm, and J. Voges. 2009. Long-term efficacy and safety of chronic globus pallidus internus stimulation in different types of primary dystonia. *Stereotactic and Functional Neurosurgery* 87:8–17.

Mehta, P. H., and J. Beer. 2010. Neural mechanisms of the testosterone–aggression relation: The role of orbitofrontal cortex. *Journal of Cognitive Neuroscience* 22 (10): 2357–2368.

Meixner, J. B., and J. P. Rosenfeld. 2011. A mock terrorism application of the P300-based concealed information test. *Psychophysiology* 48:149–154.

Meloni, R., J. Mallet and N. Faucon Biguet. 2010. Brain gene transfer and brain implants. *Studies in Ethics, Law, and Technology* 4 (3): Article 3.

Mendelsohn, D., N. Lipsman, and M. Bernstein. 2010. Neurosurgeons' perspectives on psychosurgery and neuroenhancement: A qualitative study at one center. *Journal of Neurosurgery* 113 (6): 1212–1218.

Mendez, M. F. 2009. The neurobiology of moral behavior: Review and neuropsychiatric implications. *CNS Spectrums* 14 (11): 608–620.

Menzler, K., M. Belke, E. Wehrmann, U. Lengler, A. Jansen, et al. 2011. Men and women are different: Diffusion tension imaging reveals sexual dimorphism in the microstructure of the thalamus, corpus callosum and cingulum. *NeuroImage* 54 (4): 2557–2562.

Mercadillo, R. E., and N. A. Arias. 2010. Violence and compassion: A bioethical insight into their cognitive bases and social manifestations. *International Social Science Journal* 61:221–232.

Meyer-Lindenberg, A., J. W. Buckholtz, B. Kolachana, R. Hariri, L. Pezawas, et al. 2006. Neural mechanisms of genetic risk for impulsivity and violence in humans. *Proceedings of the National Academy of Sciences of the United States of America* 103:6269–6274.

Meyer-Lindenberg, A., K. K. Nicodemus, M. F. Egan, J. H. Callicott, V. Mattay, and D. R. Weinberger. 2008. False positives in imaging genetics. *NeuroImage* 40:655–661.

Miller, F. G., M. M. Mello, and S. Joffe. 2008. Incidental findings in human subjects research: What do investigators owe research participants? *Journal of Law, Medicine & Ethics* 36 (2): 271–279.

Miller, G. 2010. fMRI lie detection fails a legal test. *Science* 328 (5984): 689–690.

Miller, F. G., and H. Brody. 2005. Enhancement technologies and professional integrity. *American Journal of Bioethics* 5 (3): 15–17.

Miller, F. G., and R. D. Truog. 2008. Rethinking the ethics of vital organ donations. *Hastings Center Report* 38 (6): 38–46.

Miller, I. N., and A. Cronin-Golomb. 2010. Gender differences in Parkinson's disease: Clinical characteristics and cognition. *Movement Disorders* 25 (16): 2695–2703.

Miller, J. 1992. Trouble in mind. *Scientific American* 267 (3): 180.

Milstein, A. C. 2008. Research malpractice and the issue of incidental findings. *Journal of Law, Medicine & Ethics* 36 (2): 356.

Mirsky, A. F., and M. H. Orzack. 1977. Final report on psychosugery pilot study. In *Appendix: Psychosurgery*. Washington, D.C.: GPO.

Mobbs, D., H. C. Lau, O. D. Jones, and C. D. Frith. 2007. Law, responsibility, and the brain. *PLoS Biology* 5:693–700.

Modinos, G., J. Ormel, and A. Aleman. 2010. Individual differences in dispositional mindfulness and brain activity involved in reappraisal of emotion. *Social Cognitive and Affective Neuroscience* 5 (4): 369–377.

Moffitt, T. E. 2005. The new look of behavioral genetics in developmental psychopathology: Gene–environment interplay in antisocial behaviors. *Psychological Bulletin* 131:533–554.

Mokdad, A. H., J. S. Marks, D. Stroup, and J. L. Gerberding. 2004. Actual causes of death in the United States, 2000. *Journal of the American Medical Association* 291:1238–1245.

Moll, J., and R. de Oliveira-Souza. 2007. Moral judgments, emotions and the utilitarian brain. *Trends in Cognitive Sciences* 11 (8): 319–321.

Moll, J., R. de Oliveira-Souza, and R. Zahn. 2008. The neural basis of moral cognition: Sentiments, concepts, and values. *Annals of the New York Academy of Sciences* 1124:161–180.

Moll, J., R. Zahn, R. de Oliveira-Souza, F. Krueger, and J. Grafman. 2005. The neural basis of human moral cognition. *Nature Reviews. Neuroscience* 6:799–809.

Mooney, A. 2012. Solving the mystery of the American psycho. http://sites.duke.edu/dukeresearch/2012/04/23/solving-the-mystery-of-the-american-psycho.

Moreno, J. D. 2004. Bioethics and the national security state. *Journal of Law, Medicine & Ethics* 32:198–208.

Moreno, J. D. 2006. *Mind Wars: Brain Research and National Defense*. New York: Dana Press.

Moreno, J. D. 2008. Using neuropharmacology to improve interrogation techniques. *Bulletin of the Atomic Scientists* 15 (July). http://www.thebulletin.org/print/web-edition/roundtables/the-military-application-of-neuroscience-research.

Moreno, J. D. 2009. The future of neuroimaged lie detection and the law. *Akron Law Review* 42:717.

Moretto, G., E. Ladavas, F. Mattioli, and G. di Pellegrino. 2010. A psychophysiological investigation of moral judgment after ventromedial damage. *Journal of Cognitive Neuroscience* 22 (8): 1888–1899.

Müller, J. L. 2010. Psychopathy: An approach to neuroscientific research in forensic psychiatry. *Behavioral Sciences & the Law* 28:129–147.

Müller, S., and H. Walter. 2010. Reviewing autonomy: Implications of the neurosciences and the free will debate for the principle of respect for the patient's autonomy. *Cambridge Quarterly of Healthcare Ethics* 19:206–217.

Müller, J. L., M. Sommer, K. Döhnel, T. Weber, T. Schmidt-Wilcke, and G. Hajak. 2008. Disturbed prefrontal and temporal brain function during emotion and cognition interaction in criminal psychopathy. *Behavioral Sciences & the Law* 26 (1):131–150.

Murphy, E. R., J. Illes, and P. B. Reiner. 2008. Neuroethics of neuromarketing. *Journal of Consumer Behaviour* 7:293–302.

Nair-Collins, M. 2010. Death, brain death, and the limits of science: Why the whole-brain death concept of death is a flawed public policy. *Journal of Law, Medicine & Ethics* 38 (3): 667–683.

Nakamura, M., P. G. Nestor, J. J. Levitt, A. S. Cohen, T. Kawashima, M. E. Shenton, and R. W. McCarley. 2008. Orbitofrontal volume deficit in schizophrenia and thought disorder. *Brain* 131 (1):180–195.

National Conference of Commissioners of Uniform State Laws. 1980. Uniform Definition of Death. http://pntb.org/wordpress/wp-content/uploads/Uniform -Determination-of-Death-1980_5c.pdf.

National Institute on Drug Abuse. 2010. Impulsive-antisocial personality traits linked to a hypersensitive brain reward system. *Mental Health Weekly Digest*, April 12, 31.

National Institute of Neurological Disorders and Stroke. 2005. *Report of t-PA Review Committee*. Washington, D.C.: National Institutes of Health.

Nature. 2007. Enhancing, not cheating. *Nature* 450:320.

Nature. 2011. First do no harm: Simple tools to diagnose mental illness should not be offered without sound supporting evidence. *Nature* 469:132. doi:10.1038/469132a .

Nelson, C. A. 2008. Incidental findings in magnetic resonance imaging (MRI) brain research. *Journal of Law, Medicine & Ethics* 36 (2): 315–329.

New, A. S., E. A. Hazlett, M. S. Buchsbaum, M. Goodman, S. A. Mitelman, R. Newmark, R. Trisdorfer, et al. 2007. Amygdala–prefrontal disconnection in borderline personality disorder. *Neuropsychopharmecology* 32:1629–1640.

Newson, A. J., and R. E. Ashcroft. 2005. Whither authenticity? *American Journal of Bioethics* 5 (3): 53–55.

Ngun, T. C., N. Ghahramani, F. J. Sanchez, S. Bocklandt, and E. Vilian. 2010. The genetics of sex differences in brain and behaviour. *Frontiers in Neuroendocrinology*. doi:10.1016/j.yfrne.2010.10.001.

Nicholson, J. 2011. Drugs used for temporary chemical castration. http://www .ehow.com/about_5394778_drugs-used-temporary-chemical-castration.html.

Nierenberg, A.A., A.C. Leon, L.H. Price, R.C. Sheldon and M.H. Trivedi. 2011. Crisis of confidence: Antidepressant risk versus benefit. *Journal of Clinical Psychiatry* 72 (3): e.11.

O'Connell, G., J. De Wilde, J. Haley, K. Shuler, B. Schafer, P. Sandercock, and J. M. Wardlaw. 2011. The brain, the science and the media: The legal, corporate, social and security implications of neuroimaging and the impact of media coverage. *EMBO Reports* 12 (7): 30–36.

O'Donovan, M. C., and M. J. Owen. 2009. Genetics and the brain: Many pathways to enlightenment. *Human Genetics* 126 (1): 91–99.

Office of Technology Assessment. 1990. *Neural Grafting: Repairing the Brain and Spinal Cord*. Washington, D.C.: GPO.

Olsson, A., K. I. Nearing, and E. A. Phelps. 2007. Learning fears by observing others: The neural systems of social fear transmission. *Social Cognitive and Affective Neuroscience* 2 (1): 3–11.

Pabst-Battin, M. 1994. *The Least Worst Death: Essays in Bioethics at the End of Life*. New York: Oxford University Press.

Pacherie, E. 2008. The phenomenology of action: A conceptual framework. *Cognition* 107 (1):179–217.

Pacholczyk, A. 2011. DBS makes you feel good! Why some of the ethical objections to the use of DBS for neuropsychiatric disorders and enhancement are not convincing. *Frontiers in Integrative Neuroscience* 5. doi:10.3389/fnint. 2011.000014.

Parens, E. 2002. How far will the treatment/ enhancement distinction get us as we grapple with new ways to shape ourselves? In *Neuroethics: Mapping the Field*, ed. S. J. Marcus, 152–158. New York: Dana Press.

Parsey, R. V. 2010. Serotonin receptor imaging: Clinically useful? *Journal of Nuclear Medicine* 51 (10): 1495–1499.

Patin, A., and R. Hurlemann. 2011. Modulating amygdala responses to emotion: Evidence from pharmacological fMRI. *Neuropsychologia* 49:706–717.

Patrick, C. J. 2008. Psychophysiological correlates of aggression and violence: An integrative review. *Philosophical Transactions of the Royal Society of London. Series B, Biological Sciences* 363:2543–2555.

Paus, T. 2010. Sex differences in the human brain: A developmental perspective. *Progress in Brain Research* 186:13–28.

Pelchat, M. L. 2009. Food addiction in humans. *Journal of Nutrition* 139: 620–622.

Penney, D., and G. McGee. 2005. Chemical trust: Oxytocin oxymoron? *American Journal of Bioethics* 5 (3): 1–2.

Pentland, A. 2007. On the collective nature of human intelligence. *Adaptive Behavior* 15:189–198.

Pereira, M. G., L. de Oliveira, F. S. Erthal, M. Joffily, et al. 2010. Emotion affects action: Midcingulate cortex as a pivotal node of interaction between negative emotion and motor signals. *Cognitive, Affective & Behavioral Neuroscience* 10 (1): 94–106.

Perlin, M. L. 2010a. Good and bad, I defined these terms, quite clear no doubt somehow. *Behavioral Sciences & the Law* 28:671–689.

Perlin, M. L. 2010b. Too stubborn to ever be governed by enforced insanity: Some therapeutic jurisprudence dilemmas in the representation of criminal defendants in incompetency and insanity cases. *International Journal of Law and Psychiatry* 33:475–481.

Perry, B. D. 2002. Childhood experience and the expression of genetic potential: What childhood neglect tells us about nature and nurture. *Brain and Mind* 3:79–100.

Pessoa, L. 2009. How do emotion and motivation direct executive control? *Trends in Cognitive Sciences* 13 (4): 160–166.

Pessoa, L. 2010a. Emergent processes in cognitive-emotional interactions. *Dialogues in Clinical Neuroscience* 12 (4): 433–448.

Pessoa, L. 2010b. Emotion and cognition and the amygdala: From "What is it?" to "What's to be done?" *Neuropsychologia* 48 (12): 3416–3429.

Pessoa, L., and R. Adolphs. 2010. Emotion processing and the amygdala: From a "low road" to "many roads" of evaluating biological significance. *Nature Reviews. Neuroscience* 11:773–783.

Pessoa, L., and J. B. Engelmann. 2010. Embedding reward signals into perception and cognition. *Frontiers in Neuroscience* 4 (article 17): 1–8.

Peters, J., and C. Büchel. 2010. Episodic future thinking reduces reward delay discounting through an enhancement of prefrontal-mediotemporal interactions. *Neuron* 66 (1): 138–148.

Petersen, G. N. 1994. Regulation of electroconvulsive therapy: The California experience. In *Psychiatric Practice under Fire*, ed. H. I. Schwartz. Washington, D.C.: American Psychiatric Press.

Pettigrew, T. F., and L. Tropp. 2006. A meta-analytic test of intergroup contact theory. *Journal of Personality and Social Psychology* 90:751–783.

Pettit, M. 2007. FMRI and BF meet FRE: Brain imaging and the federal rules of evidence. *American Journal of Law & Medicine* 33 (2/3): 319–343.

Pezaris, J. S., and E. N. Eskandar. 2009. Getting signals into the brain: Visual prosthetics through thalamic microstimulation. *Neurosurgical Focus* 27 (1): E6.

Phelps, E. A. 2001. Faces and races in the brain. *Nature Neuroscience* 4 (8): 775–776.

Phelps, E. A., C. J. Cannistraci, and W. A. Cunningham. 2003. Intact performance on an indirect measure of race bias following amygdala damage. *Neuropsychologia* 41:204–206.

Phelps, E. A., K. J. O'Connor, W. A. Cunningham, E. S. Funayma, J. C. Gatenby, J. C. Gore, and M. R. Banaji. 2000. Performance on indirect measures on race evaluation predicts amygdala activation. *Journal of Cognitive Neuroscience* 12 (5): 729–738.

Phillips, H. 2004. Private thought, public property. *New Scientist* 31 (July): 38–41.

Piech, R. M., M. McHugo, S. D. Smith, M. S. Dukic, et al. 2011. Fear-enhanced visual search persists after amygdala lesions. *Neuropsychologia* 49:596–601.

Pinker, S. 2002. *The Blank Slate: The Modern Denial of Human Nature*. New York: Viking.

Plassmann, H., J. O'Doherty and A. Rangel. 2007. Orbitofrontal cortex encodes willingness to pay in everyday economic transactions. *Journal of Neuroscience* 27: 9984–88.

Plassmann, H., J. O'Doherty, B. Shiv, and A. Rangel. 2008. Marketing actions can modulate neural representations of experienced pleasantness. *Proceedings of the National Academy of Sciences of the United States of America* 105: 1050–1054.

Poldrack, R. A. 2006. Can cognitive processes be inferred from neuroimaging data? *Trends in Cognitive Sciences* 10 (2):59–63.

Poldrack, R. A. 2011. Inferring mental states from neuroimaging data: From reverse inference to large-scale decoding. *Neuron* 72:692–697.

President's Commission for the Study of Ethical Problems in Medicine and Biomedical and Behavioral Research. 1981. *Defining Death*. Washington, DC: U.S. Government Printing Office.

Preston, S. 2011. The empathy gap. *Nature* 472:416.

Price, J. L., and W. C. Drevets. 2010. Neurocircuitry of mood disorders. *Neuropsychopharmacology* 35 (1):192–216.

Quednow, B. B. 2010. Ethics of neuroenhancement: A phantom debate. *Biosocieties* 5:153–156.

Quenqua, D. 2011. Rethinking addiction's roots, and its treatment. *New York Times*, July 10.

Qureshi, I. A., and M. F. Mehler. 2010. Genetic and epigenetic underpinnings of sex differences in the brain and in neurological and psychiatric disease susceptibility. *Progress in Brain Research* 186:77–95.

Raafat, R. M., N. Chater, and C. Firth. 2009. Herding in humans. *Trends in Cognitive Sciences* 13 (10): 420–428.

Racine, E. 2010. *Pragmatic Neuroethics: Improving Treatment and Understanding of the Mind-brain*. Cambridge, MA: MIT Press.

Racine, E., O. Bar-Ilan, and J. Illes. 2006. Neuroimaging: A decade of coverage in the print media. *Science Communication* 28:122–143.

Racine, E., S. Waldman, J. Rosenberg, and J. Illes. 2010. Contemporary neuroscience in the media. *Social Science & Medicine* 71:725–733.

Raichle, M. E., and M. A. Mintun. 2006. Brain work and brain imaging. *Annual Review of Neuroscience* 29:449–476.

Raine, A. 2002. Biosocial studies of antisocial and violent behavior in children and adults: A review. *Journal of Abnormal Child Psychology* 30 (4):311–326.

Raine, A. 2008. From genes to brain to antisocial behavior. *Current Directions in Psychological Science* 17:323–328.

Raine, A., and Y. Yang. 2006. Neural foundations to moral reasoning and anti-social behavior. *Social Cognitive and Affective Neuroscience* 1 (3): 203–213.

Raine, A., Y. Yang, K. L. Narr, and A. W. Toga. 2011. Sex differences in orbito-frontal gray as a partial explanation for sex differences in antisocial personality. *Molecular Psychiatry* 16 (2): 227–236.

Ramírez, J. M. 2010. The usefulness of distinguishing types of aggression by function. *International Social Science Journal* 61:263–272.

Ramos, B. P., and A. F. Arnsten. 2007. Adrenergic pharmacology and cognition: Focus on the prefrontal cortex. *Pharmacology & Therapeutics* 113:523–536.

Rangel, A., C. Camerer, and P. R. Montague. 2008. A framework for studying the neurobiology of value-based decision making. *Nature Reviews. Neuroscience* 9:545–556.

Rasmussen, K. G. 2011. Some considerations in choosing electroconvulsive therapy versus transcranial magnetic stimulation for depression. *Journal of ECT* 27 (1): 51–54.

Reidpath, D. D., D. Crawford, L. Tilgner, and C. Gibbons. 2002. Relationship between body mass index and the use of healthcare services in Australia. *Obesity Research* 10:526–531.

Reeves, R. R., and R. L. Panguluri. 2011. Neuropsychiatric complications of traumatic brain injury. *Journal of Psychological Nursing and Mental Health Services* 49 (3): 42–51.

Reis, J., H. M. Schambra, L. G. Cohen, E. R. Buch, B. Fritsch, and E. Zarahn. 2009. Noninvasive cortical stimulation enhances motor skill acquisition over multiple days through an effect on consolidation. *Proceedings of the National Academy of Sciences of the United States of America* 106 (5): 1590–1595.

Renvoisé, P., and C. Morin. 2007. *Neuromarketing: Understanding the Buy Buttons in Your Customer's Brain*. New York: Thomas Nelson.

Repantis, D., P. Schlattmann, O. Laisney, and I. Heuser. 2010. Modafinil and methylphenidate for neuroenhancement in healthy individuals: A systematic review. *Pharmacological Research* 62:187–206.

Restak, R. M. 1994. *Receptors*. New York: Bantam Books.

Reuters. 2005a. Expert sees obesity hitting U.S. life expectancy. http://mgm supplement.blogspot.com/2005/02/expert-sees-obesity-hitting-us-life.html.

Reuters. 2005b. Panel to review pacemaker device for brain. http://www.black herbals.com/fda_approves_implant_for_depress.htm.

Richeson, J. A., A. R. Todd, S. Trawalter, and A. A. Baird. 2008. Eye-gaze direction modulates race-related amygdala activity. *Group Processes & Intergroup Relations* 11:233–246.

Rilling, J. K., A. L. Glenn, M. R. Jairam, G. Pagnoni, D. R. Goldsmith, H. A. Elfenbein and S. O. Lilienfeld. 2007. Neural correlates of social cooperation and non-cooperation as a function of psychopathy. *Biological Psychiatry* 61 (11): 1260–1271.

Robert, J. S. 2005. Human dispossession and human enhancement. *American Journal of Bioethics* 5 (3): 27–29.

Roberts, A. J., and G. F. Koob. 1997. The neurobiology of addiction: An overview. *Alcohol Health Res World* 21:101–106.

Rodrigues, S. M., J. E. LeDoux, and R. M. Sapolsky. 2009. The influence of stress hormones on fear circuitry. *Annual Review of Neuroscience* 32: 289–313.

Rodriguez de Fonseca, F., M. Rocio, A. Carrera, et al. 1997. Activation of corticotropin-releasing factor in the limbic system during cannabinoid withdrawal. *Science* 276:2050–2054.

Rolls, E. T. 2000. Memory systems in the brain. *Annual Review of Psychology* 51:599–630.

Ronquillo, J., T. F. Denso, B. Lickel, Z.-L. Lu, A. Nandy, and K. B. Maddox. 2007. The effects of skin tone on race-related amygdala activity: an fMRI investigation. *Social Cognitive and Affective Neuroscience* 2 (1): 39–44.

Rosack, J. 2005. New data show declines in antidepressant prescribing. *Psychiatric News* 40 (17): 1.

Roskies, A. L. 2010. How does neuroscience affect our conception of volition? *Annual Review of Neuroscience* 33:109–130.

Rose, S. P. R. 2002. "Smart drugs": Do they work? Are they ethical? Will they be legal? *Nature Reviews. Neuroscience* 3:975–977.

Rose, S. P. R. 2010. The art of medicine: Brain-based ethics? *Lancet* 371 (February 2): 380–381.

Rossiter, J. R., R. B. Silberstein, P. G. Harris, and G. Nield. 2001. Brain-imaging detection of visual scene encoding in long-term memory for TV commercials. *Journal of Advertising Research* 41:13–22.

Royal, J. M., and B. S. Peterson. 2008. The risks and benefits of searching for incidental findings in MRI research scans. *Journal of Law, Medicine & Ethics* 36 (2): 305–313.

Rubenstein, L., C. Pross, F. Davidoff, and V. Iacopino. 2005. Coercive US interrogation policies: A challenge to medical ethics. *Journal of the American Medical Association* 294 (12): 1544–1549.

Rule, N. O., J. B. Freeman, J. M. Moran, J. D. E. Gabrieli, R. B. Adams, and N. Ambady. 2010. Voting behavior is reflected in amygdala response across cultures. *Social Cognitive and Affective Neuroscience* 5 (2–3): 349–355.

Ryan, C. J. 2010. One flu over the cuckoo's nest: Comparing legislated coercive treatment for mental illness with that for other illnesses. *Journal of Bioethical Inquiry*. doi:10.1007/s11673-010-9270-2.

Safer, D. J., and M. Malever. 2000. Stimulant treatment in Maryland Public Schools. *Pediatrics* 106:553–559.

Said, C. P., R. Dotsch, and A. Todorov. 2011. The amygdala and FFA track both social and non-social face dimensions. *Neuropsychologia* 49:630–639.

Salerno, J. M., and B. L. Bottoms. 2009. Emotional evidence and juror's judgments: Neuroscience for informing psychology and law. *Behavioral Sciences & the Law* 27:273–296.

Salzman, C. D., and S. Fusi. 2010. Emotion, cognition, and mental state representation in amygdala and prefrontal cortex. *Annual Review of Neuroscience* 33:173–202.

Sanchez, F. J., and E. Vilain. 2010. Genes and brain sex differences. *Progress in Brain Research* 186:65–76.

Sanfey, A. G. 2009. Expectations and social decision-making: Biasing effects of prior knowledge on ultimatum responses. *Mind and Society* 8:93–107.

Sapara, A., M. Cooke, D. Fannon, and A. Francis. 2007. Prefrontal cortex and insight in schizophrenia: A volumetric MRI study. *Schizophrenia Research* 89 (1–3):22–34.

Sapolsky, R. M. 2004. The frontal cortex and the criminal justice system. *Neuroscience and the Legal System* 359:1787–1796.

Sato, W., T. Kochiyama, K. Matsuda, K. Usui, Y. Inoue, and M. Toichi. 2011. Rapid amygdala gamma oscillations in response to fearful facial expressions. *Neuropsychologia* 49:612–617.

Savic, I., A. Garcia-Falgueras, and D. F. Swaab. 2010. Sexual differentiation of the human brain in relation to gender identity and sexual orientation. *Progress in Brain Research* 186:41–62.

Savic, I., and P. Lindström. 2008. PET and MRI show differences in cerebral asymmetry and functional connectivity between homo- and heterosexual subjects. *Proceedings of the National Academy of Sciences of the United States of America* 105:9403–9408.

Savulescu, J. 2009. Functional neuroimaging and withdrawal of life-sustaining treatment from vegetative patients. *Journal of Medical Ethics* 35 (8): 508.

Schauer, F. 2010. Neuroscience, lie-detection, and the law: Contrary to the prevailing view, the suitability of brain-based lie-detection for courtroom or forensic use should be determined according to legal and not scientific standards. *Trends in Cognitive Sciences* 14:101–103.

Schermer, M. 2011. Ethical issues in deep brain stimulation. *Frontiers in Integrative Neurosci.* 5:17. doi:10.3389/fnint.2011.00017.

Schlaepfer, T. E., and J. Fins. 2010. Deep brain stimulation and the neuroethics of responsible publishing, when one is not enough. *Journal of the American Medical Association* 303:775–776.

Schlaepfer, T. E., S. H. Lisanby, and S. Pallanti. 2010. Separating hope from hype: Some ethical implications of the development of deep brain stimulation in psychiatric research and treatment. *CNS Spectrums* 15 (5): 285–287.

Schleim, S., T. M. Springer, S. Erk, and H. Walter. 2011. From moral to legal judgment: The influence of normative context in lawyers and other academics. *Social Cognitive and Affective Neuroscience* 6 (1): 48–57.

Schmetz, M. K., and T. Heinemann. 2010. Ethical aspects of deep brain stimulation in the treatment of psychiatric disorders. *Fortschritte der Neurologie-Psychiatrie* 78 (5): 269–278.

Schmook, M. T., P. C. Brugger, M. Weber, G. Kasprian, et al. 2010. Forebrain development in fetal MRI: evaluation of anatomical landmarks before gestational week 27. *Neuroradiology Heidelberg* 52 (6): 495–499.

Schneiderman, L. 2005. The perils of hope. *Cambridge Quarterly of Healthcare Ethics* 14:235–239.

Schoenbaum, G., M. P. Saddoris, and T. A. Stalnaker. 2007. Reconciling the roles of orbitofrontal cortex in reversal learning and the encoding of outcome expectancies. *Annals of the New York Academy of Sciences* 1121:320–335.

Scholz, J., M. C. Klein, T. E. J. Behrens, and H. Johansen-Berg. 2009. Training induces changes in white-matter architecture. *Nature Neuroscience* 12:1370–1371.

Schumann, C. M., M. D. Bauman, and D. G. Amaral. 2011. Abnormal structure or function of the amygdale is a common component of neurodevelopmental disorders. *Neuropsychologia* 49:745–759.

Schwartz, P. H. 2005. Defending the distinction between treatment and enhancement. *American Journal of Bioethics* 5 (3): 17–19.

Scott, A. 1995. *Stairway to the Mind: The Controversial New Science of Consciousness*. New York: Copernicus.

Scott, S. H. 2006. Converting thoughts into action. *Nature* 442:141–142.

Selkoe, D. J. 1992. Aging brain, aging mind. *Scientific American* 267 (September): 135–142.

Seligman, R., and R. A. Brown. 2010. Theory and method at the intersection of anthropology and cultural neuroscience. *Social Cognitive and Affective Neuroscience* 5 (2–3): 130–137.

Sell, A., J. Tooby, and L. Cosmides. 2009. Formidability and the logic of human anger. *Proceedings of the National Academy of Sciences of the United States of America* 106 (35): 15073–15078.

Sen, A. N., P. G. Campbell, S. Yadla, J. Jallo, and A. D. Sharan. 2010. Deep brain stimulation in the management of disorders of consciousness: A review of physiology, previous reports, and ethical considerations. *Neurosurgical Focus* 29 (2): E14.

Seo, D., C. J. Patrick, and P. J. Kennealy. 2008. Role of serotonin and dopamine system interactions in the neurobiology of impulsive aggression and its comorbidity with other clinical disorders. *Aggression and Violent Behavior* 13:383–395.

Sergerie, K., C. Chochol, and J. L. Armony. 2008. The role of the amygdala in emotional processing: A quantitative meta-analysis of functional neuroimaging studies. *Neuroscience and Biobehavioral Reviews* 32:811–830.

Settle, J. E., C. T. Dawes, and J. H. Fowler. 2009. The heritability of partisan attachment. *Political Research Quarterly* 62 (3): 601–613.

Shamoo, A. E. 2010. Ethical and regulatory challenges in psychophysiology and neuroscience-based technology for determing behaviour. *Accountability in Research* 17:8–29.

Shaywitz, B. A., S. E. Shaywitz, K. R. Pugh, et al. 1995. Sex differences in the functional organization of the brain for language. *Nature* 373:607–609.

Shewmon, D. A. 2009. Brain death: Can it be resuscitated? *Hastings Center Report* 39 (2): 18–24.

Shibasaki, H. 2008. Human brain mapping: Hemodynamic response and electrophysiology. *Clinical Neurophysiology* 119:731–743.

Shirtcliff, E. A., M. J. Vitacco, A. R. Graf, A. J. Gostisha, J. L. Merz, and C. Zahn-Waxler. 2009. Neurobiology of empathy and callousness: Implications for the development of antisocial behavior. *Behavioral Sciences & the Law* 27:137–171.

Shiv, B., S. J. Grant, A. P. McGraw, et al. 2005. Decision neuroscience. *Marketing Letters* 16 (3/4): 375–386.

Shuman, D. W., and L. H. Gold. 2008. Without thinking: Impulsive aggression and criminal responsibility. *Behavioral Sciences & the Law* 26:723–734.

Sides, J., K. Lipsitz, and M. Grossmann. 2010. Do voters perceive negative campaigns as informative campaigns? *American Politics Research* 38 (3): 502–530.

Siegel, A., and J. Douard. 2011. Who's flying the plane: Serotonin levels, aggression and free will. *International Journal of Law and Psychiatry* 34 (1): 20–29.

Siever, L. J. 2008. Neurobiology of aggression and violence. *American Journal of Psychiatry* 165:429–442.

Silver, J. M., R. M. Hales, and S. C. Yudofsky. 2008. Neuropsychiatric aspects of traumatic brain injury. In *The American Psychiatric Publishing Textbook of Neuropsychiatry and Behavioral Neurosciences*. 5th ed., ed. S. C. Yudofsky and R. E. Hales, 595–647. Washington, DC: American Psychiatric Press.

Silveri, M. M., J. Rogowska, A. McCaffrey, and D. A. Yurgelun-Todd. 2011. Adolescents at risk for alcohol abuse demonstrate altered frontal lobe activation during Stroop Performance. *Alcoholism, Clinical and Experimental Research* 35:218–228.

Siminoff, L. A., R. M. Arnold, and M. Sear. 1996. Death. In *Encyclopedia of Biomedical Policy*, ed. R. H. Blank. Westport, CT: Greenwood Press.

Simpson, J. R. 2008. Functional MRI lie detection: Too good to be true? *Journal of American Psychiatry Law* 36 (4): 491–498.

Singer, E. 2004. They know what you want. *New Scientist* 31 (July): 36–37.

Singer, P. 1995. *Rethinking Life and Death: The Collapse of Our Traditional Ethics*. Oxford: Oxford University Press.

Singer, T., and C. Lamm. 2009. The social neuroscience of empathy. *Annals of the New York Academy of Sciences* 1156:81–96.

Singer, T., B. Seymour, J. P. O'Doherty, H. Kaube, R. J. Dolan, and C. D. Frith. 2004. Empathy for pain involves the affective but not sensory components of pain. *Science* 303 (5661): 1157–1162.

Singer, T., B. Seymour, J. P. O'Doherty, K. E. Stephan, R. J. Dolan, and C. D. Frith. 2006. Empathic neural responses are modulated by the perceived fairness of others. *Nature* 439:466–469.

Singh, I. 2005. Will the "real boy" please behave: Dosing dilemmas for parents of boys with ADHD. *American Journal of Bioethics* 5 (3): 34–47.

Skeem, J., P. Johansson, H. Andershed, M. Kerr and J.E. Louden. 2007. Two subtypes of psychopathic violent offenders that parallel primary and secondary variants. *Journal of Abnormal Psychology* 116 (2): 395–409.

Skuban, T., K. Hardenacke, C. Wopen, and J. Kuhn. 2011. Informed consent in deep brain stimulation: Ethical considerations in a stress field of pride and prejudice. *Frontiers in Integrative Neuroscience* 5:21. doi:10.3389/fnint.2011. 00007.

Slavich, G. M., T. Thornton, L. D. Torres, S. M. Monroe, and I. H. Gotlib. 2009. Targeted rejection predicts hastened onset of major depression. *Journal of Social and Clinical Psychology* 28:223–243.

Smith, K. 2009. Brain imaging measures more than we think: Anticipatory brain mechanism may be complicating MRI studies. *Nature*. doi:10.1038/news. 2009.48.

Smith, K. B., D. R. Oxley, M. V. Hibbing, J. R. Alford, and J. R. Hibbing. 2011. Linking genetics and political attitudes: Reconceptualizing political ideology. *Political Psychology* 32 (3): 369–397.

Snead, O. C. 2007. Neuroimaging and the "complexity" of capital punishment. *New York University Law Review* 82 (5): 1265–1339.

Snead, O. C. 2008. Neuroimaging and capital punishment. *New Atlantis* 19:35–64.

Society for Neuroscience. 2006. *Brain Facts: A Primer on the Brain and Nervous System*. Washington, D.C.: Society for Neuroscience.

Sogg, S. 2007. Alcohol misuse after bariatric surgery: Epiphenomenon or "Oprah" phenomenon? *Surgery for Obesity and Related Diseases* 3:366–368.

Soliman, A., R. M. Bagby, A. A. Wilson, L. Miler, M. Clark, P. Rusjan, J. Sacher, S. Houle, and J. H. Meyer. 2010. Relationship of monoamine oxidase A binding to adaptive and maladaptive personality traits. *Psychological Medicine* 1:1–10.

Sommer, M., C. Rothmayr, K. Döhnel, J. Meinhardt, J. Schwerdtner, et al. 2010. How should I decide? The neural correlates of everyday moral reasoning. *Neuropsychologia* 48:2018–2036.

Spence, S. A. 2008. Playing devil's advocate: The case against fMRI lie detection. *Leg Crim Psychologie* 13:11–25.

Spence, S. A., M. D. Hunter, and G. Harpin. 2002. Neuroscience and the will. *Current Opinion in Psychiatry* 15 (5): 519–526.

Spezio, M. L., and R. Adolphs. 2007. Emotional processing and political judgment: Toward integrating political psychology and decision neuroscience. In *The Affect Effect: Dynamics of Emotion in Political Thinking and Behavior*, ed.

W. R. Neuman, G. E. Marcus, A. N. Crigler, and M. MacKuen, 71–95. Chicago: University of Chicago Press.

Spezio, M. L., A. Rangel, R. M. Alvarez, J. P. O'Doherty, K. Mattes, A. Todorov, H. Kim, and R. Adolphs. 2008. A neural basis for the effect of candidate appearance on election outcomes. *Social Cognitive and Affective Neuroscience* 3:344–352.

Spiers, H. J., and E. A. Maguire. 2007. Neural substrates of driving behaviour. *NeuroImage* 36:245–255.

Sreedhar, R., and T. Sanjeev. 2009. Brain death and the apnea test. *Annals of Indian Academy of Neurology* 12 (3):201–203.

Stanton, S. J., J. C. Beehner, E. K. Saini, C. M. Kuhn, and K. S. LaBar. 2009. Dominance, politics, and physiology: Voters' testosterone changes on the night of the 2008 United States presidential election. *PLoS ONE* 4 (10): e7543. doi:10.1371/journal.pone.0007543.

Stein, M. B., P. R. Goldin, J. Sareen, L.T. Eyler Zorrilla and G.G. Brown. 2002. Increased amygdala activation to angry and contemptuous faces in generalized social phobia. *Archives of General Psychiatry* 59 (11): 1027–1034.

Stenziok, M., F. Krueger, G. Deshpande, R. K. Lenroot, E. van der Meer, and J. Grafman. 2010. Fronto-parietal regulation of media violence exposure in adolescents: A multi-method study. *Social Cognitive and Affective Neuroscience.* doi:10.1093/scan/nsq079.

Stenziok, M., F. Krueger, A. Heinecke, et al. 2011. Developmental effects of aggressive behavior in male adolescents assessed with structural and functional brain imaging. *Social Cognitive and Affective Neuroscience* 6 (1): 2–11.

Stevens, D., J. Sullivan, B. Allen, and D. Alger. 2008. What's good for the goose is bad for the gander: Negative political advertising, partisanship, and turnout. *Journal of Politics* 70:527–541.

Stoller, S. E., and P. R. Wolpe. 2007. Emerging neurotechnologies for lie detection and the Fifth Amendment. *American Journal of Law & Medicine* 33 (2–3): 359–375.

Sukel, K. 2008. Sex differences offer new insight into psychiatric disorders. Dana Foundation. http://www.dana.org/news/features/detail.aspx?id=13514.

Swaab, D. F. 2008. Sexual orientation and its basis in brain structure and function. *Proceedings of the National Academy of Sciences of the United States of America* 105:10273–10274.

Swaab, D. F., and M. A. Hofman. 1990. An enlarged suprachiasmatic nucleus in homosexual men. *Brain Research* 537:141–148.

Synofzik, M., and T. E. Schlaepfer. 2008. Stimulating personality: Ethical criteria for deep brain stimulation in psychiatric patients and for enhancement purposes. *Biotechnology Journal* 3 (12): 1511–1520.

Synofzik, M., and T. E. Schlaepfer. 2011. Electrodes in the brain: Ethical criteria for research and treatment with deep brain stimulation for neuropsychiatric disorders. *Brain Stimulation* 4 (1):7–16.

Szabo, C. A., J. Xiong, J. L. Lancaster, L. Rainey, and P. Fox. 2001. Amygdalar and hippocampal volumetry in control participants: Differences regarding handedness. *AJNR. American Journal of Neuroradiology* 22:1342–1345.

Tairyan, K., and J. Illes. 2009. Imaging genetics and the power of combined technologies: A perspective from bioethics. *Neuroscience* 164 (1): 7–15.

Takagi, M. 2009. Safety and neuroethical consideration of deep brain stimulation as a psychiatric treatment. *Brain Nerve* 61 (1): 33–40.

Talbot, M. 2009. Brain gain: The underworld of "neuroenhancing" drugs. *New Yorker,* April 17, 32–43.

Tallis, R. 2010. What neuroscience cannot tell us about ourselves. *New Atlantis* 29:3–26.

Tancredi, L. R., and J. D. Brodie. 2007. The brain and behavior: Limitations in the legal use of functional magnetic resonance imaging. *American Journal of Law & Medicine* 33 (2–3): 271–294.

Tanda, G., F. E. Pontieri, and G. DiChiara. 1997. Cannabinoid and heroin activation of mesolimbic dopamine transmission by a common u1 opioid receptor mechanism. *Science* 276:2048–2050.

Taylor, V. H., C. M. Curtis, and C. Davis. 2010. The obesity epidemic: The role of addiction. *Canadian Medical Association Journal* 182 (4): 327–328.

Teicher, M. H. 2002. Scars that won't heal: The neurobiology of child abuse. *Scientific American* 286 (3): 68–75.

Teten, A. L., L. A. Miller, S. D. Bailey, N. J. Dunn, and T. A. Kent. 2008. Empathic deficits and alexithymia in trauma-related impulsive aggression. *Behavioral Sciences & the Law* 26:823–832.

Teuber, H. L., S. H. Corkin, and T. E. Twitchell. 1977. Study of Cingulotomy in Man: A summary. In *Neurosurgical Treatment in Psychiatry, Pain and Epilepsy,* ed. W. H. Sweet, et al. Baltimore: University Park Press.

The Whole Brain Atlas. 2012. http://www.med.harvard.edu/aanlib/home.html.

Thompson, S. K. 2005. Note: The legality of the use of psychiatric neuroimaging in intelligence interrogation. *Cornell Law Review* 90:1601.

Thompson, S. K. 2007. A brave new world of interrogation jurisprudence? *American Journal of Law & Medicine* 33 (2–3): 341–357.

Tingley, D. 2006. Neurological imaging as evidence in political science: A review, critique, and guiding assessment. *Social Sciences Information. Information Sur les Sciences Sociales* 45 (1): 5–33.

Tingley, D. 2007. Evolving political science: Biological adaptation, rational action, and symbolism. *Politics and the Life Sciences* 25 (1/2): 23–41.

Tingley, D. 2011. Neurological imaging and the evaluation of competing theories. In *Biology and Politics: The Cutting Edge,* ed. S. Peterson and A. Somit, 187–204. Bingley, UK: Emerald Books.

Todd, R. M., J. W. Evans, D. Morris, M. D. Lewis, and M. J. Taylor. 2011. The changing face of emotion: Age-related patterns of amygdala activation to salient faces. *Social Cognitive and Affective Neuroscience* 6 (1): 12–23.

Todorov, A., and A. D. Engell. 2008. The role of the amygdala in implicit evaluation of emotionally neutral faces. *Social Cognitive and Affective Neuroscience* 3:303–312.

Todorov, A., A. N. Mandisodza, A. Goren, and C. C. Hall. 2005. Inferences of competence from faces predict election outcomes. *Science* 308:1623–1626.

Todorov, A., M. Pakrashi, and N. N. Oosterhof. 2009. Evaluating faces on trustworthiness after minimal time exposure. *Social Cognition* 27 (6): 813–833.

Tooby, J., and L. Cosmides. 2010. Groups in mind: The coalitional roots of war and morality. In *Human Morality and Sociality: Evolutionary and Comparative Perspective*, ed. H. Høgh-Olesen, 191–234. New York: Palgrave Macmillan.

Tovino, S. A. 2007. Imaging body structure and mapping brain function: A historical approach. *American Journal of Law & Medicine* 33 (2–3): 193–228.

Tovino, S. A. 2009. Neuroscience and health law: An integrative approach? *Akron Law Review* 42:469–517.

Trachtman, H. 2005. A man is a man is a man. *American Journal of Bioethics* 5 (3): 31–33.

Tracy, J. L. 2011. Emotions of inequality. *Science* 333 (6040): 289–290.

Tranel, D., and A. Bechara. 2009. Sex-related functional asymmetry of the amygdala: Preliminary evidence using a case-matched lesion approach. *Neurocase* 15 (3):217–234.

Tranel, D., G. Gullickson, M. Koch, and R. Adolphs. 2006. Altered experience of emotion following bilateral amygdala damage. *Cognitive Neuropsychiatry* 11 (3): 219–232.

Truog, R. D. 1997. Is it time to abandon brain death? *Hastings Center Report* 27 (1): 29–37.

Truog, R. D. 2005. Organ Donation without Brain Death? *Hastings Center Report* 35 (6): 3.

Tsukiura, T., and R. Cabeza. 2011. Shared brain activity for aesthetic and moral judgments: Implications for the beauty-is-good stereotype. *Social Cognitive and Affective Neuroscience* 6 (1): 138–148.

Turner, R., and C. Whitehead. 2008. How collective representations can change the structure of the brain. *Journal of Consciousness Studies* 15 (10–11): 43–57.

Uddin, L. Q., I. Molnar-Szakacs, E. Zaidel, and M. Iacoboni. 2006. rTMS to the right inferior parietal lobule disrupts self-other discrimination. *Social Cognitive and Affective Neuroscience* 1:65–71.

Van Bavel, J. J., D. J. Packer, and W. A. Cunningham. 2008. The neural substrates of in-group bias: A functional magnetic resonance imaging investigation. *Psychological Science* 19 (11):1131–1139.

Van Den Eeden, S. K., C. M. Tanner, A. L. Bernstein, R. D. Fross, A. Leimpeter, D. A. Bloch, and L. M. Nelson. 2003. Incidence of Parkinson's disease: Variation by age, gender, and race/ethnicity. *American Journal of Epidemiology* 157: 1015–1022.

Van der Gaag, C., R. B. Minderaa, and C. Keysers. 2007. The BOLD signal in the amygdala does not differentiate between dynamic facial expressions. *Social Cognitive and Affective Neuroscience* 2 (2): 93–103.

Van der Plas, E. A. A., A. D. Boes, J. A. Wemmie, D. Tranel, and P. Nopoulos. 2010. Amygdala volume correlates positively with fearfulness in normal healthy girls. *Social Cognitive and Affective Neuroscience* 5 (4): 424–431.

Valenstein, E. S., ed. 1986. *Great and Desperate Cures: The Rise and Decline of Psychosurgery and Other Radical Treatments for Mental Illness.* New York: Basic Books.

Vastag, B. 2002. Decade of work shows depression is physical. *Journal of the American Medical Association* 287 (14):1787–1788.

Veatch, R. M. 1993. The impending collapse of the whole-brain definition of death. *Hastings Center Report* 23 (4): 18–24.

Veatch, R. M. 2009. The evolution of death and dying controversies. *Hastings Center Report* 39 (3): 16–19.

Venables, N. C., C. J. Patrick, J. R. Hall, E. M. Bernat. 2011. Clarifying relations between dispositional aggression and brain potential response: Overlapping and distinct contributions of impulsivity and stress reactivity. *Biological Psychology* 86:279–288.

Verheijde, J. L., M. Y. Rady, and J. L. McGregor. 2009. Brain death, states of impaired consciousness, and physician-assisted death for end-of-life organ donation and transplantation. *Medicine, Health Care, and Philosophy* 12:409–421.

Viding, E., D. E. Williamson, and A. R. Hariri. 2006. Developmental imaging genetics: Challenges and promises for translational research. *Development and Psychopathology* 18:877–892.

Vincent, N. A. 2009. Neuroimaging and responsibility assessments. *Neuroethics* 4 (1): 35–49.

Vocks, S., M. Busch, D. Grönemeyer, D. Schulte, S. Herpertz, and B. Suchan. 2010. Neural correlates of viewing photographs of one's own body and another woman's body in anorexia and bulimia nervosa: An fmri study. *Journal of Psychiatry & Neuroscience* 35 (3): 163–176.

Volkow, N. D., and C.P. O'Brien. 2007. Issues for DSM-V: Should obesity be included as a brain disorder? *American Journal of Psychiatry* 164:708–710.

Volkow, N. D., and J. M. Swanson. 2008. The action of enhancers can lead to addiction. *Nature* 451:520.

Volkow, N. D., D. Tomasi, G.-J. Wang, J. S. Fowler, F. Telang, et al. 2011. Reduced metabolism in brain 'control networks' following cocaine-cues exposures in female cocaine abusers. *PLoS ONE* 6 (2): e16593.

Voytek, B., L. Secundo, A. Bidet-Caulet, D. Scabini, S. I. Stiver, et al. 2010. Hemicraniectomy: A new model for human electrophysiology with high spatio-temporal resolution. *Journal of Cognitive Neuroscience* 22 (11): 2491–2502.

Vrecko, S. 2010. "Civilizing technologies" and the control of deviance. *Biosocieties* 5 (1): 36–51.

Vrij, A. 2008. *Detecting Lies and Deceit: Pitfalls and Opportunities.* 2nd ed. West Sussex: Wiley.

Vul, E., C. Harris, P. Winkielman, and H. Pashler. 2009. Puzzlingly high correlations in fMRI studies of emotion, personality, and social cognition. *Perspectives on Psychological Science* 4 (3): 274–290.

Vytal, K., and S. Hamann. 2010. Neuroimaging support for discrete neural correlates of basic emotions: A voxel-based meta-analysis. *Journal of Cognitive Neuroscience* 22 (12): 2864–2885.

Wahlund, K., and M. Kristiansson. 2009. Aggression, psychopathy and brain imaging--review and future recommendations. *International Journal of Law and Psychiatry* 32:266–271.

Wallis, J. D. 2007. Orbitofrontal cortex and its contribution to decision-making. *Annual Review of Neuroscience* 30:31–56.

Walker, P. M., L. Silvert, M. Hewstone, and A. C. Nobre. 2008. Social contact and other-race face processing in the human brain. *Social Cognitive and Affective Neuroscience* 3 (1):16–25.

Walker, P. M., and J. W. Tanaka. 2003. An encoding advantage for own-race versus other-race faces. *Perception* 32 (9): 1117–1125.

Walker, P. M., L. Silvert, M. Hewstone, and A. C. Nobre. 2008. Social contact and other-race face processing in the human brain. *Social Cognitive and Affective Neuroscience* 3 (1): 16–25.

Warneken, F., and M. Tomasello. 2009. Cognition for culture. In *Cambridge Handbook of Situated Cognition*, ed. P. Robbins and M. Aydede, 467–479. Cambridge: Cambridge University Press.

Weaver, J. 2010. Amygdala at the centre of your social network. *Nature.* doi:10.1038/news.2010.699.

Weber, M. J., and S. L. Thompson-Schill. 2010. Functional neuroimaging can support causal claims about brain function. *Journal of Cognitive Neuroscience* 22 (11): 2415–2416.

Weiss, E., C. M. Siedentopfc, A. Hofera, E. A. Deisenhammerd, M. J. Hoptman, et al. 2003. Sex differences in brain activation pattern during a visuospatial cognitive task: A functional magnetic resonance imaging study in healthy volunteers. *Neuroscience Letters* 344:169–172.

Westen, D. 2007. *The Political Brain.* Jackson, TN: Public Affairs.

Westen, D., P. S. Blagov, K. Harenski, C. Kilts, and S. Hamann. 2006. Neural bases of motivated reasoning: An fMRI study of emotional constraints on partisan political judgment in the 2004 U.S. Presidential election. *Journal of Cognitive Neuroscience* 18 (11): 1947–1958.

Wheeler, M. T., and S. T. Fiske. 2006. Controlling racial prejudice and stereotyping: Social-cognitive goals affect amygdala and stereotype activation. *Psychological Science* 16 (1): 56–63.

Whelan, R., P. J. Conrod, J-B. Poline, A. Lourdusamy, T. Banaschewski, G. J. Barker, M. A. Bellgrove, C. Büchel, M. Byrne, T. D. R. Cummins, M. Fauth-Bühler,

H. Flor et al. 2012. Adolescent impulsivity phenotypes characterized by distinct brain networks. *Nature Neuroscience* 15: 920-925. doi:10.1038/nn.3092.

White, A. E. 2010. The lie of fMRI: An examination of the ethics of a market in lie detection using functional magnetic resonance imaging. *HEC Forum.* doi:10.1007/s10730-010-9141-6.

White, G. B. 2005. Splitting the self: The not-so-subtle consequences of medicating boys for ADHD. *American Journal of Bioethics* 5 (3): 57–59.

Wijdicks, E. F. M. 2002. Brain death worldwide: Accepted fact but no global consensus in diagnostic criteria. *Neurology* 58 (1): 20–25.

Wijdicks, E. F. M. 2010. The case against confirmatory tests for determining brain death in adults. *Neurology* 75:77–83.

Wijdicks, E. F. M., P. N. Varelas, G. S. Gronseth, and D. M. Greer. 2010. Evidence-based guideline update: Determining brain death in adults. *Neurology* 74 (3): 1911–1918.

Wild, J. 2005. Brain imaging ready to detect terrorists, say neuroscientists. *Nature* 437:457.

Wilens, T. E., J. Biederman, A. Kwon, R. Chase, L. Greenberg, E. Mick, and T. J. Spencer. 2003. A systematic chart review of the nature of psychiatric adverse events in children and adolescents treated with selective serotonin reuptake inhibitors. *Journal of Child and Adolescent Psychopharmacology* 13 (2): 143–152.

Williams, L. M., B. J. Liddell, A. H. Kemp, R. A. Bryant, R. A. Meares, A. S. Peduto, and E. Gordon. 2006. Amygdala-prefrontal dissociation of subliminal and supraliminal fear. *Human Brain Mapping* 27:652–661.

Wilson, M. 2010. The re-tooled mind: How culture re-engineers cognition. *Social Cognitive and Affective Neuroscience* 5 (2–3): 180–187.

Windle, M. 2010. A multilevel developmental contextual approach to substance use and addiction. *Biosocieties* 5 (1): 124–136.

Wolf, S. M. 2008. Introduction: The challenge of incidental findings. *Journal of Law, Medicine & Ethics* 36 (2):216–218.

Wolf, S. M., F. P. Lawrenz, C. A. Nelson, J. P. Kahn, M. K. Cho, E. W. Clayton, J. G. Fletscher, et al. 2008. Managing incidental findings in human subjects research: Analysis and recommendations. *Journal of Law, Medicine & Ethics* 36 (2): 219–248.

Wolf, S. M., J. Paradise, and C. Caga-anan. 2008. The law of incidental findings in human subjects research: Establishing researchers' duties. *Journal of Law, Medicine & Ethics* 36 (2): 361–383.

Wolpe, P. R. 2003. Neuroethics of enhancement. *Brain and Cognition* 50:387–395.

Wolpe, P. R., K. R. Foster, and D. D. Langleben. 2005. Emerging neurotechnologies for lie-detection: Promises and perils. *American Journal of Bioethics* 5 (2): 39–49.

Wright, R., S. Houston, M. Ellis, S. Holloway, and M. Hudson. 2003. Crossing racial lines: Geographies of mixed race partnering and multiraciality in the United States. *Progress in Human Geography* 27:457–474.

Yeomans, M. R., and R. W. Gray. 1997. Effects of naltrexone on food intake and changes in subjective appetite during eating: Evidence for opioid involvement in the appetizer effect. *Physiology and Behavior* 62:15–21.

Yin, H. H., S. B. Ostlund, and B. W. Balleine. 2008. Reward guided learning beyond dopamine in the nucleus accumbens: The integrative functions of cortico-basal ganglia networks. *European Journal of Neuroscience* 28:1437–1448.

Young, S. 2006. *Designer Evolution: A Transhumanist Manifesto*. Amherst, MA: Prometheus Books.

Zahn, R., R. de Oliveira-Souza, I. Brameti, G. Garrido, and J. Moll. 2009. Subgenual cingulate activity reflects individual differences in empathic concern. *Neuroscience Letters* 457 (2):107–110.

Zahn-Waxler, C., E. A. Shirtcliff, and K. Marceau. 2008. Disorders of childhood and adolescence: Gender and psychopathology. *Annual Review of Clinical Psychology* 4:275–303.

Zangen, A. 2010. Novel perspectives on drug addiction and reward. *Neuroscience and Biobehavioral Reviews* 35:127–128.

Zaretsky, M., A. Mendelsohn, M. Mintz and T. Hendler. 2010. In the eye of the beholder: Internally driven uncertainty of danger recruits the amygdala and dorsomedial prefrontal cortex. *Journal of Cognitive Neuroscience* October 22 (10): 2263–2275.

Ziauddeen, H., I. S. Farooqi, and P. C. Fletcher. 2012. Obesity and the brain: How convincing is the addiction model? *Nature Reviews. Neuroscience* 13 (4): 279–286.

Zielinski, P. B. 2011. Brain death, the pediatric patient, and the nurse. *Pediatric Nursing* 37 (1):17–21, 38.

Zink, C. F., G. Pagnoni, J. Chappelow, M. Martin-Skurski, and G. S. Berns. 2006. Human striatal activation reflects degree of stimulus saliency. *NeuroImage* 29:977–983.

Index

Note: Page numbers in italics refer to tables.

Basic Bioethics
Arthur Caplan, editor

Books Acquired under the Editorship of Glenn McGee and Arthur Caplan

Peter A. Ubel, *Pricing Life: Why It's Time for Health Care Rationing*

Mark G. Kuczewski and Ronald Polansky, eds., *Bioethics: Ancient Themes in Contemporary Issues*

Suzanne Holland, Karen Lebacqz, and Laurie Zoloth, eds., *The Human Embryonic Stem Cell Debate: Science, Ethics, and Public Policy*

Gita Sen, Asha George, and Piroska Östlin, eds., *Engendering International Health: The Challenge of Equity*

Carolyn McLeod, *Self-Trust and Reproductive Autonomy*

Lenny Moss, *What Genes Can't Do*

Jonathan D. Moreno, ed., *In the Wake of Terror: Medicine and Morality in a Time of Crisis*

Glenn McGee, ed., *Pragmatic Bioethics, 2d edition*

Timothy F. Murphy, *Case Studies in Biomedical Research Ethics*

Mark A. Rothstein, ed., *Genetics and Life Insurance: Medical Underwriting and Social Policy*

Kenneth A. Richman, *Ethics and the Metaphysics of Medicine: Reflections on Health and Beneficence*

David Lazer, ed., *DNA and the Criminal Justice System: The Technology of Justice*

Harold W. Baillie and Timothy K. Casey, eds., *Is Human Nature Obsolete? Genetics, Bioengineering, and the Future of the Human Condition*

Robert H. Blank and Janna C. Merrick, eds., *End-of-Life Decision Making: A Cross-National Study*

Norman L. Cantor, *Making Medical Decisions for the Profoundly Mentally Disabled*

Margrit Shildrick and Roxanne Mykitiuk, eds., *Ethics of the Body: Post-Conventional Challenges*

Alfred I. Tauber, *Patient Autonomy and the Ethics of Responsibility*

David H. Brendel, *Healing Psychiatry: Bridging the Science/Humanism Divide*

Jonathan Baron, *Against Bioethics*

Michael L. Gross, *Bioethics and Armed Conflict: Moral Dilemmas of Medicine and War*

Karen F. Greif and Jon F. Merz, *Current Controversies in the Biological Sciences: Case Studies of Policy Challenges from New Technologies*

Deborah Blizzard, *Looking Within: A Sociocultural Examination of Fetoscopy*

Ronald Cole-Turner, ed., *Design and Destiny: Jewish and Christian Perspectives on Human Germline Modification*

Holly Fernandez Lynch, *Conflicts of Conscience in Health Care: An Institutional Compromise*

Mark A. Bedau and Emily C. Parke, eds., *The Ethics of Protocells: Moral and Social Implications of Creating Life in the Laboratory*

Jonathan D. Moreno and Sam Berger, eds., *Progress in Bioethics: Science, Policy, and Politics*

Eric Racine, *Pragmatic Neuroethics: Improving Understanding and Treatment of the Mind-Brain*

Martha J. Farah, ed., *Neuroethics: An Introduction with Readings*

Jeremy R. Garrett, ed., *The Ethics of Animal Research: Exploring the Controversy*

Books Acquired under the Editorship of Arthur Caplan

Sheila Jasanoff, ed., *Reframing Rights: Bioconstitutionalism in the Genetic Age*

Christine Overall, *Why Have Children? The Ethical Debate*

Yechiel Michael Barilan, *Human Dignity, Human Rights, and Responsibility: The New Language of Global Bioethics and Bio-Law*

Tom Koch, *Thieves of Virtue: When Bioethics Stole Medicine*

Timothy F. Murphy, *Ethics, Sexual Orientation, and Choices about Children*

Daniel Callahan, *In Search of the Good: A Life in Bioethics*

Robert H. Blank, *Intervention in the Brain: Politics, Policy, and Ethics*